CURRENT TOPICS IN
ATOMIC, MOLECULAR AND OPTICAL PHYSICS

CURRENT TOPICS IN ATOMIC, MOLECULAR AND OPTICAL PHYSICS

Invited Lectures delivered at the Conference on Atomic
Molecular and Optical Physics (TC2005)
13th – 15th December, 2005
Indian Association for the Cultivation Of Science
Kolkata, India

edited by

CHANDANA SINHA

SHIBSHANKAR BHATTACHARYYA

Indian Association for the Cultivation of Science, India

World Scientific

NEW JERSEY • LONDON • SINGAPORE • BEIJING • SHANGHAI • HONG KONG • TAIPEI • CHENNAI

Published by

World Scientific Publishing Co. Pte. Ltd.
5 Toh Tuck Link, Singapore 596224
USA office: 27 Warren Street, Suite 401-402, Hackensack, NJ 07601
UK office: 57 Shelton Street, Covent Garden, London WC2H 9HE

British Library Cataloguing-in-Publication Data
A catalogue record for this book is available from the British Library.

CURRENT TOPICS IN ATOMIC, MOLECULAR AND OPTICAL PHYSICS

ISBN-13 978-981-270-379-8
ISBN-10 981-270-379-9

Printed in Singapore by Mainland Press

Preface

Since the early days of modern physics, study of the dynamics, structure and interactions of atoms and molecules (between themselves and with external fields) has repeatedly changed our theoretical perception of the external world. In this sense, Atomic Molecular and Optical (AMO) Science has shown amazing resilience and freshness. In recent years, trapping of single atoms and ions with the concomitant development of atom optics, the realization of sub femtosecond lasers, successful implementation of BEC and fundamental breakthroughs in the experimentally realizable approaches to quantum computation, have all contributed to a new revolution in this established field. Huge theoretical as well as experimental programmes are now being taken up to understand, interpret and model these newly discovered phenomena and to predict and discover other novel ones. We are now in a stage where many more spectacular scientific results and technological applications of these new fields are well in sight.

With these thoughts in mind, a Topical Conference on Atomic, Molecular and optical Physics (TC- 2005) was held in Indian Association for the Cultivation of Science, Kolkata between 13th-15th December, 2005. The aim was the dissemination of the specialized knowledge acquired by the experts in various groups and the sharing of their expertise and experience. Accordingly, the talks in the conference were tentatively categorized as belonging to one or the other of a few emerging or currently emerged areas in the process of reaching maturity. Field-matter interaction and quantum control, the application of these techniques to quantum computation and communication and properties of Bose Einstein Condensates were discussed by various speakers. The conventional sub-fields of collision and structures continue to be of fundamental importance and were also included among the topics discussed. These talks, taken together, not only demonstrated the remarkable diversity of the range of problems currently under the purview of AMO physics, but also gave a fascinating glimpse of their underlying unity.

The present volume is primarily based on the talks delivered at the conference. We believe that this book will give the reader a snapshot of the present status of some important aspects of the topics mentioned above - both from a theoretical and from an experimental point of view. At the same time some of the articles can be read as topical reviews highlighting recent achievements. The volume contains articles on theoretical studies of control of molecules and nanostructures by intense fields, optical control of information transfer, theoretical and experimental studies of atomic Bose Einstein Condensates, quantum mechanical electronic structure theory, ion atom collision, optically nonlinear material and quantum chaos. We hope that an audience consisting of physicists in various disciplines will appreciate the broad coverage and find the book useful.

We gratefully acknowledge financial assistance obtained from various Government Agencies and research institutes to the conference TC- 2005 and towards the publication of the present volume. The project has been sponsored by Indian Association for the Cultivation of Science (IACS), Department of Science and Technology (DST), Board of Research in Nuclear Sciences (BRNS), Council of Scientific and Industrial Research (CSIR), Government of West Bengal (Higher Education Department) and West Bengal DST. We also obtained funding from S. N. Bose National Centre for Basic Sciences (SNBNCBS), Saha Institute of Nuclear Physics (SNIP), Department of Theoretical Physics and Department of Materials Science, IACS. We thank all of them for their generous help.

We are greatly indebted to all the authors who sent manuscripts of their talks for inclusion in this volume. It is a pleasure to acknowledge their cooperation. Finally we would also like to record our deep and grateful appreciation of the work of our students without whose help the organization of this conference would have been impossible.

Kolkata

Chandana Sinha
S. S. Bhattacharyya

CONTENTS

Ultrafast Dynamics of Nano and Mesoscopic Systems Driven by Asymmetric Electromagnetic Pulses

A. Matos-Abiague, A. S. Moskalenko, and J. Berakdar

Max-Planck-Institut für Mikrostrukturphysik, Weinberg 2, 06120 Halle, Germany

This work provides an overview on the theoretical description of the electron dynamics in nano and mesoscopic semiconductor structures driven by short asymmetric electromagnetic pulses. For double quantum well structures we show how the electron can be steered within picoseconds into certain spatial regions and discuss ways to maintain in time this non-stationary situations. We also show how charge polarization and charge current can be swiftly generated in mesoscopic rings when irradiated with electromagnetic pulses. We also envisage the possibility of pulse-induced electron removal from quantum dots.

Keywords: 78.67.-n, 42.65.Ky, 42.65.Re

1. Introduction

The everlasting development in the generation and engineering of short laser pulses [1] is giving rise to an increasing number of their utilization in tracing in time the response and the transitions of matter between various states. Even subfemto second resolution has become available which opens the way for exploring the dynamics of new fundamental physical, chemical, and biological processes [2]. Another exciting development in fast optical probes is the generation of strongly asymmetric monocycle linearly (or circularly) polarized electromagnetic pulses [3], often called half-cycle (HCPs) for the reason which will become obvious below. The time structure of the electric filed amplitude of an HCP is shown in Fig.1. It consists of a very short, strong half-cycle (it is in fact this part which is referred to as an HCP), followed by a second long and a much weaker half-cycle of an opposite polarity (the tail of the HCP) [4].

A method for generating HCPs is the following (cf. Fig.1): a wafer of biased gallium-arsenide (GaAs) semiconductor is irradiated by a short pulse from a Ti:Sapphire chirped-pulse amplifier. The wafer is photoconductive with a band gap of $\sim$ 1.4 eV. Upon the illumination of one side of the

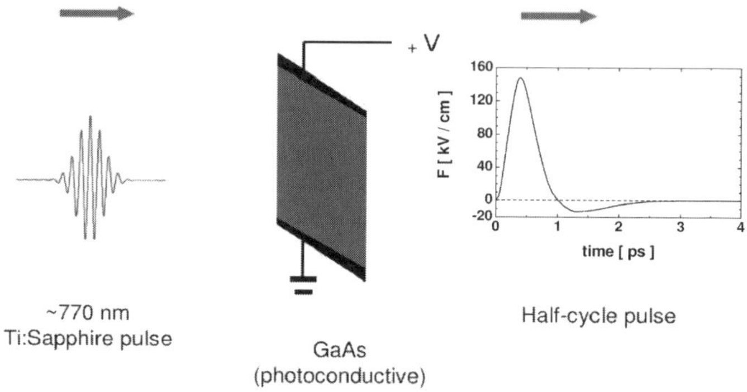

Fig. 1. The method for the generation of half-cycle pulses, as reported in [3].

wafer with the ca. 770 nm laser pulse the GaAs wafer turns conductive and the electrons are quickly accelerated and radiate a short unipolar coherent electromagnetic pulse. The polarization axis is in the direction of the bias field. The strength of the HCP depends linearly on the bias field strength. The relaxation of GaAs wafer to the insulating ground state following the excitation process occurs on a much longer time scale (hundreds of picoseconds) which leads to the extended tail of the HCP (with opposite polarity). Experimentally realized HCPs have an amplitude asymmetry ratio of 13 : 1 between the HCP peak field and peak tail. [3]. Reported HCPs possess peak fields of up to several hundreds of kV/cm and have a duration in the range between nanosecond and subpicoseconds.

2. Theoretical considerations

The purpose of this work is to expose the nature of the electron dynamics driven by HCPs. To highlight the features akin to the interaction of an HCP with charged particles let us consider a general system described by the Hamiltonian $H^{(0)}$ which is subjected at $t = t_1$ to an electromagnetic pulse. The system propagates as prescribed by the time-evolution operator $U(t, 0)$ which satisfies the equation of motion

$$i\hbar \frac{\partial U(t,0)}{\partial t} = [H^{(0)} + V(t)]U(t,0) \quad . \tag{1}$$

$V(t)$ describes the interaction of the pulse with the system.

For the time-evolution operator the following relations apply

$$U(t,0) = U_0(t,t_1)U(t,t_1,0)U_0(t_1,0), \tag{2}$$

$$U(t,t_1,0) = U_0^\dagger(t,t_1)U(t,0)U_0^\dagger(t_1,0), \quad (t > t_1). \tag{3}$$

Here $U_0(t,t_1)$ is the evolution operator of the undriven system. From Eqs. (1) and (2) we deduce that

$$U(t,t_1,0) = \hat{\mathrm{T}}\exp\left[-\frac{i}{\hbar}\int_{-t_1}^{t-t_1} e^{iH^{(0)}t'/\hbar}V(t'+t_1)e^{-iH^{(0)}t'/\hbar}dt'\right], \tag{4}$$

where $\hat{\mathrm{T}}$ is the time-ordering operator. In what follows we will be dealing with systems with a relevant characteristic time scale being much longer than the duration of the HCP. In such a case, i.e. for a short interaction time, the time ordering in Eq. (4) becomes irrelevant, and the propagator is cast as (this approximation amounts to the first order in the Magnus expansion of the exponential [5])

$$U(t,t_1,0) = \exp\left[-\frac{i}{\hbar}\int_{-t_1}^{t-t_1} e^{iH^{(0)}t'/\hbar}V(t'+t_1)e^{-iH^{(0)}t'/\hbar}dt'\right]. \tag{5}$$

Noting that

$$e^{-A}Be^{A} = B + [B,A] + \frac{1}{2!}[(B,A),A] + \dots\ , \tag{6}$$

we derive the following expansion for the propagator

$$U(t,t_1,0) = \exp\left[-\frac{i}{\hbar}V_0 + \frac{i}{\hbar^2}[V_1, iH^{(0)}] + \frac{i}{2!\hbar^3}\{[V_2,H^{(0)}],H^{(0)}\} + \dots\right], \tag{7}$$

$$V_n = \int_0^t (t'-t_1)^n V(t')dt' \quad , \quad n = 0,1,2,\dots \quad t > t_1\ . \tag{8}$$

2.1. *The impulsive approximation*

Let us assume the interaction potential $V(t)$ to have, in the configuration space, the form

$$V(t) = \mathbf{r}.\mathbf{e}Fa(t-t_1),$$

where $\mathbf{r}$ is the position coordinate of the driven charge and $\mathbf{e}$ is the external-field polarization vector. F is the peak amplitude of the field. The time envelope of the pulse we denote by $a(t-t_1)$. E.g., if the pulse has a Gaussian form we write

$$a(t-t_1) = \exp[-(t-t_1)^2/(2\sigma^2)].$$

For pulses strongly peaked at t_1 we conclude that

$$V_0 = \mathbf{r}.\mathbf{e}F\sigma\sqrt{2\pi} \ , \ \ V_1 = 0 \ , \ \ V_2 = V_0\sigma^2 \ . \tag{9}$$

The relevance of the various terms in the exponential appearing in Eq. (7) depends decisively on the parameters of the external fields. For example, if the duration of the HCP, quantified by σ is much smaller than the characteristic time of the system τ_c, estimated by $\Delta\epsilon/\hbar \sim \tau_c^{-1}$ (where $\Delta\epsilon$ is the level spacing near the ground state) then the first term of the expansion in Eq. (7) dominates and the propagator (Eq. (7)) reduces to the form

$$U(t, t_1, 0) = \exp\left[\frac{i}{\hbar}\mathbf{r}.\mathbf{p}\right] \ . \tag{10}$$

Here the vector $\mathbf{p}$ plays the role of a momentum and is evaluated as the time-integral over the pulse, i.e.

$$\mathbf{p} = -\mathbf{e}F\int_{-\infty}^{\infty} a(t')dt' \ . \tag{11}$$

Within this sudden-excitation limit (also called the impulsive approximation (IA)) the evolution operator has then the final form

$$U(t, 0) = U_0(t, t_1)e^{\frac{i}{\hbar}\mathbf{r}.\mathbf{p}}U_0(t_1, 0) \ . \tag{12}$$

According to this finding the pulse delivers an instantaneous *kick* to the system accompanied by a transfer of the momentum $\mathbf{p}$. The particle's wave functions just before ($t = t_1^-$), and right after ($t = t_1^+$) the interaction with the pulse are interrelated via the condition

$$\Psi(\mathbf{r}, t_1^+) = e^{\frac{i}{\hbar}\mathbf{r}.\mathbf{p}}\Psi(\mathbf{r}, t_1^-) \ . \tag{13}$$

This relation tells that the system evolves in a field-free manner in between the kicks. This does not mean however that the system upon the pulse is in a stationary state, as shown in full details below.

3. Electron localization and dynamical control in semiconductor double quantum wells

As an illustration let us consider the dynamics of the charge carriers confined in a double quantum well (DQW) as typically formed in $Al_xGa_{1-x}As$ structures. Within the parabolic band and the effective mass approximations (the effective mass $m^* = 0.067m_0$) the particle moves in an effective potential V_{conf} which has the form plotted in Fig. 2 (a). The particle is then subjected to a sequence of HCPs as schematically shown in Fig. 2 (b), i.e.

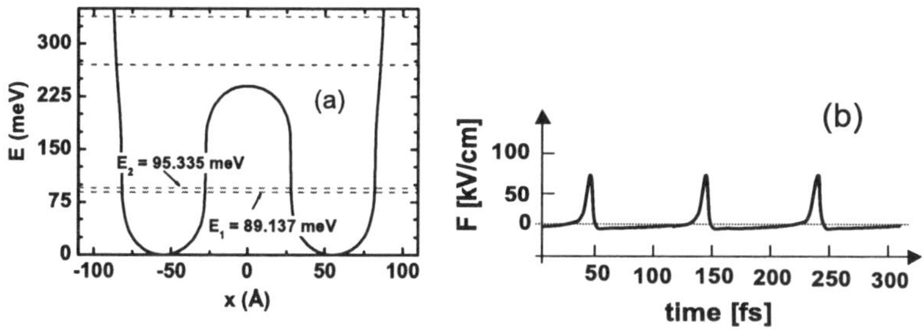

Fig. 2. (a) The confining potential V_{conf} of the charge carrier, as used in the text. The central barrier height is $\sim$ 240 meV. The dashed lines point to the first lowest energy levels. (b) The electric field amplitude vs. time for the sequence of HCPs acting on the particle in (a).

$$V(x,t) = x \sum_{k=0}^{N_p-1} F_k a(t - t_p - t_k) \ , \tag{14}$$

where

$$a(t) = \begin{cases} \exp\left[-\frac{t^2}{2\sigma^2}\right] \cos \Omega t & \text{if } -\frac{\pi}{2\Omega} \leqslant t < T - \frac{\pi}{2\Omega} \\ 0 & \text{otherwise} \end{cases} . \tag{15}$$

Here F_k stands for the peak field of the k-th pulse, $t_p = \frac{\pi}{2\Omega}$ corresponds to the time at which the positive tail of the first applied pulse is centered, T is the time between consecutive pulses, N_p is the number of applied pulses, and σ characterizes the width of the pulses. The parameter $\Omega = \frac{\pi}{3\sigma\sqrt{\ln 2}}$ in Eq. (15) guarantees a ratio 8:1 between the peak amplitudes of the positive and negative tails of the pulses. The duration d of the positive tail of each pulse is given by $d = 3\sigma\sqrt{\ln 2}$ (in the calculations presented below we use $\sigma = 20$ fs).

In the ground state the particle is delocalized, i.e. it has the same probability to be in the left or in the right well. Here we are interested in the possibility of localizing the electron in one of the wells upon the application of the HCPs. The localization can be quantified by the time-dependent probability

$$P_L(t) = \int_{-\infty}^{0} \Psi^*(x,t)\Psi(x,t)dx \tag{16}$$

which is the probability of finding the particle in the left well. We apply at the time $t = t_p$ a pulse with a strength F_{aux} that creates a time-dependent coherent state. We then pose the question of which kind of sequence of pulses is needed to localize the particle in one of the well upon the time period $t = t_p + t_{loc}$. Here t_{loc} is the localization time.

For an insight into the possible solution to this task let us consider excitations restricted to the two lowest-energy levels only, i.e. we reduce at first the system to a two-level system. Furthermore, we note that the characteristic time $\tau_c = 2\pi/\omega_c$ where ω_c is the frequency corresponding to the energy difference between the ground and the first excited states of the field-free quantum well, is on the range of $\tau_c \approx 665$ fs. On the other hand the pulses we will apply have a duration of ~ 80 fs. As explained in the preceding section we can utilize in this case the impulsive approximation for the time evolution upon excitation.

The wave function of the system is cast in terms of the wave functions $\Psi_n^{(0)}(x)$ $(n = 1, 2)$ of the two unperturbed lowest levels, i.e.

$$\Psi(x,t) = \sum_{n=1}^{2} C_n(t)\Psi_n^{(0)}(x) \ . \tag{17}$$

Here the expansion coefficients $C_n(t)$ are expressible in terms of a two-dimensional spinor $\mathbf{C}(t) = (C_1(t), C_2(t))^T$ whose dynamics is governed by the time-dependent Schrödinger equation

$$i\hbar\frac{\partial\mathbf{C}(t)}{\partial t} = \left[-(\hbar\omega_c/2)\sigma_z + \mu_{12}\sum_{k=0}^{N_{p-1}} F_k a(t - t_p - t_k)\sigma_x\right]\mathbf{C}(t) \ . \tag{18}$$

σ_x and σ_z are Pauli matrices and the transition dipole is introduced as $\mu_{12} = \langle\Psi_1^{(0)}(x)|x|\Psi_2^{(0)}(x)\rangle$.

From Eqs. (16) and (17) we deduce that

$$P_L(t) = \frac{1}{2} + \mathrm{Re}[C_1^*(t)C_2(t)] \ . \tag{19}$$

Once the pulse is applied at $t = t_p$ the spinor $\mathbf{C}(t) = (C_1(t), C_2(t))^T$ transforms in time as

$$\mathbf{C}(t) = U_0(t, t_p)U(t, t_p, 0)U_0(t_p, 0)\mathbf{C}(0) \ , \tag{20}$$

$$U_0(t, t') = \begin{pmatrix} e^{\frac{i}{2}\omega_c(t-t')} & 0 \\ 0 & e^{-\frac{i}{2}\omega_c(t-t')} \end{pmatrix} \ , \tag{21}$$

$$U(t, t_p, 0) = e^{\frac{i}{\hbar}p_0\sigma_x} = \begin{pmatrix} \cos\left(\frac{\mu_{12}p_0}{\hbar}\right) & i\sin\left(\frac{\mu_{12}p_0}{\hbar}\right) \\ i\sin\left(\frac{\mu_{12}p_0}{\hbar}\right) & \cos\left(\frac{\mu_{12}p_0}{\hbar}\right) \end{pmatrix} \ , \tag{22}$$

$$p_0 = p_{aux} = F_{aux} \int_{-\infty}^{\infty} a(t - t_p)dt \ . \tag{23}$$

If we start with the initial condition $\mathbf{C}(t) = (1,0)^T$, i.e. from the ground state, Eqs. (19) - (22) tell us that if the pulse has a peak amplitude

$$\mu_{12} p_0 / \hbar = \pi/4 \tag{24}$$

the probability to be in the left well becomes

$$P_L(t_p + \tau_c/4) = 1.$$

This means $\tau_c/4$ is the time it takes for the electron to localize in the left well (within the IA and the two-level approximation). The localization peak field is then determined by Eqs. (23) and (24) This localization phenomenon is not sustainable without the application of additional, appropriately designed fields. To achieve time sustainability one may apply for example a train of HCPs with a period T at the time $t_1 = t_p + \tau_c/4 + \gamma$, $\gamma \ll \tau_c/4$ (here we assume $F_k = F$, $k = 1, 2, ..., N_p - 1$). The system evolution at stroboscopic times $t = t_1 + (k-1)T$; $k = 1, ..., N_p - 1$ follows the relation

$$\mathbf{C}(t_1 + (k-1)T) = [U(t_1 + T, t_1)]^{k-1} \mathbf{C}(t_1) \quad ; \quad k = 1, ..., N_p - 1 \ , \tag{25}$$

$$\begin{aligned} U(t_1 + T, t_1) &= U_0(t_1 + T, t_1) U(t_1 + T, t_1, t_1) \\ &= \begin{pmatrix} e^{\frac{i}{2}\omega_c T} \cos\left(\frac{\mu_{12} p}{\hbar}\right) & i e^{\frac{i}{2}\omega_c T} \sin\left(\frac{\mu_{12} p}{\hbar}\right) \\ i e^{-\frac{i}{2}\omega_c T} \sin\left(\frac{\mu_{12} p}{\hbar}\right) & e^{-\frac{i}{2}\omega_c T} \cos\left(\frac{\mu_{12} p}{\hbar}\right) \end{pmatrix} \ , \end{aligned} \tag{26}$$

$$p = F \int_{-\infty}^{\infty} a(t - t_1)dt \ . \tag{27}$$

Now we drive the system periodically such that it undergoes a cyclic evolution, for the spinor vectors this means that

$$\mathbf{C}(t_1 + l\mathcal{T}) = e^{i\phi_l} \mathbf{C}(t_1) \quad ; \quad l = 0, 1, 2, ... \ . \tag{28}$$

The real phase ϕ_l is gained by the wave function upon the lth evolution cycle that has a duration of $\mathcal{T}$. The key point is that for periodic cyclic evolution $P_L(t)$ turns periodic with period $\mathcal{T}$

$$P_L(t_1 + l\mathcal{T}) = P_L(t_1) \quad ; \quad l = 0, 1, 2, ... \ . \tag{29}$$

So if $P_L(t_1) \approx 1$ the particle localizes also in the left well at any time $t = t_1 + l\mathcal{T}$. From Eqs. (25)-(28) we conclude that for peak amplitudes such that

$$\mu_{12} p / \hbar = (2n+1)\pi/2; \ n \in \mathcal{Z} \tag{30}$$

the system undergoes a cyclic evolution with [6]

$$\mathcal{T} = T \quad \text{if} \quad \gamma = T/2, \tag{31}$$

$$\text{or} \quad \mathcal{T} = 2T \quad \text{if} \quad \gamma \neq T/2. \tag{32}$$

The peak amplitudes $F_k = F$ suitable for the suppression of the electron tunnelling are then determined by Eqs. (27) and (30).

3.1. *Numerical illustrations*

Here we present two types of calculations: One type of calculations is based on the above analytical results for the two-level systems and the second derives from the full-numerical solution (including all levels). Figs. 3(a) and (b) show the time dependence of P_L whereas Fig. 3(c) shows the average probability $\langle P_L \rangle_{\tau_s} = (1/\tau_s) \int_0^{\tau_s} P_L(t)dt$ for $\tau_s = 2$ ps as a function of the pulse amplitude. Fig. 3(c) evidences that the localized particle can be displaced in controllable way in between the wells by tuning the pulse amplitudes.

Once the particle has been localized in one of the well we can, as we explained above, sustain this localization by applying a train of HCPs with a period $T \ll \tau_c/4$. An example of such a case is shown in Fig. 4 where a strong and a sustainable localization is achieved for the field parameters displayed on the figures.

4. Driven electron dynamics in mesoscopic rings

A further example that we present here is the dynamics of phase-coherent electrons confined in mesoscopic rings (MRs). Such systems have served for the demonstration of a variety of fundamental phenomena [7–13] such as the Aharonov-Bohm effect [14] and the appearance of persistent currents [15] when pierced by a magnetic flux. A number of studies have dealt with the electron-electron interaction [16–18], the impurity scattering [16] and disorder effects [11,17] on the persistent currents. Experimentally several measurements on persistent currents [15,19–21] have been conducted successfully. Furthermore, dynamical processes induced by a *time-dependent* electric field acting on a MR threaded by a static magnetic field have been investigated [22–25] in which case a stationary non-equilibrium current arises. This direct non-equilibrium current caused by nonlinear effects is however an odd function of the static magnetic flux and vanishes thus if the static magnetic flux is zero [23–25]. Studies on MRs subjected to external CW laser fields are reported in [26–28].

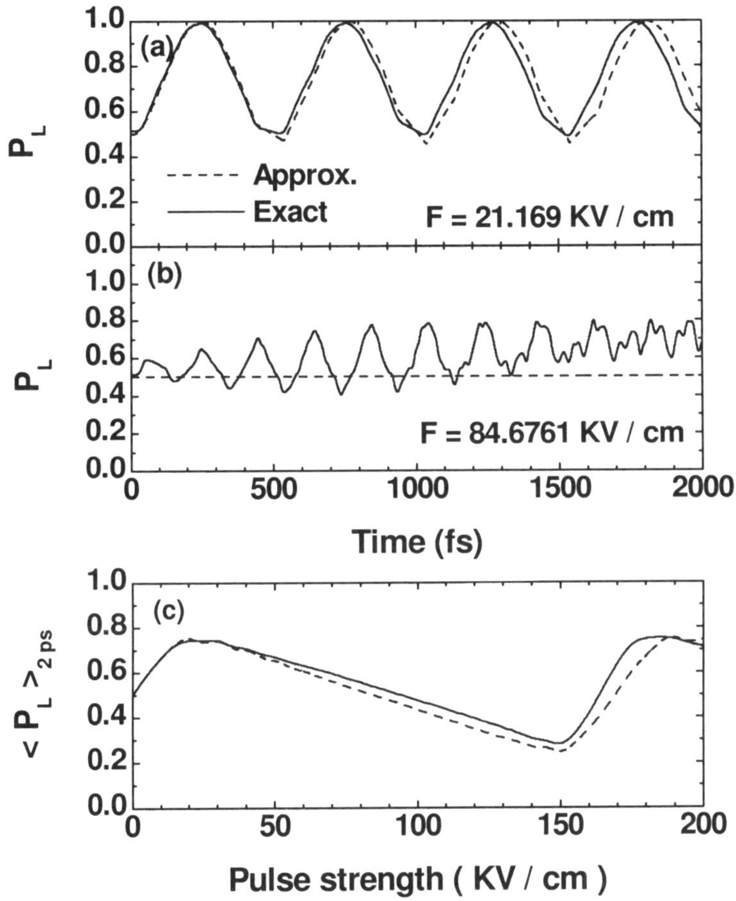

Fig. 3. (a) $P_L(t)$ for a pulse amplitude appropriate for the localization in one well. (b) Same as in (a) however the pulse amplitude is chosen such that the particle oscillates between the two wells. (c) P_L averaged over 2 ps as function of the pulse strength. Results of two-level approximation (dashed lines) and the results of full numerical calculations involving all the levels (solid lines) are shown.

Here we will be interested in the generation of nonequilibrium charge currents and charge polarization by means of HCPs [29]. Fig. 5 shows a schematics of the system to be studied here: An isolated ring with a width d and a radius $\rho_0 \gg d$ confines N charge carriers. The ring is subjected at the time $t = t_1$ to one or a train of HCPs that have an amplitude $F_1(t)$. As we will see below this pulse induces a non-equilibrium charge redistribution

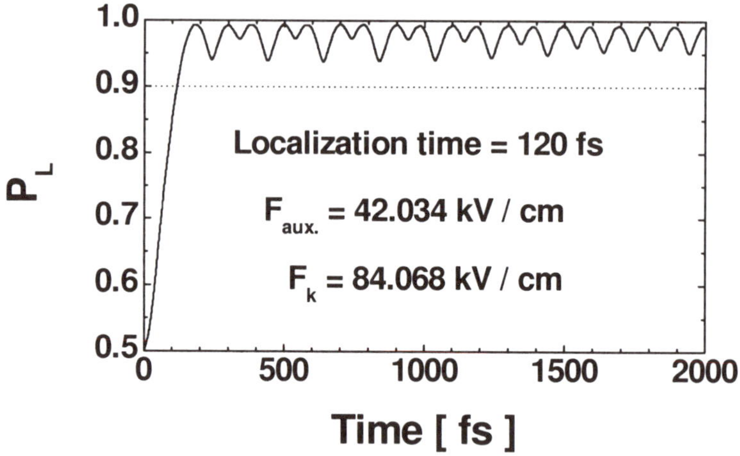

Fig. 4. P_L as a function of the time . After a pulse with $F_{aux} \approx 42$ kV/cm by a time delay $\tau = t_1 - t_p = 220$ fs we apply a train of HCPs with a period of $T = 100$ fs and peak amplitudes $F_k \approx 84.1$ kV/cm

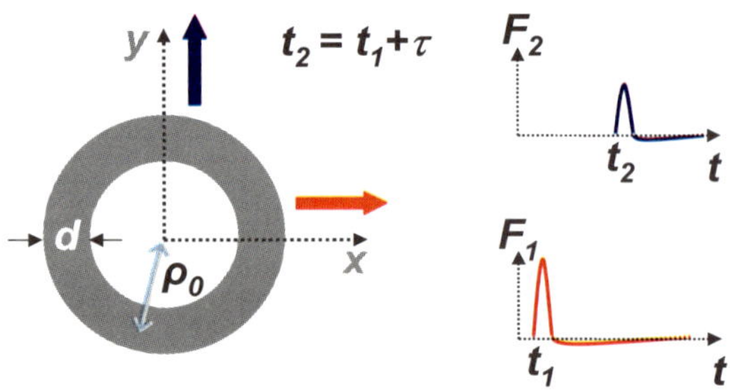

Fig. 5. A ballistic ring with a width d and a radius $\rho_0 \gg d$ subjected to a sequence of two crossed time-asymmetric pulses applied at $t = t_1$ and $t = t_2 = t_1 + \tau$. The pulses are linearly polarized along the perpendicular x and y axes, respectively.

in the ring, generating thus a charge polarization. However, it does not destroy the clockwise-anticlockwise symmetry of the charge density. We demonstrate further that charge currents can be generated if the ring is irradiated upon a time delay τ after the pulse F_1 by a second pulse $F_2(t)$. In this way clockwise-anticlockwise symmetry is broken and a time-dependent current $I(t)$ and an associated magnetization are induced.

4.1. *Charge polarization buildup*

For simplicity, we consider here a 1D ring. The influence of the radial channels can be incorporated in the model without much effort [29]. We concentrate on the case where the round trip time (few hundreds picoseconds) of the charge carriers in the MR is much longer than the pulse duration τ_d (picoseconds) in which case the IA is applicable. Thus, the wave function of the charge carrier with a charge q before and after the application of the HCP (that has the polarization axis along the x axis) are related via

$$\Psi(\theta, t=0^+) = \Psi(\theta, t=0^-) e^{i\alpha\cos\theta} \quad , \tag{33}$$

where $\alpha = q\rho_0 p/\hbar$ and θ identifies the angular position of the charge carriers with respect to the x axis.

The function $\Psi_{m_0}(\theta, t)$ describes the dynamics of the particle that has started from the initial state labelled by orbital quantum number m_0. $\Psi_{m_0}(\theta, t)$ can be expanded on the ring stationary eigenstates as

$$\Psi_{m_0}(\theta, t) = \frac{1}{\sqrt{2\pi}} \sum_{m=-\infty}^{\infty} C_m(m_0, t) e^{im\theta} e^{-i\frac{E_m t}{\hbar}} \quad . \tag{34}$$

E_m are the stationary orbital energies of the unperturbed states, i.e.

$$E_m = \frac{\hbar^2 m^2}{2m^* \rho_0^2} \quad , \quad m = 0, \pm 1, \pm 2, \cdots \ . \tag{35}$$

From Eq. (33) we conclude for the expansion coefficients the relations

$$C_m(m_0, t) = \begin{cases} \delta_{m,m_0} & \text{for } t \leqslant 0 \\ i^{m_0 - m} J_{m-m_0}(\alpha) & \text{for } t > 0 \end{cases} , \tag{36}$$

where $J_l(x)$ are the Bessel functions [30]. The energy associated with a particle initially in the m_0th state evolves as

$$\begin{aligned} E_{m_0}(t) &= \langle \Psi_{m_0}(\theta, t) | H | \Psi_{m_0}(\theta, t) \rangle, \\ &= i\hbar \left\langle \Psi_{m_0}(\theta, t) \left| \frac{\partial}{\partial t} \right| \Psi_{m_0}(\theta, t) \right\rangle \quad . \end{aligned} \tag{37}$$

From Eqs. (34) - (36, 37) we find thus for $t > 0$

$$E_{m_0}(t>0) = \frac{\hbar^2}{2m^* \rho_0^2} \sum_{m=-\infty}^{\infty} \left[m J_{m_0 - m}(\alpha) \right]^2 \quad , \tag{38}$$

$$E_{m_0}(t) = \begin{cases} \frac{\hbar^2 m_0^2}{2m^* \rho_0^2} & \text{for } t \leqslant 0 \\ \frac{\hbar^2}{2m^* \rho_0^2} \left(m_0^2 + \frac{\alpha^2}{2} \right) & \text{for } t > 0 \end{cases} . \tag{39}$$

Recalling that $\alpha = q\rho_0 p/\hbar$ we arrive at the formula

$$E_{m_0}(t>0) = E_{m_0}(t<0) + \frac{q^2}{2}\frac{p^2}{2m^*}. \tag{40}$$

This relation indicates that the HCP shifts the unperturbed energy spectrum by an amount which scales quadratically with the pulse strength independently of the ring size. $E_{m_0}(t<0)$ increases quadratically with m_0 so that the ring energy is virtually unchanged if $m_0^2 \gg \alpha^2$. To explore the possibility of creating charge polarization in the ring we study the quantity

$$\langle \cos\theta \rangle_{m_0}(t) = \int_0^{2\pi} |\Psi_{m_0}(\theta,t)|^2 \cos\theta d\theta \ , \tag{41}$$

which characterizes the charge localization in the direction of the pulse polarization and it varies in the interval $[-1,1]$. When the extremal values -1 and 1 are reached then a perfect localization at the angles $\theta=\pi$ and $\theta=0$, respectively, is achieved. The dipole moment μ_{m_0} along the x axis corresponding to a particle initially in the m_0th stationary state is proportional to $\langle \cos\theta \rangle_{m_0}(t)$, i.e.

$$\mu_{m_0}(t) = q\rho_0 \langle \cos\theta \rangle_{m_0}(t) \ . \tag{42}$$

Making use of Eqs. (34) - (36) and (41) we derive the relations

$$\langle \cos\theta \rangle_{m_0}(t) = \Theta(t)\alpha h(\Omega) \sin\left[\frac{2\pi t}{\tau_p}\right] \cos\left[\frac{4\pi m_0 t}{\tau_p}\right] \ , \tag{43}$$

$$\Omega = \alpha\sqrt{2 - 2\cos[4\pi t/\tau_p]} \quad ; \quad \tau_p = \frac{4\pi m^* \rho_0^2}{\hbar} \ , \tag{44}$$

and

$$h(\Omega) = J_0(\Omega) + J_2(\Omega) \ . \tag{45}$$

The total dipole moment induced along the x axis by application of an HCP is

$$\mu(t) = \sum_{m_0,\sigma} f(m_0,t)\mu_{m_0}(t) \ . \tag{46}$$

Here σ is the spin of the particle, f is the non-equilibrium distribution function, and $\mu_{m_0}(t)$ is given by Eq. (42).

4.1.1. *Numerical results*

For a demonstration of the HCP-induced polarization let us consider a ballistic GaAs-AlGaAs ring similar to that used in the experiment reported in Ref. [15]. We assume the ring width is $d \ll \lambda_F$ so that only the lowest radial channel is relevant. The ring radius is $\rho_0 = 1.35\ \mu$m, and the electron effective mass is $m^* = 0.067 m_e$, and $N = 1400$. Zero temperature is assumed.

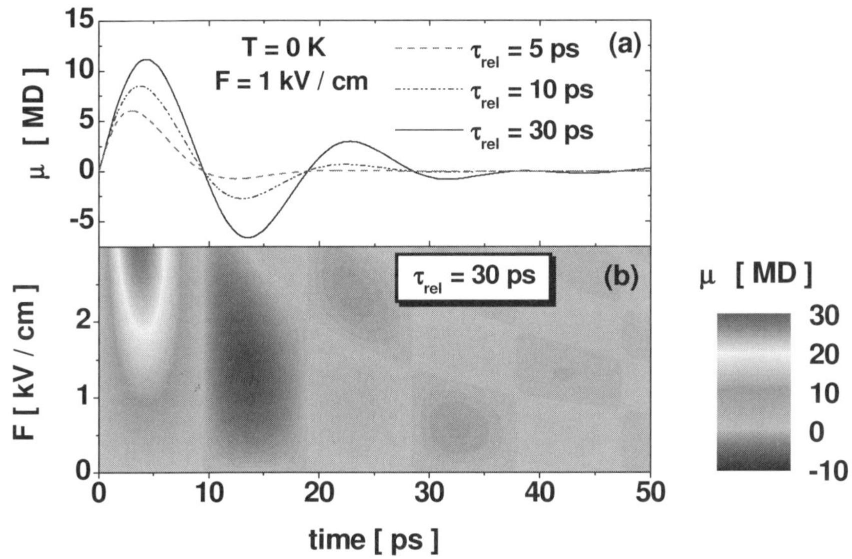

Fig. 6. Time dependence of the dipole moment μ corresponding to the case of spin $\frac{1}{2}$ particles for different values of the relaxation time τ_{rel} (a) and with varying the pulse strength F (b).

Using the relaxation time approximation [31] for the evaluation of the non-equilibrium distribution function we calculate the time dependence of the total dipole moment μ^σ for different values of the relaxation time τ_{rel} Figs. 6 (a). In Figs. 6 (b) the dipole moment is displayed as a density plot as function of the time and the pulse field amplitude $F = F_1$. The duration of the pulse is 1 ps so that the dynamics of the polarization occurs in a field-free manner. Furthermore, the largest polarization is achieved with a magnitude being in the range of several 10^7 D. As to be expected, the induced dipole moment increases with the pulse field strength, even though the time within which the polarization is formed decreases for stronger fields [see Fig. 6 (b)]. The modulation observed in Fig. 6 (b) point to the possibility of tuning the polarization by an appropriate choice of the applied pulses.

4.2. *Pulse induced charge currents*

If two linearly polarized pulses are applied to the ring in the way show in Fig. (5) a net time-dependent current $I(t)$ and an associated magnetization is created

$$I(t) = \sum_{l_0, m_0, \sigma} f(l_0, m_0, \sigma, N, T, t) I_{l_0, m_0}(t).$$

Here the radial and angular quantum numbers l_0 and m_0 as well as the spin σ label the initial state from which the current starts and I_{l_0, m_0} is the associated partial current which in its turn is obtained from the time dependent wave function, derived in a similar manner as described in the preceding section (although now we are also including the influence of the radial channels).

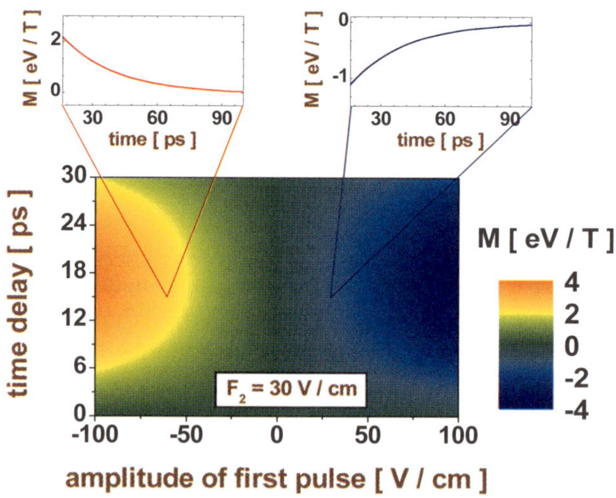

Fig. 7. Induced magnetization M as a fuhction of the time delay τ and the amplitude F_1 of the first pulse. The amplitude of the second pulse is set to the value $F_2 = 30$ V/cm. Positive and negative values of F_1 refer to pulse polarizations in the x and $-x$ directions, respectively. The upper left and right insets show for the time delay $\tau = 15$ ps the time dependence of the magnetization for respectively the field amplitudes of $F_1 = -60$ V/cm and $F_1 = 30$ V/cm. The ring radius is $\rho - 0 = 1350$ nm and its width is $d = 10$ nm. The induced current is inferred by noting that for this ring a magnetization of 1 eV/Tesla corresponds to a current of 8.9 nA.

The HCP induced current results in a time-dependent field-free magnetization $M(t)$. For rings with $d \ll \rho_0$ this dynamic magnetization is given by the classical formula $M(t) \approx \pi\rho_0^2 I(t)$. Fig. (7) shows a prototypical example of the the pulses-induced magnetization in the ring as a function of the delay time τ between the two pulses and as a function of the field amplitude strength F_1 of the first pulse (F_2 = 30 V/cm). A net current I of 8.9 nA corresponds in Fig. 7 to a ring magnetization of 1 eV/Tesla. Thus, the induced peak magnetization is an order of magnitude larger than the magnetization measured in mesoscopic rings threaded by a magnetic flux [15] in which case one measures a persistent current of $\sim$ 4 nA. Here we should stress however, that in contrast to persistent currents, the HCP generated currents are non-equilibrium states that generally decay in time, as demonstrated by the insets of Fig. 7 for two peak magnetizations. Nevertheless, as evident from these numerical illustrations the current lasts for almost 100 ps. To maintain the current beyond this time period one should utilize a train of appropriate pulses, as done above for the case of a double quantum well.

The induced magnetization direction is changed by changing the polarity of the first pulse (cf. Fig. 7). Positive or negative polarity of F_1 leads respectively to a charge polarization either along the x or $-x$ directions and this eventually results in a different magnetization direction. In addition to being able to tune the magnetization direction, the peak magnetization value can also be controlled by changing the time lag τ between the pulses.

The numerical results shown in Fig. 8 demonstrate how collective magnetic phases in an artificially structured planar array of rings [21] can be generated and controlled by means of HCPs: The magnetic moment of the individual rings in the array can be varied for changing for example the radii ρ_0 or the particle number N [see Figs. 8(a)-(d)]. In the example shown in Fig. 8 we have however, not yet included the dipole-dipole coupling between the magnetization of the rings, for we assume the distance between the rings to be large for the dipolar interaction to be effective. For closer packed rings such interaction effects are expected to be quite important. A study of this case is however still outstanding.

5. Pulse-induced electron removal from a quantum dot

As a final example we consider the electron removal from a semiconductor quantum dot upon irradiation with a half-cycle pulse. A quantum dot, also called a nanocrystal or artificial atom, is a semiconductor crystal having a size on the scale of the carriers wave length and surrounded by another

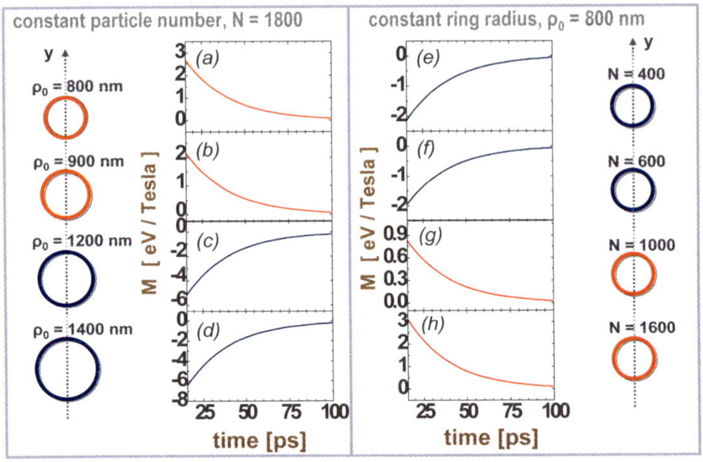

Fig. 8. The time-dependence at $T = 0$ of the magnetization induced by 1 ps pulses in an array of four non-interacting rings with a fixed particle number N and varying radii ρ_0 [(a)-(d)] or with fixed ρ_0 and varying N [(e)-(h)]. All rings are 100 nm wide, the field amplitude strengths are $F_1 = F_2 = 80$ V/cm and the delay time between the pulses is $\tau = 15$ ps.

semiconductor material with a broader band gap. Methods and procedures for the fabrication of quantum dots as well as further physical properties are found in Ref. [32]. The boundary of a quantum dot can be smooth or abrupt. Here we consider the case of an abrupt boundary, modelled by a rectangular energetic wells with finite barrier heights for electrons and holes in all three dimensions. Usually, a quantum dot like an atom has a series of discrete bound energy levels and a continuum band of infinite number of energy levels. Using appropriate semiconductor materials and fabrication procedures, the size and the barrier heights of the quantum dots can be controlled. Quantum dots have two types of carriers, electrons and holes which can recombine with each other. The carriers at the discrete energy levels are spatially confined inside the dot with some tunnelling tails penetrating to the region outside the dot. Here we investigate the probability of transferring the confined carrier which initially resides in a discrete level to the continuum part of the spectrum by the application of a half-cycle pulse.

For simplicity, we consider a spherical quantum dot containing a single discrete electron level (the condition for such a situation will be clarified below) which is occupied before the pulse application. We also assume that electrons of the semiconductor materials inside and outside the quantum dot have simple parabolic isotropic bands determined by the same effective mass

m^*. The radius of the quantum dot will be denoted as R and the energy barrier height as U_e. To render the following equations more transparent we introduce normalized radial coordinate $\tilde{r} = r/R$ and normalized energy barrier height $u = U_e/E_{\rm n}$ where the normalization energy is given by $E_{\rm n} = \frac{\hbar^2}{2m^* R^2}$. Solving the Schrödinger equation with the corresponding boundary conditions one finds the equation

$$\lambda j_1(\lambda) k_0(\kappa) - \kappa j_0(\lambda) k_1(\kappa) = 0, \tag{47}$$

determining all discrete energy levels $E = \lambda^2 E_{\rm n}$ of the system. Here $\kappa = \sqrt{u - \lambda^2}$, $j_l(x)$ denote the spherical Bessel functions of the first kind and $k_l(x)$ denote the modified spherical Bessel functions of the third kind [30] [explicitly, we have $j_0(x) = \sin(x)/x$, $j_1(x) = \sin(x)/x^2 - \cos(x)/x$, and $k_0(x) = \frac{\pi}{2}\exp(-x)/x$, $k_1(x) = \frac{\pi}{2}(x+1)\exp(-x)/x^2$]. Depending on the normalized barrier height u Eq. (47) can have no real solutions or have finite number of real solutions. Generally, the number of discrete levels increases making the values of the confining potential U_e or the dot radius R larger. Both of these factors are combined in the normalized energy barrier height u. One finds that for $u \in \big((\pi/2)^2, (3\pi/2)^2\big)$ there is exactly one discrete electron level in the dot. The wave function is given by

$$\begin{aligned} \psi(\tilde{r}) &= C_{\rm in} j_0(\lambda \tilde{r}), \quad & \tilde{r} \leqslant 1; \\ \psi(\tilde{r}) &= C_{\rm out} k_0(\kappa \tilde{r}), \quad & \tilde{r} > 1; \end{aligned} \tag{48}$$

where $C_{\rm out} = C_{\rm in} j_0(\lambda)/k_0(\kappa)$ and

$$C_{\rm in} = \left[\frac{2\pi}{\lambda^2}\left(1 - j_0(2\lambda)\right) + \frac{2\pi}{\kappa} j_0^2(\lambda)\right]^{-1/2}. \tag{49}$$

When the half-cycle pulse duration σ satisfies the following condition

$$\sigma \ll \frac{\hbar}{(u - \lambda^2)E_{\rm n}} \tag{50}$$

the impulsive approximation can be applied and we find the wave function right after the pulse application using the matching given by Eq. (13). The probability P_0 for an electron to stay at the level inside the dot after the excitation is given by the absolute value square of the projection of the wave function after the pulse application onto the unperturbed wave function of the quantum dot level,

$$P_0(\tilde{p}) = \left|\int {\rm d}^3\tilde{r}\ |\psi(\tilde{r})|^2\, {\rm e}^{i\vec{\tilde{p}}\vec{\tilde{r}}}\right|^2 = \left|\frac{4\pi}{\tilde{p}}\int_0^\infty r{\rm d}\tilde{r}\ |\psi(\tilde{r})|^2 \sin(\tilde{p}\tilde{r})\right|^2, \tag{51}$$

where $\tilde{p} = pR/\hbar$ with p being the transferred momentum. The probability of quantum dot ionization is $P_{\rm ion} = 1 - P_0$. The analytic expression for

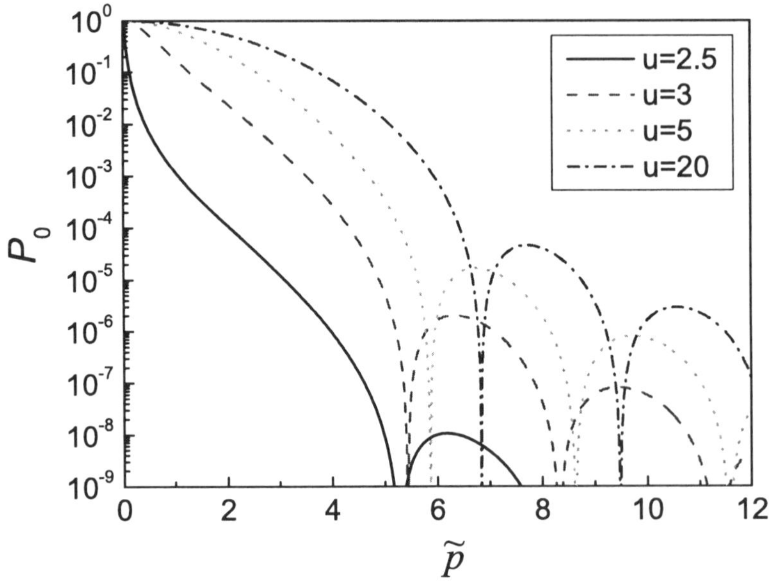

Fig. 9. Dependence of the probability P_0 to stay at the discrete level after the excitation by a half-cycle pulse on the normalized transferred momentum $\tilde{p}$.

$P_{\rm ion}(\tilde{p})$ is too cumbersome, therefore, we do not give it here but illustrate this dependence in Fig. 9 for several values of the barrier height. We see from this figure that for efficient ionization of shallow levels, weaker pulses are required than for the case of deeper levels. Adjusting the pulse strength it is possible to achieve complete ionization of the quantum dot level.

6. conclusions

We discussed in this work the use of short asymmetric electromagnetic pulses for tracing and controlling the electron dynamics in double quantum well structures and mesoscopic rings. Using analytical and numerical analysis we demonstrate how charge polarization and charge currents can be created in the rings on a picosecond time scale and how these physical phenomena can be modified in a controllable way by changing the parameters of the external driving fields. We also discuss the possibility of HCP-induced electron removal from quantum dots.

References

1. T. Brabec and F. Krausz, Rev. Mod. Phys. 72, 545 (2000)

2. J.-C. Diels, W. Rudolph, Ultrashort Laser Pulse Phenomena (Academic Press, New York, 1996).
3. R. R. Jones, D. You, and P. H. Bucksbaum, Phys. Rev. Lett. **70**, 1236 (1993); R.R. Jones, *ibid* **76**, 3927 (1996); N.E. Tielking, R.R. Jones, Phys. Rev. A **52**, 1371 (1995); J. G. Zeibel, R. R. Jones, *ibid* **68**, 023410 (2003); D. You, R. R. Jones, and P. H. Bucksbaum, Opt. Lett. **18**, 290 (1993); C. O. Reinhold, J. Burgdörfer, M. T. Frey, and F. B. Dunning, Phys. Rev. Lett. **79**, 5226 (1997); M. T. Frey, F. B. Dunning, C. O. Reinhold, S. Yoshida, and J. Burgdörfer, Phys. Rev. A **59**, 1434 (1999); S. Yoshida, C. O. Reinhold, and J. Burgdörfer, Phys. Rev. Lett. **84**, 2602 (2000); B. E. Tannian, C. L. Stokely, F. B. Dunning, C. O. Reinhold, S. Yoshida, and J. Burgdörfer, Phys. Rev. A 62 (2000) 043402.
4. N. E. Tielking, T. J. Bensky, and R. R. Jones, Phys. Rev. A **51**, 3370 (1995).
5. N. E. Henriksen, Chem. Phys. Lett. **312**, 196 (1999).
6. A. Matos-Abiague and J. Berakdar, Appl. Phys. Lett. **84**, 2346 (2004); Phys. Rev. B **69** 155304 (2004).
7. Y. Imry, *Introduction to mesoscopic physics*, 2nd. edition (University press, Oxford, 2002).
8. M. Büttiker, Y. Imry, and R. Landauer, Phys. Lett. A **96**, 365 (1983).
9. R. Landauer and M. Büttiker, Phys. Rev. Lett. **54**, 2049 (1985).
10. H. F. Cheung, Y. Gefen, E. K. Riedel, and W. H. Shih, Phys. Rev. B **37**, 6050 (1988).
11. J. F. Weisz, R. Kishore, and F. V. Kusmartsev, Phys. Rev. B **49**, 8126 (1994).
12. W. C. Tan and J. C. Inkson, Phys. Rev. B **60**, 5626 (1999).
13. D. Loss and P. Goldbart, Phys. Rev. B **43**, 13762 (1991).
14. S. A. Washburn and R. A. Webb, Add. Phys. **35**, 375 (1986).
15. D. Mailly, C. Chapelier, and A. Benoit, Phys. Rev. Lett. **70**, 2020 (1993).
16. A. Müller-Groeling and H. A. Weidenmüller, Phys. Rev. B **49**, 4752 (1994).
17. G. Bouzerar, D. Poilblanc, and G. Montambaux, Phys. Rev. B **49**, 8258 (1994).
18. T. Chakraborty and P. Pietiläinen, Phys. Rev. B **50**, 8460 (1994).
19. L. P. Lévy, G. Dolan, J. Dunsmuir, and H. Bouchiat, Phys. Rev. Lett. **64**, 2074 (1990).
20. V. Chandrasekhar, R. A. Webb, M. J. Brady, M. B. Ketchen, W. J. Gallagher, and A. Kleinsasser, Phys. Rev. Lett. **67**, 3578 (1991).
21. W. Rabaud, L. Saminadayar, D. Mailly, K. Hasselbach, A. Benoît, and B. Etienne, Phys. Rev. Lett. **86**, 3124 (2001).
22. K. B. Efetov, Phys. Rev. Lett. **66**, 2794 (1991).
23. V. E. Kravtsov and V. I. Yudson, Phys. Rev. Lett. **70**, 210 (1993).
24. O. L. Chalaev and V. E. Kravtsov, Phys. Rev. Lett. **89**, 176601 (2002).
25. P. Kopictz and A. Völker, Eur. Phys. J. B **3**, 397 (1998).
26. M. Moskalets and M. Büttiker, Phys. Rev. B **66**, 245321 (2002).
27. K. Yakubo and J. Ohe, Physica E **18**, 97 (2003).
28. G. M. Genking and G. A. Vugalter, Phys. Lett. A **189**, 415 (1994).
29. A. Matos-Abiague and J. Berakdar,Phys. Rev. Lett. **94**, 166801/1-4 (2005); Europhysics Letters **69**, 277-283 (2005); Phys. Rev. B **70**, 195338/1-10

(2004).
30. M. Abramowitz and I. Stegun (Eds.), Handbook of Mathematical functions, Dover Publications, New York, 1972.
31. J. M. Ziman, *Principles of the Theory of Solids*, Scnd. edition (University press, Cambridge, 1998); W. Jones and N. H. March, *Theoretical Solid State Physiscs*, Vol. 2 (Dover Publications, New York, 1985).
32. P. Harrison, Quantum Wells, Wires and Dots, Wiley, New York, 2001.

ONE-DIMENSIONAL NON-LINEAR OSCILLATORS AS MODELS FOR ATOMS AND MOLECULES UNDER INTENSE LASER FIELDS

AMITA WADEHRA* AND B.M.DEB†

1. Introduction

In recent years, there has been considerable interest in the area of intense laser-matter interactions. This is mainly due to the nonlinearity of the problem as well as the discovery of counter-intuitive and novel multiphoton phenomena. Both experimental realization and theoretical understanding have revealed a wealth of information about these processes such as high harmonics generation (HHG), above-threshold ionization (ATI), above-threshold dissociation (ATD), stabilization in superintense fields, alignment with external fields, dissociation through bond-softening and Coulomb explosion[1-9]. These phenomena not only provide interesting insights into the fundamental physics and chemistry of the interactions but also pave the way for further advancement of laser science and technology. For example, HHG, in which an atom or molecule exposed to intense laser field absorbs n photons of the incident laser frequency and emits photons with multiples of this frequency, has been exploited to make attosecond lasers[10,11]. Such ultrashort femtosecond and attosecond pulses are required to observe, control and manipulate reaction dynamics at the quantum level. Other possible applications point towards the generation of XUV and X-ray lasers, compact accelerators, high precision machining of dielectrics with minimum collateral damage, etc. Therefore, even after extensive theoretical and experimental research over particularly the last two decades, this area still has numerous possibilities for exploration.

---

* Department of Physics, Ohio State University, Columbus, OH 43210, U.S.A.
† S. N. Bose National Centre for Basic Sciences, JD Block, Sector III, Salt Lake, Kolkata 700098, India and Jawaharlal Nehru Centre for Advanced Scientific Research, Bangalore 560064, India.

An adequate theoretical description of the interaction of intense electromagnetic fields with atoms and molecules requires a non-perturbative approach. A complete and consistent treatment should involve the simultaneous description of the electromagnetic field coupled to the atomic and molecular systems, which are described by the time-dependent Schrödinger equation (TDSE). The advantage of relying on the numerical integration of TDSE is that solutions can be obtained for all regimes of frequency and intensity with no restrictions on the type of laser pulse, without using basis sets. However, an important issue concerning the applicability of theoretical techniques is the expensive nature of such calculations that renders the treatment of even simple atomic and molecular systems quite formidable. The necessity to save excessive computational labour has encouraged the pursuit of model systems with reduced dimensionality that not only makes the computations easier but also provides considerable freedom to conduct numerical experiments by varying the grid and laser parameters. Thus, although one-dimensional calculations might be regarded as an oversimplification of the problem, yet many interesting results have been obtained. Several simplified models have been introduced for this purpose, such as the one-dimensional approximations in a soft-core Coulomb potential[12,13], and even simpler models such as two-level systems in a strong field[14,15]. However, one-dimensional nonlinear oscillators have been the most popular for such studies[16-28].The most commonly studied is the Morse oscillator, the time evolution of which has been studied by a number of methods[17-24]. Next comes the simplest centrosymmetric system, the quartic oscillator that has been exploited to understand certain aspects of HHG[25]. Another interesting class of nonlinear oscillators with extensive applications for modeling various physicochemical and biological systems is a double-well oscillator (DWO). The interaction of a low-depth symmetric DWO with lasers of very low intensities has been studied[27] as well as driven quantum tunneling in quartic DWOs[28].

The main purpose of this article is to review recent efforts in our laboratory towards studying the intense laser-matter nonlinear interactions by numerically solving the TDSE for two interesting classes of one-dimensional nonlinear oscillators. These are symmetric as well as asymmetric DWOs and Morse oscillators. However, the launching of the numerical method for solving the TDSE requires the solution at zero time as an input which, for the sake of internal consistency, should be calculated by the same method. This is a nontrivial exercise in which the same method is to be employed for calculating probability densities and expectation values at zero time and nonzero times. Efforts in this direction have led to a numerical TD method for obtaining highly accurate ground and excited states of atoms[29], molecules[30] as well as one dimensional and two-dimensional nonlinear oscillators of various types[16, 31-36].

In Section 2 we briefly describe the method for calculating the ground and excited states of one-dimensional nonlinear oscillators. Section 3 discusses the responses of an electron moving in DWOs under intense laser fields. Section

4 discusses the multiphoton vibrational interaction of NO molecule, modeled as a Morse oscillator, with intense far-infrared lasers. Section 5 makes a few concluding remarks.

2. Time-dependent calculation of ground and excited states of one-dimensional nonlinear oscillators

Consider the Hamiltonian (atomic units employed throughout)

$$H = (-1/2)\ d^2/dx^2 + V(x) \tag{1}$$

The following cases have been considered.

(a) *Anharmonic oscillators*[31] :

$$V(x) = (1/2)\,\omega^2 x^2 + \sum_{i=2}^{s} a_i x^i\,;\quad a_i \text{ real},\ s \le 16 \tag{2}$$

(b) *Double-well oscillators*[31] :

$$V(x) = -(1/2)\,\omega^2 x^2 + \sum_{i=3}^{s} a_i x^i\,;\quad a_i \text{ real},\ s \le 14 \tag{3}$$

(c) *Multiple-well oscillators*[32] :

$$V(x) = \pm (1/2)\,\omega^2 x^2 + \sum_{i=1}^{s} a_{2i} x^{2i}\,;\quad a_i \text{ real},\ s \le 5 \tag{4}$$

(d) *Self-interacting oscillators*[33] :

$$V(x) = (1/2)\,\omega^2 x^2 + \lambda < x^{2r} > x^{2s}\,;\quad 1 \le r, s \le 6 \tag{5}$$

where λ is a real parameter denoting self-interaction, i.e., a situation where a system and its environment influence each other (*feedback*). Such oscillators of various types serve as important models for understanding numerous chemical, physical and biological phenomena[31-33].

The TDSE is

$$H\,\Psi(x,t) = i\ \partial\Psi(x,t)/\partial t \tag{6}$$

where H is given by Eq. (1). Assuming the validity of Eq. (1) in *imaginary time* τ, one first writes it in τ and then replaces τ by -it, where t is real time. As a result, the TDSE is transformed into a diffusion equation,

$$HR(x,t) = -\,\partial R(x,t)/\partial t \tag{7}$$

where the diffusion function R(x,t) replaces Ψ(x,t) in Eq. (6). Within a diffusion quantum Monte Carlo approach, the evolution of Eq.(7) up to a sufficiently long time eventually yields a stationary ground state corresponding to the global minimum of the expectation value $< R(x,t) \mid H \mid R(x,t) >$, $R^2(x,t)$ being the probability density (see ref.[37] for a detailed discussion on this approach). For an excited eigenstate, the energy eigenvalue is obtained by requiring the state to be orthogonal to all lower states. For the different types of V(x), given by Eqs. (2) – (5), Eq. (7) was numerically solved by a split-operator, finite-difference method which employs a modified Thomas algorithm [31]. *The method is exact in principle.* Fig. (1) depicts the potential energy and probability density for a symmetric DWO with $V(x) = -29x^2 + 0.08x^4 + 0.01x^6 + 0.008x^8 + 0.0008x^{10} + 0.0005x^{12}$. For this system, the ground and the first excited state are "pseudodegenerate"[31] up to 12 significant figures while the second and third excited states are "pseudodegenerate" up to 10 significant figures. A similar "pseudodegeneracy" was found for the multiple-well oscillator $V(x) = 13.5x^2 - 1.5x^4 + 0.035x^6$ (see Fig. 2). Such a situation was found to be quite prevalent for symmetric potentials with deep minima.

Fig. 3 depicts energy eigenvalues for the self-interacting potential $V(x) = 0.5x^2 + <x^{2r}> x^{2s}$. It is worthwhile to note that, for $r \geq 3$, the eigenvalues

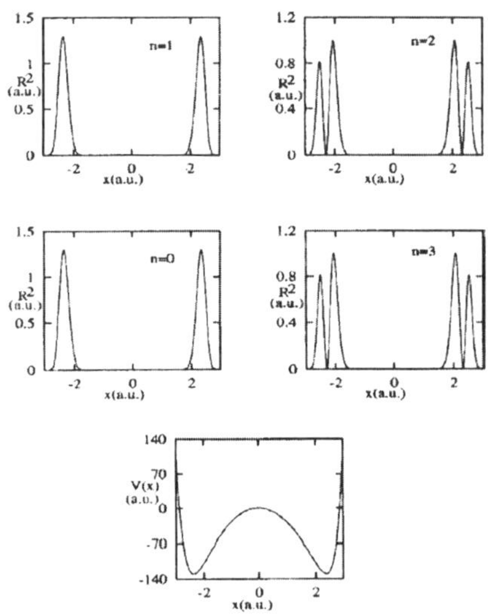

Fig. 1 Potential energy and probability density plots for the symmetric DWO given by $V(x) = -29x^2 + 0.08x^4 + 0.01x^6 + 0.008x^8 + 0.0008x^{10} + 0.0005x^{12}$. "n" is the vibrational quantum number (Reproduced from [31] by permission from the American Physical Society).

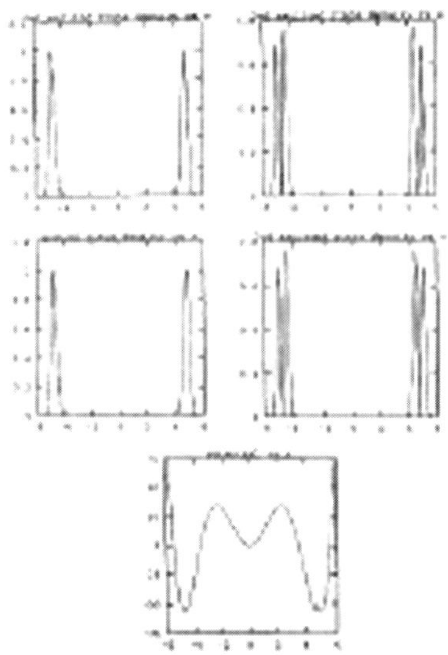

Fig. 2 Potential energy (bottommost curve and probability density plots against x, in a.u., for the symmetric three-well oscillator given by $V(x)=13.5x^2-1.5x^4 + 0.035x^6$ (Reproduced from [32] by permission from the Indian Academy of Sciences).

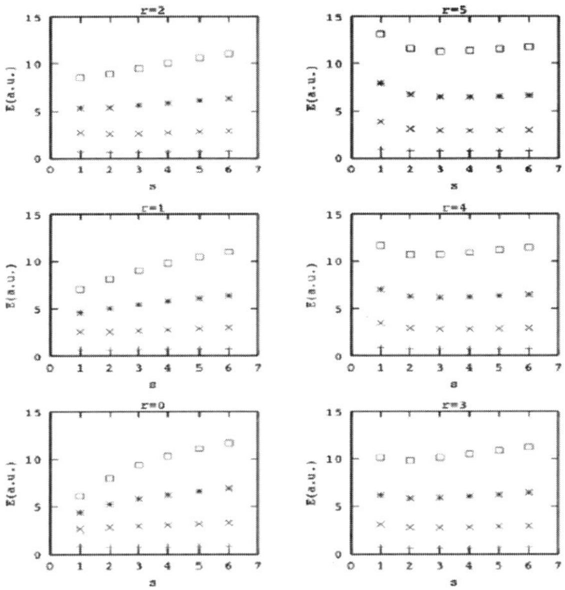

Fig. 3 Energy eigenvalues (a.u.) of ground and first three excited states of self-interacting oscillators gover- - ned by $V(x) = 0.5\,x^2 + \langle x^{2r}\rangle\, x^{2s}$, keeping r constant and varying s for each plot ; r = 0 corresponds to non-self-interacting oscillators (Reproduced from [33] by permission from John Wiley and Sons).

pass through a minimum with respect to s, the minimum deepening with higher excited states. This points to a threshold for real systems and phenomena where feedback might play a determining role.

The same numerical methodology, mentioned above, was employed for evolving the TDSE in real time, as discussed below.

3. Responses of an electron in symmetric and asymmetric DWOs under intense laser fields

The potentials employed for symmetric (V_s) and asymmetric (V_a) DWO were[38]

$$V_s(x) = -7.0x^2 + x^4 + 0.5x^6 \tag{8}$$

$$V_a(x) = -2.5x^2 + x^3 + x^4 \tag{9}$$

The responses of an electron moving in these potentials were analyzed under intense and superintense laser fields by numerically solving the TDSE and evolving the systems for 96 fs in 1064 nm laser at intensities $10^{15} - 10^{22}$ W cm^{-2}. The initial state of the system is the ground state, obtained by solving TDSE in imaginary time, as discussed in the previous Section. The principal objective of this work was to determine the suitability of DWOs to study such nonlinear interactions and compare their behaviour with that of atoms and molecules under similar conditions. Our detailed investigation of HHG and ATI processes gave clear evidence of strong qualitative parallelism between the responses of

DWOs and real atomic/molecular systems. In order to justify the origin of the HHG and ATI spectra in our infinite-barrier potentials which do not favour ionization, a mechanism was proposed in terms of a "pseudocontinuum". According to this, a classical limit is presumably reached for very high vibrational quantum numbers where energy levels crowd together approximating a "continuum". The HHG spectra (Fig.4a) showed characteristic features of a typical atomic spectrum, viz., an initial rapid decrease in harmonic intensity, a plateau region of harmonics with comparable intensities and then a sharp cut-off. The spectra showed odd, even and shifted harmonics due to the potential of asymmetric DWO and due to the almost degenerate states of symmetric DWO. For symmetric oscillator, the shifted even harmonics coalesce to generate pure even harmonics at I = 5 x 10^{18} Wcm^{-2} (Fig.4b). This results from violation of parity conservation and has been earlier related to the accidental degeneracy of Floquet quasienergy states for a symmetric DWO in less intense laser fields[27]. The plateau region showed an increase in the number of harmonics at higher intensities subsequently leading to the appearance of double and triple plateaus (Fig.4b), as observed for molecular ions. A very interesting trend towards stabilization was observed at intensities above 10^{20} Wcm^{-2} (Fig.4c) similar to that calculated for atoms in the superintense regime. In this context, we had predicted and provided an explanation for "staircase-type" HHG spectrum (Fig.4d) that is still waiting to be discovered experimentally! The ATI-like spectra exhibit similar features to HHG, viz. peaks are separated by the laser frequency (photon energy in a.u.), increase with increasing laser intensity and show multiplet structure at higher intensities

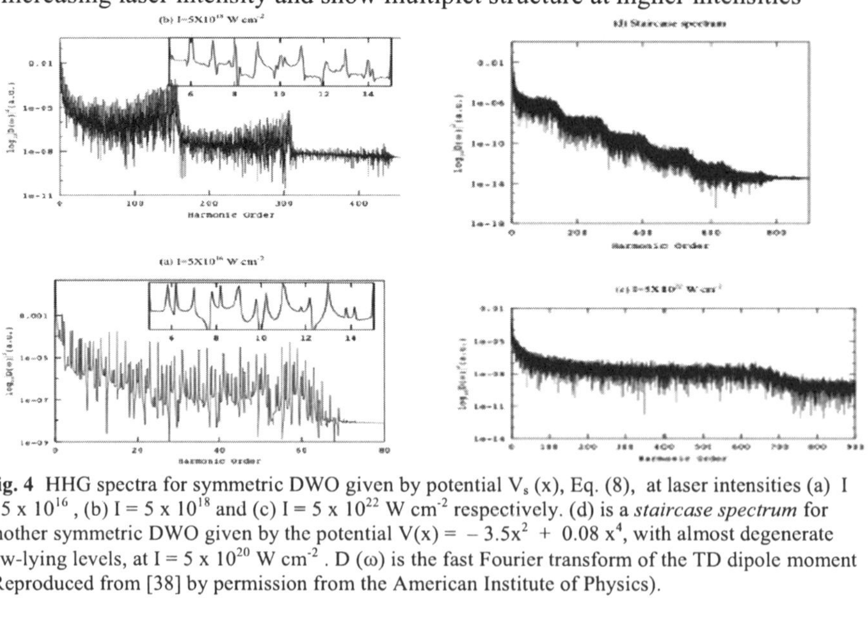

Fig. 4 HHG spectra for symmetric DWO given by potential V_s (x), Eq. (8), at laser intensities (a) I = 5 x 10^{16} , (b) I = 5 x 10^{18} and (c) I = 5 x 10^{22} W cm^{-2} respectively. (d) is a *staircase spectrum* for another symmetric DWO given by the potential V(x) = $-$ 3.5x^2 + 0.08 x^4, with almost degenerate low-lying levels, at I = 5 x 10^{20} W cm^{-2} . D (ω) is the fast Fourier transform of the TD dipole moment (Reproduced from [38] by permission from the American Institute of Physics).

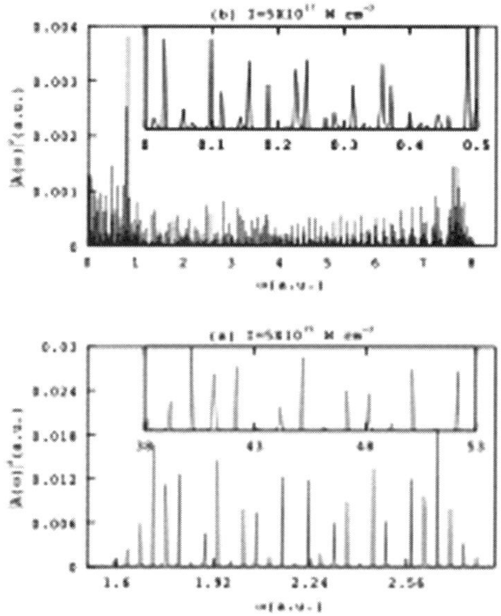

Fig. 5 ATI spectra for asymmetric DWO given by potential $V_a(x)$, Eq. (9), at laser intensities (a) I = 5 x 10^{15} and (b) I = 5 x 10^{17} W cm^{-2} respectively. $A(\omega)$ is the fast Fourier transform of the TD autocorrelation function (Reproduced from [38] by permission from the American Institute of Physics).

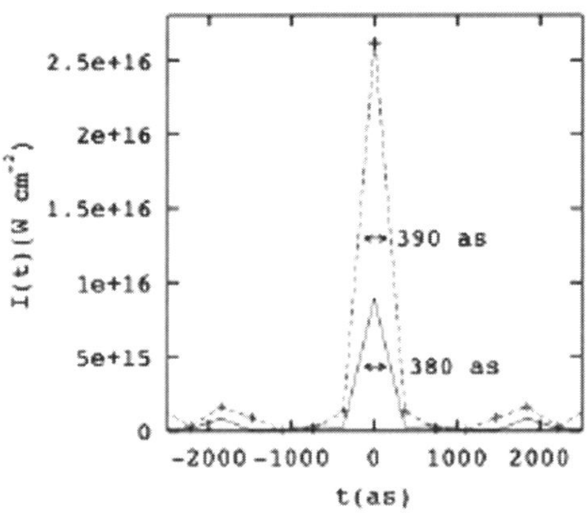

Fig. 6 Attosecond pulse intensity I(t) in W cm^{-2} plotted against time(as) for both symmetric DWO (dotted line) given by Eq. (8) and asymmetric DWO(solid line) given by Eq. (9) (Reproduced from [38] by permission from the American Institute of Physics).

(Fig.5). We also examined the applicability of the high harmonics for generation of attosecond pulses (Fig. 6). An inverse Fourier synthesis performed over a window of spectral components near the middle or end of the plateau region of the harmonics for both asymmetric and symmetric DWOs resulted in the generation of 380 and 390 *as* pulses respectively. The above results demonstrated that not only suitably designed one-dimensional DWOs are prospective model systems representing real atoms and molecules but also that any quantum system possessing a continuum or a pseudocontinuum is a suitable candidate for such processes (see ref.[38] for details).

4. Multiphoton vibrational dynamics of Morse oscillators in intense laser fields

The above results encouraged us to employ another interesting nonlinear oscillator, the Morse oscillator, which is a popular model system for diatomic molecules. Various studies have been reported on the excitation and dissociation dynamics of molecules (modeled by Morse potential), e.g. NO, HF and H_2 under infrared and far-infrared lasers[17-24]. However, the multiphoton interactions between vibrational motions and intense femtosecond far-infrared lasers seemed not to have been paid any attention. Our study[39] focused on studying such interactions of the NO molecule which is abundant in living systems and is most commonly studied for interaction with living bodies, laser fields and also for adsorption on solids. The main objective was to generate the HHG and ATD (above-threshold dissociation) spectra by employing a far-infrared laser with different intensities and disregarding any explicit interaction between electrons and the laser field. The laser frequency was chosen such that two photons were necessary to excite the NO molecule from the ground to the first excited vibrational state and 68 photons were necessary to just lift the molecule from the ground state to the lowest edge of the vibrational continuum.

In presence of the laser field, the TD Hamiltonian may be written as

$$H = H_0 + d(x)\, f(t)\, E_0 \sin(\omega_L t) \tag{10}$$

where the unperturbed Hamiltonian H_0 is given by

$$H_0 = -(1/2\mu)\, d^2/dx^2 + D_e[1 - \exp(-\alpha(x - x_{eq}))]^2 \tag{11}$$

involving the Morse potential and μas the reduced mass of the molecule. In Eq. (10), E_0 is the peak field strength ($E_0 = (8\pi I/c)^{1/2}$; I is the laser intensity), ω_L is the laser frequency, f(t) is a linear ramp function and d(x) is the dipole moment operator given by

$$d(x) = x \exp[-x / 0.2515\, x_{eq}] \tag{12}$$

In Eq. (11), x is the internuclear distance, x_{eq} is its equilibrium value, D_e is the dissociation energy and α is a positive constant.

With the Hamiltonian in Eq. (10), the TDSE was solved numerically using a far-infrared laser of wavelength 10503 nm with varying intensities. The multiphoton vibrational dynamics of the laser-molecule system were probed through various TD quantities such as the probability density, dissociation probability, potential energy curve and dipole moment. It was observed from the computations that multiphoton interactions of vibrational motions with the laser field elevated the NO molecule to its vibrational continuum resulting in very rich HHG and ATD spectra (see ref.[39] for details).

Fig. 7 depicts the TD dissociation probability above a threshold laser intensity. After an initial induction period of about 9 fs, the dissociation probability increased rapidly but then gradually tapered off with time. Fig. 8 shows the variation of the TD potential with internuclear distance at two different laser intensities. Depending on the laser intensity, the potential can become repulsive at certain times whereas at other times it can become more attractive, with a deeper minimum, than the potential at zero time. This indicates that the usual selection rules for spectral transitions, which do not explicitly consider the time-dependence of the potential, may not be applicable in situations where the TD potential can change its shape drastically.

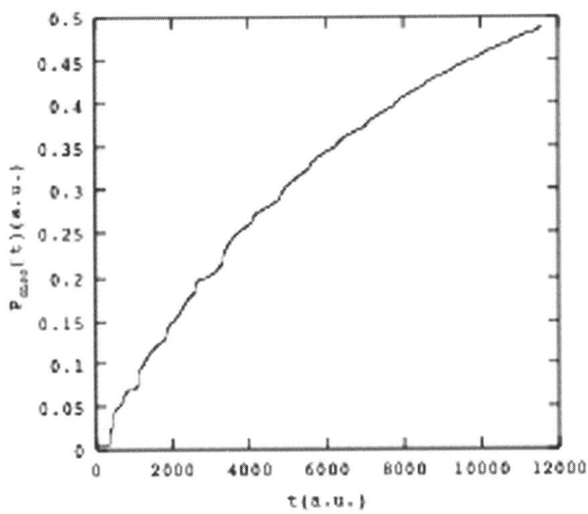

Fig. 7 Dissociation probability $P_{diss}(t)$ for the NO molecule plotted against time t, in a.u., for I=1x 10^8 W cm^{-2} (Reproduced from [39] by permission from the Indian Academy of Sciences).

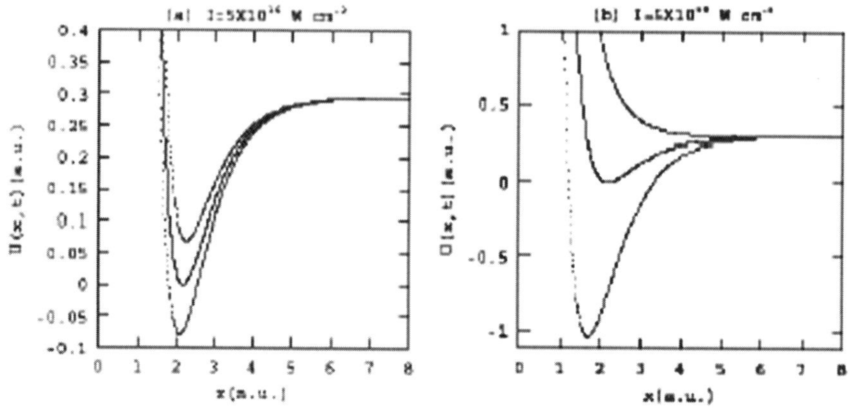

Fig. 8 TD potential U(x,t) (Morse + laser), given in Eqs. (10) and (11), of the NO molecule plotted against internuclear distance x, in a.u., for (a) I = 5×10^{16} and (b) 5×10^{18} W cm^{-2} respectively. The solid line refers to the unperturbed Morse potential, the dotted line above the Morse potential refers to the crest (electric field = E_0) and the dotted line below the Morse potential refers to the trough (electric field = $-E_0$) of the eighth optical cycle (Reproduced from [39] by permission from the Indian Academy of Sciences).

The periodicity of the TD dipole moment was *not* the same as that of the laser field in this case. This had an effect on the HHG and ATD spectra. Fig. 9 depicts the HHG spectra for two different laser intensities showing both odd, even and shifted harmonics. At a relatively lower intensity, the HHG spectrum clearly showed a "staircase spectrum" (see Section 2) with six different bumps. The first bump is associated with the ground state while the sixth bump is associated with the fifth excited state. Overall, the HHG spectra indicated a "simultaneous" occurrence of HHG and dissociation. Fig. 10 shows the ATD spectra for four laser intensities. Up to I = 1×10^{13} Wcm^{-2} (Fig. 10 a,b), the ATD spectra showed only one peak At I = 5×10^{16} Wcm^{-2} (Fig. 10c), there were several prominent peaks separated by the photon energy whereas at I = 5×10^{18} Wcm^{-2} (Fig. 10d) the spectrum was quite rich (see ref.[39] for details). It may, however, be noted that in situations where electronic motions are of considerable significance, ATI in molecules is the precursor of ATD (see, e.g., ref.[40]).

5. Conclusion

The examples discussed briefly in this article clearly establish that modeling multiphoton interactions of atoms and molecules with intense laser fields through one-dimensional nonlinear oscillators yields rich information and insight. These oscillators reproduce quite a few of the characteristic features of such interactions of real atoms and molecules. Therefore, in order to further

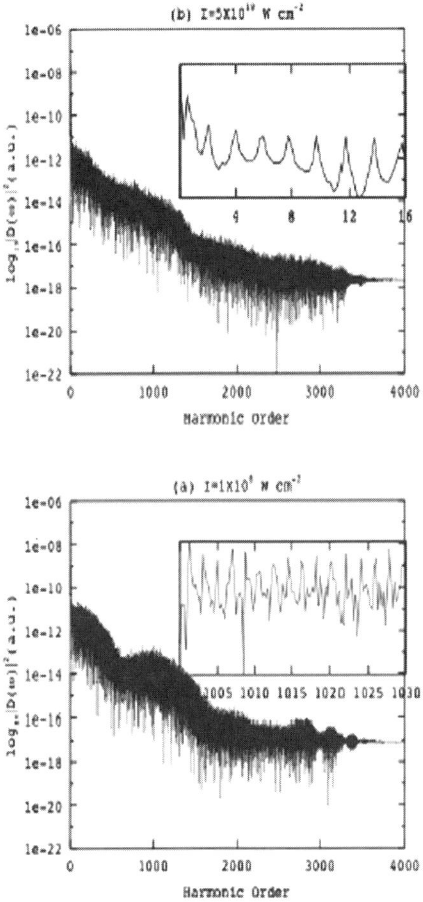

Fig. 9 Vibrational HHG spectra(a.u.) of the NO molecule at (a) I = 1×10^8 and (b) I = 5×10^{18} W cm^{-2}. D(ω) has the same meaning as described in Fig. 4. The inset of (a) shows odd, even and shifted harmonics while the inset of (b) shows the first few prominent even harmonics (Reproduced from [39] by permission from the Indian Academy of Sciences).

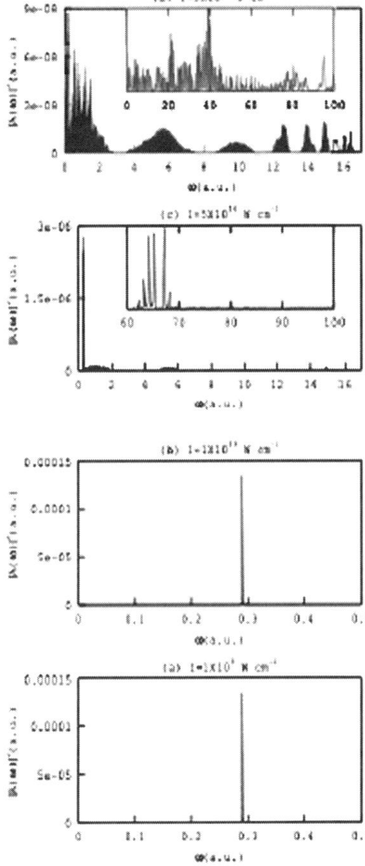

Fig. 10 Pure vibrational ATD spectra(a.u.) of the NO molecule at (a) I=1×10^8, (b) 1×10^{13} (c) 5×10^{16} and (d) 5×10^{18} W cm^{-2} respectively. A(ω) has the same meaning as described in Fig. 5. The insets of (c) and (d) show a resolution into peaks (Reproduced from [39] by permission from the Indian Academy of Sciences).

consolidate such models, one might even go beyond one dimension and examine nonlinearly coupled two- and three-dimensional nonlinear oscillators - with and without feedback – interacting with intense laser fields. An important question then comes up, viz., can such two- and three-dimensional oscillators exhibit quantum chaos under intense laser fields ? Since the answer is *yes*[35,36], it would be necessary to examine quantum chaos of atoms and molecules under intense laser fields.

References

1. *Atoms in Intense Laser Fields*, edited by M.Gavrila (Academic Press, New York, 1992).
2. M.H.Mittleman, *Introduction to the Theory of Laser-Atom Interactions* (Plenum Press, New York, 1993).
3. F.H.M. Faisal, *Theory of Multiphoton Processes* (Plenum Press, New York, 1987).
4. *Molecules in Laser Fields*, edited by A.D. Bandrauk (Dekker, New York, 1994).
5. K. Burnett, V.C. Reed, and P.L. Knight J.Phys.B **26**, 561 (1993).
6. M. Protopapas, C.H.Keitel and P.L.Knight, Rep. Prog. Phys. **60**, 389 (1997).
7. C.J. Joachain, M. Dorr and N.J. Kylstra, Adv. At. Mol. Opt. Phys. **42**, 225 (2000).
8. J.H.Eberly, R.Grobe, C.K.Law and Q.Su, Adv.At.Mol.Opt.Phys. Suppl.**1** 301 (1992).
9. J.H. Eberly and K.C.Kulander, Science **262**, 1229 (1993).
10. F.L.Klein, K.Midorikawa and A.Suda, Phys.Rev.A **58**, 3311 (1998).
11. N.A.Papadogiannis, B.Witzel, C.Kalpouzos and D.Charalambidis, Phys.Rev.Lett. **83**, 4289 (1999).
12. J.H. Eberly, Q.Su and J.Javanainen, Phys.Rev.Lett. **62**, 881 (1989).
13 V.C. Reed and K. Burnett, Phys. Rev.A **46**, 424 (1992).
14 B.Sundaram and P.W. Milonni, Phys. Rev.A **41**, 6571 (1990).
15. E.Kaplan and P.L. Shlonikov, Phys.Rev.A **49**, 1275 (1994).
16. N.Gupta, A.Wadehra, A.K.Roy and B.M.Deb, in *Recent Advances in Atomic and Molecular Physics*, edited by R.Srivastava (Phoenix Press, New Delhi, 2001).
17. C. Cerjan and R. Kosloff, J. Phys. B 20. 4441 (1987).
18. R.B. Walker and R.K. Preston, J.Chem.Phys. 67, 2017 (1977).
19. T.F. Jiang, Phys. Rev.A 48, 3995 (1993).
20. M.E. Goggin and P.W. Milonni, Phys.Rev.A 37, 796 (1988).
21. V.Averbukh and N.Moiseyev, Phys.Rev.A 57, 1345 (1998).
22. R.Heather and H. Metiu, J.Chem. Phys. 86, 5009 (1987).
23. R Heather and H. Metiu, J.Chem. Phys. 88, 5496 (1988).
24. J J-L Ting, J.Phys.B 27, 1249 (1994).

25. J J-L Ting, Phys. Rev.A **51**, 2641 (1995).
26. Ph.Balcou, A.L' Huillier and D.Escande, Phys. Rev.A **53**, 3456 (1996).
27. R. Bavli and H. Metiu, Phys.Rev.A **47**, 3299 (1993).
28. M. Grifoni and P.Hänggi, Phys.Rep. **304**, 229 (1998).
29. A.K.Roy, B.K.Dey and B.M.Deb, Chem. Phys. Lett. **308**, 523 (1999).
30. B.K.Dey and B.M.Deb, J.Chem.Phys. **110**, 6229 (1999).
31. A.K.Roy, N.Gupta and B.M.Deb, Phys. Rev.A, **65**, 012109 (2002).
32. N.Gupta, A.K.Roy and B.M.Deb, Pramana- J.Phys. **59**, 575 (2002).
33. A. Wadehra, A.K.Roy and B.M. Deb, Int.J. Quantum Chem. **91**, 597 (2003).
34. A. K. Roy, A. J. Thakkar and B. M. Deb, J.Phys. A: Math. Gen. 38, 2189 (2005).
35. N. Gupta and B. M. Deb, Chem. Phys., in press (2006).
36. N. Gupta and B. M. Deb, Pramana-J. Phys., to appear (2006).
37. B.L. Hammond, W.A. Lester Jr., and P.J.Reynolds, Monte Carlo Methods in Ab Initio Quantum Chemistry (World Scientific Press, Singapore, 1994).
38. A. Wadehra, Vikas and B.M.Deb, J.Chem.Phys. **119**, 6620 (2003).
39. A.Wadehra and B.M.Deb, Proc. Indian Acad. Sci. (Chem.Sc., C.N.R. Rao Festschrift) **115**, 349 (2003); erratum, J. Chem. Sci. **116**, 129 (2004).
40. A. Wadehra and B. M. Deb, Eur. Phys. J. D 39, 141 (2006).

Experimenting with Topological States of Bose-Einstein Condensates

Chandra Raman*

School of Physics, Georgia Institute of Technology,
Atlanta, Georgia 30332-0430 USA
**E-mail: craman@gatech.edu*
http://www.physics.gatech.edu/people/faculty/craman.html

Bose-Einstein condensed atomic gases (BECs) are a new class of quantum fluids. They are produced by cooling a dilute atomic gas to nanokelvin temperatures using laser and evaporative cooling techniques. In this paper we review the basic principles behind the experimental realization of Bose-Einstein condensation. We discuss experiments performed in our laboratory at Georgia Tech on quantized vortices formed in the condensate and on a new type of metastable BEC created in a quadrupole magnetic trap.

Keywords: Bose-Einstein condensate, laser cooling, atom laser

1. Introduction

When a gas of bosonic atoms is cooled below a critical temperature T_c, a large fraction of the atoms condenses into the lowest quantum state. This phenomenon was first predicted by Albert Einstein in 1925[1] and is a consequence of quantum statistics, as outlined schematically in Figure 1. At ordinary temperatures a gas of atoms behaves as a collection of point-like particles (Figure 1a), a state familiar to most of us. If it can be prevented from freezing, at extremely low temperatures within 1/1000th of a degree above absolute zero, the properties of the gas change dramatically. The atoms can be regarded as quantum-mechanical wavepackets which have an extent on the order of a thermal de Broglie wavelength $\lambda_{dB} = (2\pi\hbar^2/mk_BT)^{1/2}$, where T is the temperature and m the atomic mass (Figure 1b). It is a remarkable fact that experimentally, it is feasible to cool and maintain gases of atoms at such low temperatures. When T is reduced to the point that λ_{dB} is comparable to the interatomic separation (Figure 1c), the atomic wavepackets "overlap" and the indistinguishability of particles becomes important. At this temperature, bosons undergo a quantum-mechanical phase transition and form a Bose-Einstein condensate, a coherent cloud of atoms all occupying the same quantum mechanical state. The transition temperature and the peak atomic density n are related as $n\lambda_{dB}^3 \simeq 2.612$. The quest for Bose-Einstein condensation has a long history and is nicely summarized in various contributions to the 1998 Varenna summer school.[2]

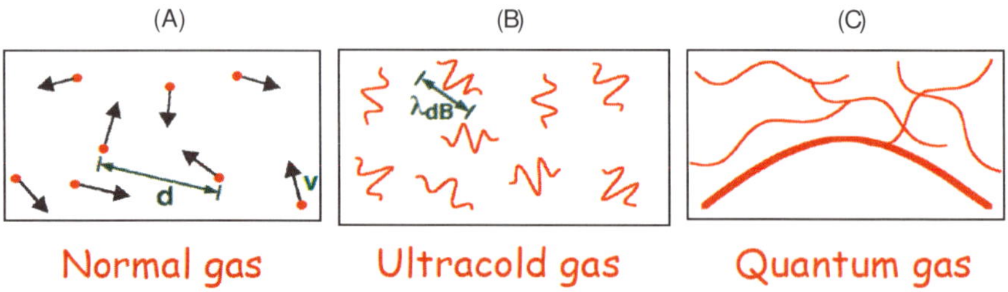

Fig. 1. (A) Classical gas at ordinary temperatures. (B) Same gas at temperatures around 1 milliKelvin where quantum effects become important. (C) Quantum degenerate gas of bosons appearing in the microKelvin to nanoKelvin temperature range.

The realization of Bose-Einstein condensation (BEC) in dilute atomic gases[3–6] achieved several long-standing goals. First, neutral atoms were cooled into the lowest energy state, thus exerting ultimate control over the motion and position of atoms, limited only by Heisenberg's uncertainty relation. Second, a coherent macroscopic sample of atoms all occupying the same quantum state was generated, leading to the realization of atom lasers, devices which generate coherent matter waves. Third, degenerate quantum gases were produced with properties quite different from the quantum liquids ^{3}He and ^{4}He. This provides a testing ground for many-body theories of the dilute Bose gas which were developed many decades ago but never tested experimentally.[7] BEC of dilute atomic gases is a macroscopic quantum phenomenon with similarities to superfluidity, superconductivity and the laser.[8] More generally, atomic Bose-Einstein condensates are a new "nanokelvin" laboratory where interactions and collisions at ultralow energy can be studied.

2. Basic techniques

Achieving Bose-Einstein condensation requires techniques to cool gases to sub-microkelvin temperatures and atom traps to confine them at high densities and to keep them away from the hot walls of the vacuum chamber. Over the last $\sim$ 20 years, such techniques were developed in the atomic physics and low-temperature communities.[2] The Georgia Tech experiment, based on the pioneering work of the MIT group,[9] uses a multistage cooling process to cool hot sodium vapor down to temperatures where the atoms form a condensate. A beam of sodium atoms is created from an oven where the density is about 10^{14} atoms per cm^{-3}, similar to the eventual density of the condensate. The gas is cooled by nine orders of magnitude from 600 K to 1 μK first by slowing the atomic beam, followed by optical trapping and laser cooling, then by magnetic trapping and evaporative cooling.

Table 1 shows how these cooling techniques together reduce the temperature of the atoms by a factor of a billion. The phase space density enhancement is almost equally distributed between laser cooling and evaporative cooling, providing about six orders of magnitude each. Bose-Einstein condensation can be regarded as "free cooling," as it increases the quantum occupancy by another factor of about a million without any extra effort. This reflects one important aspect of BEC: the fractional population of the ground state is no longer inversely proportional to the number of states with energies smaller than $k_B T$, but quickly approaches unity when the sample is cooled below the transition temperature.

	Temperature	Density (cm^{-3})	Phase-space density
Oven	500 K	10^{14}	10^{-13}
Laser cooling	50 μK	10^{11}	10^{-6}
Evaporative cooling	500 nK	10^{14}	1
BEC			10^{7}

Atom clouds are observed either by absorptive or dispersive techniques. In this paper we will focus on the absorptive method as it is relevant for the data being presented. Typically, the BEC phase transition can be observed by imaging the shadow cast by an atom cloud which expands ballistically for a time τ after suddenly switching off the magnetic trap (see Figure 2 for an example). One shines a pulse of light onto the atoms which is much shorter than τ, and is near resonant with the atomic transition from the ground $3S_{1/2}, F = 1$ hyperfine level to the excited $3P_{3/2}, F = 2$ hyperfine level. The shadow cast by the atom cloud is imaged onto a CCD camera. For long enough τ, the spatial extent of the cloud is simply proportional to the expansion velocity, which is related to the temperature of atoms before release.[10] The signature of BEC is the sudden appearance of a slow component with anisotropic expansion.[3,4] This can be regarded as observing BEC in momentum space. As one lowers the temperature further by reducing the final radiofrequency used for forced evaporative cooling, the condensate number grows and the thermal wings of the distribution become shorter. This sequence is seen in Figure 2.

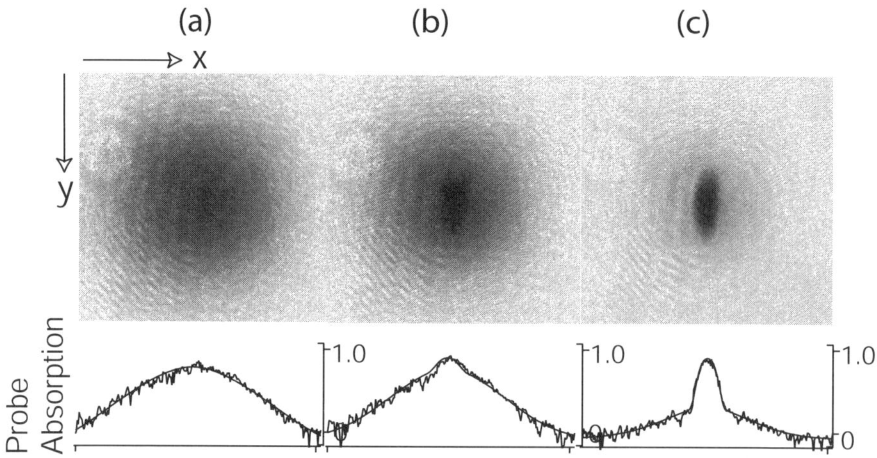

Fig. 2. Transition to BEC. Absorption images taken at final radiofrequencies of a) 0.55, b) 0.45, c) 0.30 MHz show the formation of a Bose condensate as the temperature is lowered from left to right. The condensate expands anisotropically, appearing as the central dark elliptical region in the images. The field of view in each image is 2.7 × 2.7 mm. Below each image is a horizontal slice through the absorption data as well as through a 2-dimensional bimodal fit that measures the normal and condensed components of the gas.

3. Optically Plugged Quadrupole Trap at Georgia Tech

Large volume magnetic traps are a workhorse in the field of quantum gases.[10] This is due to the fact that they can capture an entire laser cooled atom cloud, a feature difficult to achieve with optical traps,[11–13] one which greatly facilitates achieving the initial conditions for evaporative cooling. A simple configuration of coils which will trap low magnetic field seeking particles is the quadrupole trap, formed by a pair of coils running current in the opposite direction in the so-called "anti-Helmholtz" configuration. Unfortunately, by itself this trap is not so useful for evaporative cooling– within a region of $1-2$ μm radius near the magnetic field zero at the trap center, the atoms can spontaneously undergo spin flips and are lost from the trap.[14] This Majorana loss can be eliminated if one removes the field zero from the cloud, for example, using a fast, rotating bias field,[3,14] or alternately, using a Ioffe-Pritchard design which has a finite bias field at the trap minimum.[15] Our approach is based on an idea of Ketterle to use the optical dipole force of a blue-detuned laser beam to repel atoms from the region containing the hole.[4] The resulting potential energy surface depends on both laser and magnetic fields, and the minimum is displaced from the coil center so that the atoms experience a non-zero magnetic field.

At Georgia Tech we have demonstrated that such an "optically plugged" quadrupole trap (OPT) is a simple and robust method of creating a BEC.[16] Our design uses a stable, solid state "plug" laser at 532 nm requiring little or no adjustment for several weeks of operation. Moreover, the focusing of an additional, intense laser beam adds only a minor complexity to the apparatus, comparable to that required for optical confinement of BECs.[17] The quadrupole design has many advantages. These include a large capture volume–close to the physical size of the coils–and tight confinement due to the linearity of the potential. Moreover, only one pair of coils are required, and this maximizes the optical access to the atoms.

Our experimental sequence starts with a Zeeman slowed ^{23}Na atomic beam based on a "spin-flip" design whose flux is about 10^{11}atoms/s. About 10^{10} atoms are loaded in 3 seconds into a dark MOT in the F=1 hyperfine level.[18] Roughly 1/3 of the atoms (the weak-field seekers) are transferred into the OPT (the magnet and laser beam are turned on simultaneously), whose axis of symmetry is vertical. Each coil has 24 windings of 1/8" square cross-section copper tubing. The average diameter of each coil is 4 inches and their spacing is 2.25 inches. A current of 350 A flows through the tube walls, while cooling water flows through the tube itself, and the total voltage drop including a high current switch is 20 Volts. The predicted field gradient is 320 Gauss/cm at this current. Following the loading of the trap, rf evaporative cooling for 42 seconds resulted in an almost pure Bose-Einstein condensate of 10^{6-7} atoms. In order to achieve such high atom numbers, we reduced the trap current by a factor of 14 toward the end of the evaporation stage, thus lowering inelastic losses associated with high atomic density. To ensure the magnetic field zero did not move with respect to the stationary plug beam, it was imperative to carefully cancel stray magnetic fields. By observing the motion of the cloud center and adjusting 3 pairs of Helmholtz coils, we reduced stray DC fields to $\lesssim$ 20 milliGauss, resulting in $\lesssim 10\mu$m motion of the field zero, well below the plug beam diameter.

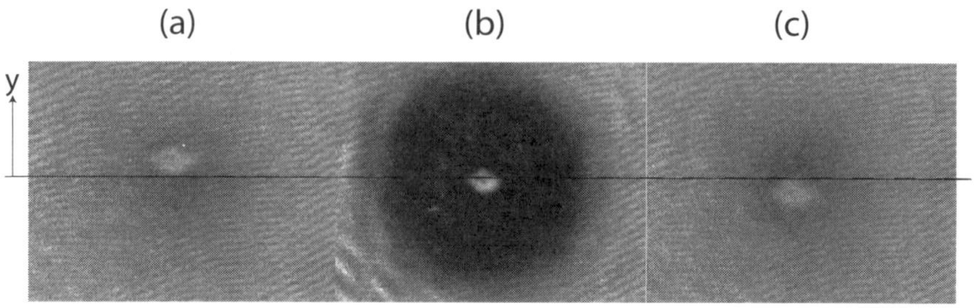

Fig. 3. Plugging the hole. Catastrophic loss results if the "plug" is not carefully aligned. In b) it is aligned precisely with the magnetic field zero (indicated by a dashed line), resulting in a large number of atoms near the end of the rf evaporation stage, while in a) and c) it is misaligned along the y-direction by $+90\mu$m and -65μm, respectively, resulting in very few atoms. The field of view in each image is 1mm × 1mm.

We explored the crucial role played by the plug during evaporation. Figure 3 demonstrates how strong Majorana losses are–when the plug beam was misaligned from the magnetic field center by more than 50μm along any direction (Figs. 3a,3c), tremendous losses ensued and very few atoms remained near the end of the evaporation ramp. However, when it was correctly aligned, as in Fig. 3b, we obtained a huge increase in probe absorption, an unmistakable signature that the hole had been successfully plugged and that we could produce a BEC.

4. Theory of the Weakly Interacting Bose Gas

A gas of bosonic atoms in the condensed state possesses an order parameter $\Psi(\vec{x},t)$ that satisfies the Gross-Pitaevskii equation[19]

$$i\hbar\frac{\partial\Psi}{\partial t} = \left(-\frac{\hbar^2}{2m}\nabla^2 + V_{ext}(\vec{x},t) + g|\Psi|^2\right)\Psi$$

where $\hbar = \frac{h}{2\pi}$ is the reduced Planck constant, m is the atomic mass and V_{ext} is the sum of all external potentials including the trapping potential. Ψ is a single-particle Schrodinger wavefunction for which $\int |\Psi|^2 d^3x = N$, the total atom number. It approximates the many body physical description in the limit of weak interactions between the atoms and very low temperatures, both of which can be satisfied in the laboratory for a wide range of conditions. Since Ψ is in general complex, it may be written as $\sqrt{n(\vec{x},t)}e^{iS(\vec{x},t)}$, where n is the superfluid density and S the phase of the wavefunction. The superfluid velocity field $\vec{v} = \frac{\hbar}{M}\nabla S$ is proportional to the gradient of the phase. Since phase is only defined modulo 2π, it is readily apparent that the circulation $\oint \vec{v} \cdot \vec{dl} = k \times \frac{h}{m}$ must be quantized, where $k = 1, 2, 3.....$ The simplest state for which this condition is satisfied is a single vortex state with $k = 1$. For large angular momentum, the superfluid typically breaks up into a lattice of N_v singly-quantized vortices that form a triangular structure.[20] An experimental image of such a lattice can be seen in Figure 6b, and the experiment is discussed in detail in a later section.

5. Metastable BEC in a linear potential

We have realized a Bose-Einstein condensate whose spin state is metastable.[21] By turning off the plug laser, we have transferred the BEC into a "linear" trap formed by only the quadrupole coils, where the magnetic field is zero at the center. Our goal was to understand how a BEC undergoes Majorana transitions to an untrapped state. We have observed a slow decay of atoms over 100-200 milliseconds due to these Majorana transitions (see Figure 4A). For a thermally excited atom one may use a semiclassical approach to understand the rate of spin flips by considering the motion of an atom along a trajectory in the presence of the field zero. However, for a BEC one must use a quantum picture, in which the spin-flip transition occurs through a coherent evolution of both spatial and spin wavefunctions. Our work was aimed at reconciling these two pictures of the process–quantum and semiclassical.

An atom moving through an inhomogeneous B-field can flip its spin through a nonadiabatic process. Such Majorana transitions are most significant when the magnitude of $\vec{B}$ is close to zero, as it is near a region of radius b near the center of a quadrupole trap. b can be determined from a calculation of the Landau-Zener tunneling rate and is of order $\sim 1\mu$m dimension.[14] If the de Broglie wavelength $\lambda \leq b$, as it is usually for thermal atoms, one may use a semiclassical picture where the atomic motion can be treated classically and the internal spin states are treated quantum-mechanically. In this model, the rate of loss Γ_M only depends on the cloud size,[14]

$$\Gamma_M \sim \frac{\hbar}{MR^2} \tag{1}$$

and is simply related to the statistical probability that an atom trapped in a cloud of radius R will pass directly through a hole of radius b located at the origin. For a gas above the transition temperature, the scaling relation

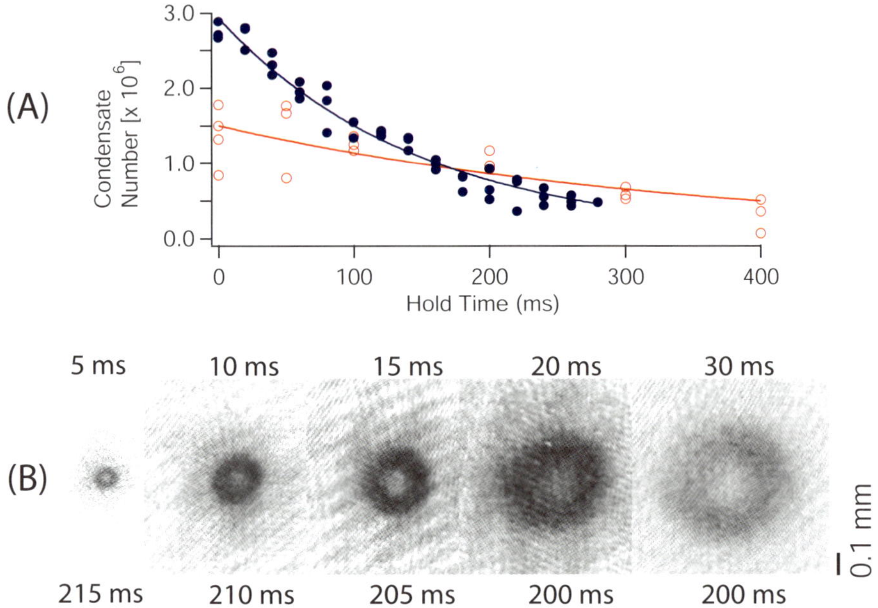

Fig. 4. Metastable BEC. Condensates could be stored for 30 oscillation periods in a purely linear potential formed by a quadrupole trap. (A) Condensate number is plotted against hold time for traps with axial magnetic field gradients of 21 G/cm (filled circles) and 13 G/cm (open circles). (B) Ring shaped expansion of a BEC is often observed when the linear potential is shut off. Shown are absorption images of the expanded condensates after various times of flight (given above each image) and various hold times (given below). Possible explanations for ring formation include the creation of vortex states and the Majorana loss dynamics. Data is taken from the reference[21].

(1) for the rate of loss of atoms from the trap was experimentally verified by Cornell's group.[14]

For a BEC one must treat the entire problem quantum mechanically since all atoms occupy a single wavefunction that extends over a region that is typically much larger than b. For atoms whose internal spin is F and a three-dimensional quadrupole magnetic field $\vec{B} = B'(x\hat{x} + y\hat{y} - 2z\hat{z})$, the resulting Hamiltonian contains the potential energy term:

$$V = -\vec{\mu} \cdot \vec{B}(\vec{r}) = \mu_B g_F B'(x\hat{F}_x + y\hat{F}_y - 2z\hat{F}_z) \quad (2)$$

where $\vec{F} = (\hat{F}_x, \hat{F}_y, \hat{F}_z)$ is the vector spin-F operator and μ_B is the Bohr magneton. Since V acts on both spin and spatial coordinates, the trapped and untrapped states are coupled to one another. In general, this process is completely coherent, and could be used as a mechanism for creating spinor condensates, as has been discussed for other inhomogeneous field geometries.[22] We note that the same coupling exists in Ioffe-Pritchard traps, but is much smaller due to the finite bias field B_z that preserves the spin orientation.

In general, one must solve the problem using a coupled channel approach similar to that involved in the theory of Feshbach resonances.[23] In this approach, the potential V from Eqn. (2) above must be included into a 3-spin component version of the Gross-Pitaevskii Eqn. (4). While such a solution is beyond the scope of this paper, we can compare the loss rate for our BEC using the semi-classical Eqn. (1), using the average Thomas-Fermi radius of the condensate for R. We concluded that the loss dynamics for both condensed and normal components are fairly similar, when the cloud sizes are scaled appropriately.[21]

While the loss rate is fairly well understood, the quantum dynamics of the trapped atoms are more complex. After hold times of 100-200 ms, the time-of-flight distribution frequently displayed a dramatic and unexpected signature–a single, clear hole in the center of the cloud (see Figure 4B). The mechanism of ring formation is currently under investigation. Possible explanations include vortex formation or a feature of the Majorana loss dynamics. The inhomogeneous magnetic field geometry in the linear potential could have application to the phase-imprinting of superfluid persistent currents.

6. Bragg Scattering from Rotating Condensates

6.1. *Technique*

Time-of-flight images such as those we have shown in Figure 2, and for vortex filled clouds such as in Figure 6b, typically measure only the column density distribution of the atoms. That is, the signal in the image is proportional to $|\Psi|^2$, and does not directly measure the phase S of the wavefunction[a]. For vortex states, in addition to the depletion of atoms at the core (the small, dark "holes" in Figure 6b), one would like to *directly* probe the phase of the wavefunction S. One technique for doing so is two photon Bragg scattering.[24,25] It is sensitive to the velocity of the atoms, which is $\propto \nabla S$. The full details of Bragg scattering process are beyond the scope of this paper, and therefore, we will concentrate only on a few key

[a]The amplitude of Ψ does depend on the phase S, however, since the two are coupled during the time-of-flight evolution.

features which are relevant to our data. More details on Bragg scattering can be found elsewhere.[26,27]

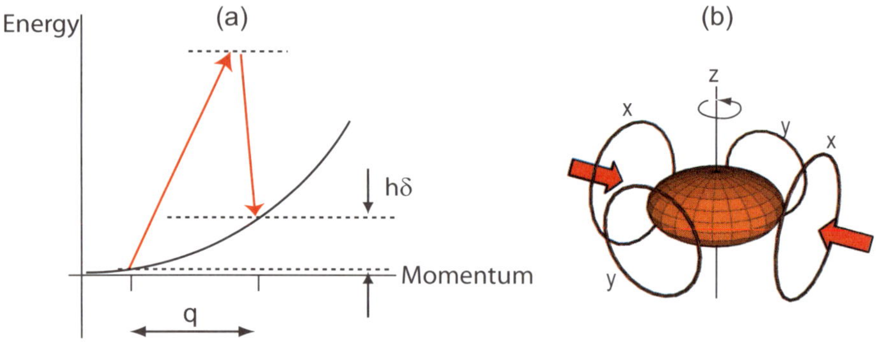

Fig. 5. (a) Energy-momentum relation for atoms in a BEC showing the momentum transfer q from the two-photon process. The Bragg resonance occurs at a frequency difference δ between the two laser beams that satisfies both energy and momentum conservation. (b) Diagram of experimental geometry. Vortices are created by rotating a BEC about the z-axis by phase control of the transverse fields produced in coil pairs x and y that control the TOP (time-orbiting potential) trap. The Bragg beam containing frequencies ω and $\omega + 2\pi\delta$ is applied along the x-direction and retroreflected.

In brief, the Bragg method employs two laser beams with frequencies ω and $\omega + 2\pi\delta$. An atom scatters a photon from one laser beam into another. The net result is to impart to it a momentum $\mathbf{q}$. For counterpropagating laser beams of wavelength λ, $q = |\mathbf{q}| = 2h/\lambda$ is twice the momentum of a single photon. The two photon process and the energy-momentum relation are shown schematically in Figure 5a. For a condensate, q/M is typically much greater than the initial velocity of the atoms, and therefore, the diffracted cloud can be easily distinguished from the non-diffracted atoms. This is because the former have traveled an additional distance $\approx q/M \times t_{tof}$ during the time-of-flight t_{tof} after the trap has been shut off. The two clouds can then be separated in the images, as we show below. There is a resonance in the scattering of light when

$$\delta = \delta_{MF} + \frac{q^2}{2Mh} + \frac{\mathbf{q} \cdot \mathbf{v}}{h} \tag{3}$$

which expresses the conservation of momentum and energy. In the above equation, the second term is the recoil energy, which must be provided by the energy difference between the two photons. For sodium atoms near the principal resonance, $\frac{q^2}{2mh} = 100$ kHz. The third term is simply the Doppler shift, which makes the Bragg technique velocity sensitive. It is this term which is of primary importance to this work, as the Bragg process selects a group of atoms with the same projection of velocity v_x along the direction of the momentum transfer $\mathbf{q} = q\hat{x}$. For a trapped BEC, one also has to consider the effects of interactions. In the mean-field and local density approximations,[27,28] this causes an extra $\delta_{MF} = \frac{4\mu}{7h}$ frequency shift[b], where μ is the chemical potential. This shift is a consequence of the Bogoliubov dispersion relation for condensate excitations, as discussed in reference.[24] Free particles have $\delta_{MF} = 0$. In our case, for a stationary condensate, the Bragg resonance is peaked at a frequency $\delta_o = \frac{q^2}{2mh} + \delta_{MF} \approx 101$ kHz. Thus $\delta = \delta_o + \frac{\mathbf{q}\cdot\mathbf{v}}{h}$.

6.2. *Experiment and Data*

In our experiments we transfer a sodium BEC with typically $1 - 3 \times 10^6$ atoms into the oblate potential of a "TOP", or time-orbiting potential magnetic trap,[14] according to the method described in reference.[29] This trap has azimuthal symmetry in the $x - y$ plane. It consists of a rapidly rotating magnetic bias field superimposed on a static quadrupole magnetic trap. The net effect is to move the magnetic field zero outside of the cloud, thus preventing nonadiabatic spin flips,[14] and resulting in a time-averaged potential $V_{ext} = \frac{1}{2}M\left(\omega_t^2(x^2 + y^2) + \omega_z^2 z^2\right)$. Parameters for the trap are a radial gradient $B'_\rho = 12$ Gauss/cm and a bias rotation of $\omega_{TOP} = 2\pi \times 5$ kHz. The measured transverse oscillation frequency is $\omega_\rho = 2\pi \times 31$ Hz, with $\omega_z = \sqrt{8}\omega_\rho$. For details of experimental techniques, the reader is referred to several excellent review articles.[10]

We produce the vortex-lattice by creating a rotating elliptical asymmetry in the horizontal x-y plane of the TOP trap.[30] The TOP trap employs a fast rotating bias field $\vec{B} = (B_x(t), B_y(t))$ at a frequency ω_{TOP} much greater than that of the atomic motion. This creates a time-averaged harmonic potential. Therefore, we can create a slowly rotating elliptical potential by superimposing slow variations upon the fast oscillation of the bias field. To produce the fields, we combine the signals of two digital frequency synthesizers operating at frequencies $\omega_1 = \omega_{TOP} + \omega_{AR}$ and $\omega_2 = \omega_{TOP} - \omega_{AR}$. These signals are each split, phase shifted and summed together to produce the two fields: $B_x(t) = B_0\cos(\omega_1 t) + \epsilon\cos(\omega_2 t)$ and $B_y(t) = B_0\sin(\omega_1 t) - \epsilon\sin(\omega_2 t)$,

[b] The inhomogeneous density distribution also causes a broadening of the resonance. For our parameters, this is discussed in.[29]

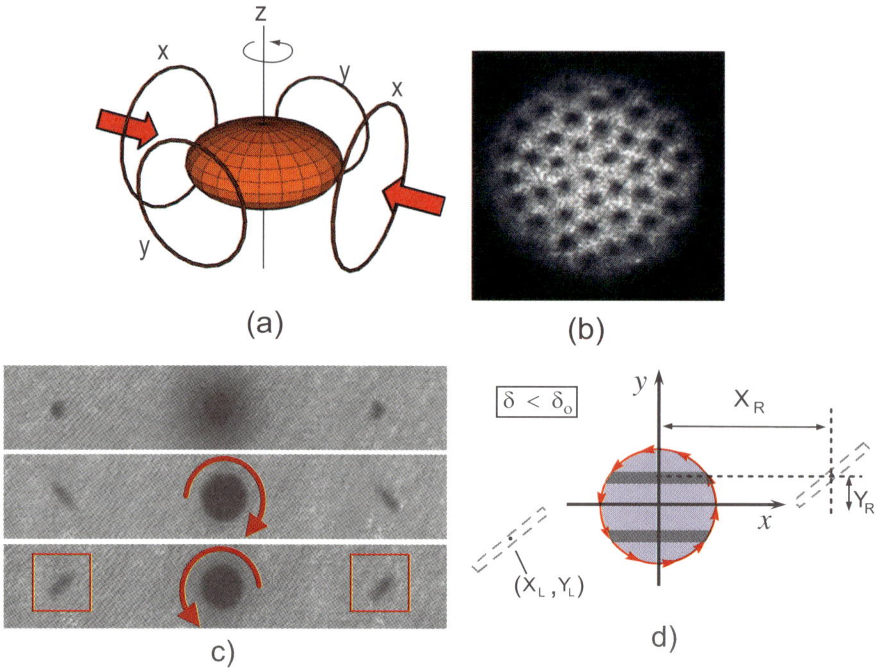

Fig. 6. Bragg scattering detects the phase of a rotating BEC. (a) Atoms in a TOP trap were made to rotate about the z-axis by phase control of the transverse fields produced in coil pairs x and y. This procedure formed a lattice of quantized vortices (b) whose density profile was imaged in time-of-flight after 40 ms. The field of view is 0.6 mm. The Bragg beam was applied along the x-direction and retroreflected, as shown in (a). (c) Time-of-flight images probed the phase of the rotating condensate, which cannot be detected through the density profile of (b). Shown are images of undiffracted (center) and Bragg-diffracted atoms (left and right) for non-rotating condensates (top), clockwise (middle) and counter-clockwise (bottom) rotating clouds. (d) A thin, horizontal band of atoms which are Doppler shifted into resonance are Bragg scattered (the dark shaded regions). As this band of atoms moves, the spread in velocities in the $x-y$ plane causes part of the band to move up while another part moves down, forming a tilted stripe after 10 ms time-of-flight, as observed in the red boxes in (c). All images were taken at $\delta = 102\text{kHz}$.

where ϵ and ω_{AR} are the amplitude and frequency of the rotating asymmetry, respectively. The two currents are individually amplified using 100 Watt car audio amplifiers, and capacitively coupled to a pair of Helmholtz coils of approximately 10 cm diameter along the x and y directions, respectively (see Figure 5b). In order to maximize the number of vortices, we chose $\omega_{AR} = 2\pi \times 22$ Hz, which is very close to the frequency $\simeq 0.7\omega_\rho$ that drives the quadrupole mode in our harmonic trap.[31] After applying the rotating asymmetry for 1.5 seconds, it was turned off and the atomic cloud allowed to equilibrate in the trap for another 1 to 1.5 seconds. This procedure reliably created vortex lattices with approximately 40 ± 10 vortices, as shown in Figure 6b.

After producing vortices and allowing the lattice to equilibrate, we pulse the Bragg diffracting beams along the x-direction (see Figure 5)b) for a time τ_B, while the atoms are still in the trap. The Bragg beams are detuned by 1.7 GHz from the $F = 1$ to $F' = 2$ resonance, and are created by back-reflecting a single beam that contains two frequencies ω_L and $\omega_L + 2\pi\delta$. δ is the difference between the frequencies of two rf synthesizers that are used to drive a single acousto-optic modulator. This creates 2 groups of diffracted atoms propagating to the left and to the right, respectively. We applied a Bragg pulse of square shape with $\tau_B = 250\mu s$, and then turned off the magnetic trap within 100μs. The atoms expanded for a variable t_{tof} before we took an absorption image using laser light resonant with the $F = 1 \rightarrow 2$ transition in a 250μs pulse. The result is shown in the 3 horizontal strips that constitute Figure 6c. The undiffracted cloud is in the center of each strip, while the two groups of diffracted atoms are on the right and left side of each strip.

We can clearly observe *spatial structures* in the outcoupled atom cloud arising from the rotation of the cloud. In the top strip of Figure 6c one can see the diffraction from an initially stationary condensate. No particular structure is visible. However, in the middle strip, we have initially prepared a vortex lattice, which causes the diffracted atoms to form a tilted, elongated spatial pattern. Moreover, when we reversed the direction of the applied rotation (by replacing $\omega_{AR} \rightarrow -\omega_{AR}$), the tilt angle with respect to the y-direction reverses, as shown in the lowest strip.

While earlier work[24,32,33] had explored the spectroscopic nature of the Bragg method for condensates, it required repeated experimental runs to obtain an entire spectrum. By contrast, the data in Figure 6c show that Bragg scattering provides a significant amount of information from a single image, i.e., from the *spatial profile* of the diffracted atoms. In particular, since the Bragg method is sensitive to the *phase* of the wavefunction, it can distinguish between clockwise and counter-clockwise rotation, whereas direct time-of-flight imaging of the density profile cannot provide this information.

To further understand our observations, we examine the coarse-grained velocity field discussed earlier: $v = \vec{\Omega} \times \vec{r}$, with $\Omega = |\vec{\Omega}|$ proportional to the number of vortices. Since the Bragg process selects a group of atoms with the same $v_x = \Omega y$, the resonance condition is given by $\delta = \delta_o + 2y\Omega/\lambda$. Therefore, for a spectrally narrow Bragg pulse, with $\delta < \delta_0$ and a counter-clockwise rotation, the resonance corresponds to a thin, horizontal band of atoms with $y > 0$ for atoms which are Bragg scattered to the right, and $y < 0$ for atoms scattered to the left (the dark shaded regions in Figure 6d). This band is identical to what is shown in Figure 6c, and within it there is a detailed microscopic structure near the vortex cores which is not resolved in our current experiment. As this band of atoms moves, the spread in velocities in the $x - y$ plane causes part of the band to move up while another part moves down. Thus it forms a tilted stripe whose angle increases with time. At long TOF the stripe should become fully stretched along the vertical axis of the images.

A calculation of the rotation frequency from the evolution of the tilt angle as function of time, $\theta(t) = \arctan(\Omega t)$, yielded $\Omega = 2\pi \times (15.4 \pm 1.1$ Hz).[29] Moreover, one can use the locations of the diffracted cloud (X_R, X_L and Y_R, Y_L in Figure 6d) along with spectral information to extract a complete picture of the two-dimensional velocity flow. This resulted in a similar value for the rotation rate.[29] It is in good agreement with the estimate based on the total quantized vorticity of the lattice. For that we note the fact that in the rigid body limit $\Omega = (hN_v)/(2m\pi R_\rho^2)$,[34] and therefore, by measuring the number of vortices N_v one can calculate the rotation rate. We used $R = 37\mu$m and N_v was determined by a manual counting of the number of vortices from several images taken at long TOF. This resulted in $N_V = 37 \pm 7$, which leads to $\Omega = 2\pi \times (13.3 \pm 2.6$ Hz).

Our Bragg technique is more general than time-of-flight imaging, which relies on the spatial scaling of the density distribution during expansion. It is an important first step toward measurement of the phase profile in situations where the time-of-flight signature could be obscure. Examples of this include bent or tangled vortex configurations and vortices in complex potentials where the time-of-flight expansion dynamics might be unknown.

7. Conclusion

After 11 years since the first discovery, Bose-Einstein condensation continues to intrigue us. It has defined a new paradigm in atomic, molecular and optical physics. The list of future challenges is long and includes deeper understanding and control over topological states, the development of practical high-brightness atom lasers and their application in atom optics and precision measurements, and the realization of analog condensed matter systems using the precision and control afforded by atomic methods.

This work has been possible through the efforts of a number of graduate and undergraduate students and postdoctoral fellows with whom we have had the fortune to work. Thanks go to Devang Naik, Bradley Kaiser, Andrew Seltzman, Sergio Muniz, and Mishkatul Bhattacharya. This work was supported by the National Science Foundation, the Army Research Office, and the Department of Energy.

References

1. A. Einstein, *Sitzungsber. Preuss. Akad. Wiss.* **Bericht 3**, 18 (1925).
2. Bose-einstein condensation in atomic gases, in *Proceedings of the International School of Physics Enrico Fermi, Course CXL*, eds. M. Inguscio, S. Stringari and C. E. Wieman (Amsterdam, 1999). See cond-mat/9904034 for the contribution by W. Ketterle, D.S. Durfee,D.M. Stamper-Kurn, cond-mat/9903109 for the contribution of E. Cornell, J.R. Ensher, and C.E. Wieman, and physics/9812038 for the contribution of D. Kleppner, T.J. Greytak, T.C. Killian, D.G. Fried, L. Willmann, D. Landhuis, and S.C. Moss.
3. M. H. Anderson, J. R. Ensher, M. R. Matthews, C. E. Wieman and E. A. Cornell, *Science* **269**, p. 198 (1995).
4. K. B. Davis, M.-O. Mewes, M. R. Andrews, N. J. v. Druten, D. S. Durfee, D. M. Kurn and W. Ketterle, *Physical Review Letters* **75**, 3969 (1995).
5. C. C. Bradley, C. A. Sackett and R. G. Hulet, *Physical Review Letters* **78**, p. 985 (1997).
6. D. G. Fried, T. C. Killian, L. Willmann, D. Landhuis, S. C. Moss, D. Kleppner and T. J. Greytak, *Physical Review Letters* **81**, p. 3811 (1998).
7. K. Huang, **II**, 3 (1964).
8. A. Griffin, D. W. Snoke and S. Stringari (1995).
9. E. W. Streed, A. P. Chikkatur, T. L. Gustavson, M. Boyd, Y. Torii, D. Schneble, G. K. Campbell, D. E. Pritchard and W. Ketterle, *Review of Scientific Instruments* **77**, p. 023106 (2006).
10. E. A. Cornell and W. Ketterle, Bose-einstein condensation in atomic gases, in *Bose-Einstein condensation in Atomic Gases, Proceedings of the International School of Physics Enrico Fermi, Course CXL*, eds. M. Inguscio, S. Stringari and C. E. Wieman, 1999) pp. 15–66.
11. M. D. Barrett, J. A. Sauer and M. S. Chapman, *Physical Review Letters* **87**, p. 010404 (2001).
12. S. R. Granade, M. E. Gehm, K. M. O'Hara and J. E. Thomas, *Physical Review Letters* **88**, p. 120405 (2002).
13. T. Weber, J. Herbig, M. Mark, H. C. Nagerl and R. Grimm, *Science* **299**, 232 (2003).
14. W. Petrich, M. H. Anderson, J. R. Ensher and E. A. Cornell, *Physical Review Letters* **74**, p. 3352 (1995).
15. M. O. Mewes, M. R. Andrews, N. J. v. Druten, D. M. Kurn, D. S. Durfee and W. Ketterle, *Physical Review Letters* **77**, 416 (1996).

16. D. S. Naik and C. Raman, *Physical Review A (Atomic, Molecular, and Optical Physics)* **71**, p. 033617 (2005).
17. D. M. Stamper-Kurn, M. R. Andrews, A. P. Chikkatur, S. Inouye, H. J. Miesner, J. Stenger and W. Ketterle, *Physical Review Letters* **80**, 2027 (1998).
18. W. Ketterle, K. B. Davis, M. A. Joffe, A. Martin and D. E. Pritchard, *Physical Review Letters* **70**, 2253 (1993).
19. L. Pitaevskii and S. Stringari, *Bose-Einstein condensation*International Series of Monographs on Physics, International Series of Monographs on Physics (Clarendon Press, Oxford, 2003).
20. J. R. Abo-Shaeer, C. Raman, J. M. Vogels and W. Ketterle, *Science* **292**, 476 (2001).
21. D. S. Naik, S. R. Muniz and C. Raman, *Physical Review A (Atomic, Molecular, and Optical Physics)* **72**, p. 051606 (2005).
22. H. Pu, S. Raghavan and N. P. Bigelow, *Physical Review A* **63**, 063603 (2001).
23. H. Friedrich, *Theoretical Atomic Physics* (Springer-Verlag, Berlin, 1990).
24. J. Stenger, S. Inouye, A. P. Chikkatur, D. M. Stamper-Kurn, D. E. Pritchard and W. Ketterle, *Physical Review Letters* **82**, 4569 (1999).
25. J. E. Simsarian, J. Denschlag, M. Edwards, C. W. Clark, L. Deng, E. W. Hagley, K. Helmerson, S. L. Rolston and W. D. Phillips, *Physical Review Letters* **85**, 2040 (2000).
26. D. M. Stamper-Kurn, A. P. Chikkatur, A. Gorlitz, S. Gupta, S. Inouye, J. Stenger, D. E. Pritchard and W. Ketterle, *International Journal of Modern Physics B* **15**, 1621 (2001).
27. P. B. Blakie, R. J. Ballagh and C. W. Gardiner, *Physical Review A* **65**, p. 033602 (2002).
28. F. Zambelli, L. Pitaevskii, D. M. Stamper-Kurn and S. Stringari, *Physical Review A* **61**, p. 063608 (2000).
29. S. R. Muniz, D. S. Naik and C. Raman, *Physical Review A (Atomic, Molecular, and Optical Physics)* **73**, p. 041605 (2006).
30. E. Hodby, G. Hechenblaikner, S. A. Hopkins, O. M. Marago and C. J. Foot, *Physical Review Letters* **88**, p. 010405 (2001).
31. K. W. Madison, F. Chevy, V. Bretin and J. Dalibard, *Physical Review Letters* **86**, p. 4443 (2001).
32. S. Richard, F. Gerbier, J. H. Thywissen, M. Hugbart, P. Bouyer and A. Aspect, *Physical Review Letters* **91**, p. 010405 (2003).
33. M. Theis, G. Thalhammer, K. Winkler, M. Hellwig, G. Ruff, R. Grimm and J. H. Denschlag, *Physical Review Letters* **93**, p. Art. No. 123001 (2004).
34. P. Nozires and D. Pines, *The Theory of Quantum Liquids* (Perseus Books, Cambridge, Massachusetts, 1999).

Laser Cooling and Trapping of Rb Atoms

S Chakraborty, A Banerjee, A Ray, B Ray , K G Manohar[1], B N Jagatap[1] and P N Ghosh*

Physics Department, University of Calcutta, 92 A. P. C. Road, Calcutta 70009.
[1] LPT Division, Bhabha Atomic Research Centre, Mumbai 400085.

Introduction

At normal temperature and pressure atoms in the gas phase move in random directions with an r. m. s. speed of a few hundred m/sec. Cooling is an attempt to reduce the speed without allowing them to condense. The term cooling or heating is associated with concept of temperature. Thermodynamic definition of temperature is a statistical or average property of a system of particles. In case of free atoms in a dilute gas the term kinetic temperature is often used as an expression of measure of atomic speeds. In case of a large number of atoms this velocity is the root mean square velocity of the gas atoms. But kinetic temperature may mean the expression of kinetic energy of the atoms.

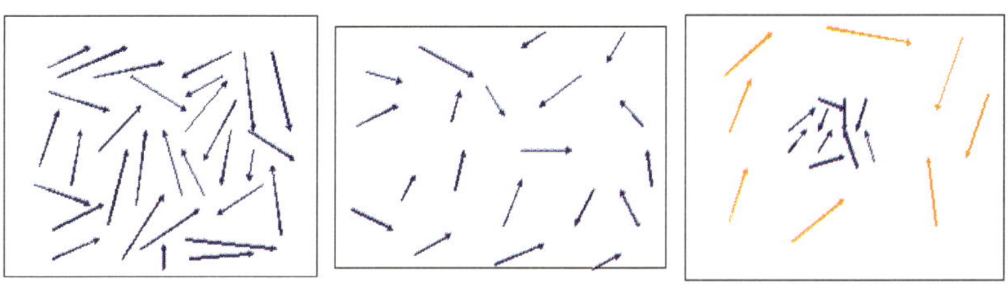

(a) (b) (c)

Fig. 1 (a) Gas phase atoms at NTP, (b) Cooled and dilute gas phase atoms and (c) Cooled and trapped atoms.

* Based on a lecture delivered at TC 2005

Cooling and Trapping

Cooling is the process of reducing the kinetic energy or velocity of atoms. Trapping is the confinement of atoms within a small region of space. Hence the process of cooling and trapping simultaneously means reducing the velocity as well as reducing their position spread. This finally leads to lower uncertainty in momentum and position. But the uncertainty product being of the order of Planck constant h, it is possible to get velocity of the order of m/sec and position confinement of less than 1 cm for the atoms.

Cooling a dilute gas of atoms

For cooling the gas atoms we need cooling them at low density, so that no nucleation occurs. Thus the process differs from the usual condensation process where the atoms or molecules come close together and a strong intermolecular interaction develops. The thermodynamic process involves release of latent heat and there is a phase change from the gas to the liquid phase. This usual concept of cooling is completely different from the cooling of atoms where the attempt is only to slow the atoms down while the interatomic separation remains large so that chemical bond formation is prevented. In order to achieve this condition the density of the atomic gas should be low enough so that three-body collisions are not allowed. If three atoms come together two of them collide while the third takes away the extra energy. Thus a molecule is formed. Molecules can spontaneously form droplets or clusters. If only two atoms participate in the collision molecules cannot be formed. This will prevent condensation. In case of a gas at very low pressure mean free path is large and collisions are usually two-particle processes. Hence laser cooling experiments are performed in a cell at a very low pressure of the order of one nano-torr or less. The atoms are confined in the centre of the cooling cell so that they do not move to the cell walls and get absorbed there.

Atomic velocity is initially reduced by the transfer of momentum from the photons that are absorbed by the atom coming from the opposite direction. Hence the laser beams coming from six mutually perpendicular opposite directions can reduce the velocity components from all directions. Doppler shift of frequency causes the atoms to absorb only the atoms coming from the opposite directions. But this method has a limit and the kinetic temperature of the atoms can be reduced to a few hundred milikelvin only. Further reduction of temperature is possible by using polarized laser beams (1).

Magneto-Optic trapping

Trapping is achieved by applying magnetic field varying linearly with position to confine the cold atoms to the center of the trap. The transitions across the magnetic sublevels rely on the polarization of light. This leads to differential absorption that exerts a restoring force on the atoms (2).

Cooling and trapping experiment at Calcutta University

Cooling and trapping of rubidium atoms is performed in a glass cell (Fig 2) made of Pyrex BK7. The cell has nine windows and a glass to metal seal for connection to the vacuum pump and for inlet of Rb atoms. The cell is fabricated at the BARC Glass Workshop. Six of the windows are used for counterpropagating cooling laser beams. The other windows are used for observation of fluorescence by CCD camera. For the purpose of trapping anti Helmholtz coils are prepared by winding Cu wires through which current is passed in opposite directions. This produces quadrupolar magnetic field with zero field at the centre of the cell with a field gradient of 10 Gauss/cm.

All six laser beams are taken from an external cavity diode laser operating at 780 nm using beam splitters and mirrors. The laser frequency is detuned from the hyperfine component $F = 3 \rightarrow F' = 4$ of ^{85}Rb D_2 transition. The frequency is locked by using a PID loop designed and fabricated in the laboratory. A repumper laser beam operating in the same frequency region is used and is locked to the transition $F = 2 \rightarrow F' = 3$. The laser beam is introduced into the cell through one of the windows used for cooling laser. Meeting point of the cooling laser beams should coincide with the zero of the magnetic field. $\lambda/4$ plates are used in front of each mirror facing the windows to obtain circularly polarized beams. The cell is evacuated to pressure of 10^{-9} Torr by using Turbo and Ion pump. The Rb atoms are kept in glass ampoule inside a flexible steel hose and are introduced into the cell through an angle valve. The pressure inside the cooling cell of the atoms is 10^{-8} Torr.

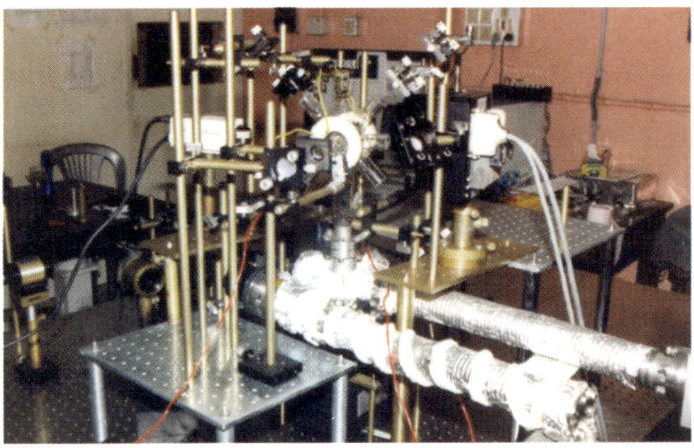

Fig. 2 Experimental arrangement for laser cooling and trapping of Rb atoms.

Fluorescence from the trapped atoms is measured by a CCD camera from Apogee. The current producing the magnetic field may be adjusted to obtain different field gradients. This causes variation in the shape and size of the observed MOT cloud. Fig 3 shows the MOT cloud at a current of 3 A. Gaussian fit of the atom cloud by using Mathmatica shows the diameter of cold atom cloud as 0.2 mm. Temperature of MOT cloud is estimated as 130 μK. The number of trapped atoms N is of the order of 10^6.

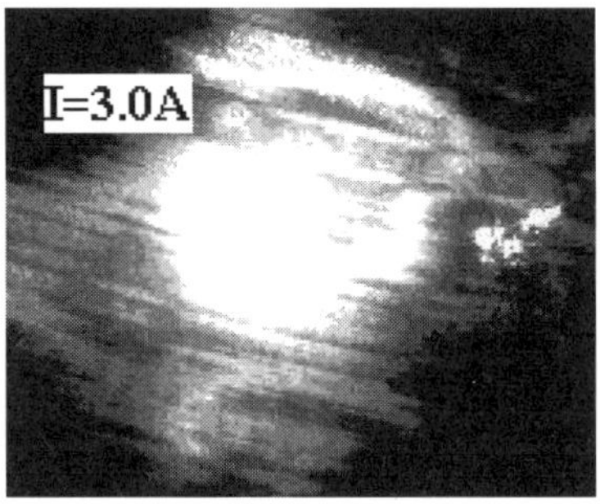

Fig 3 CCD Image of MOT cloud at a field current of 3 Amp.

One unexpected feature of the atom cloud is revealed in a 3D plot (Fig 4) with Mathmatica. It shows regularly placed modulations on the fluorescence density of the Gaussian background. Such modulations have a sinusoidal nature (Fig 4) and the spacing between the crests is approximately is much larger than $\lambda/2$. It is possible that the polarization of the beams is not fully circular. The linear component of the laser beam polarization arising from misalignment may form a standing wave with polarization gradient resulting in the formation of optical potential and variation of atom density. The fact that the laser beams from three mutually perpendicular directions are not exactly orthogonal to each other will lead to nonuniformity of the fringes and may produce the structure with larger fringe spacing. This problem has to be investigated in more details.

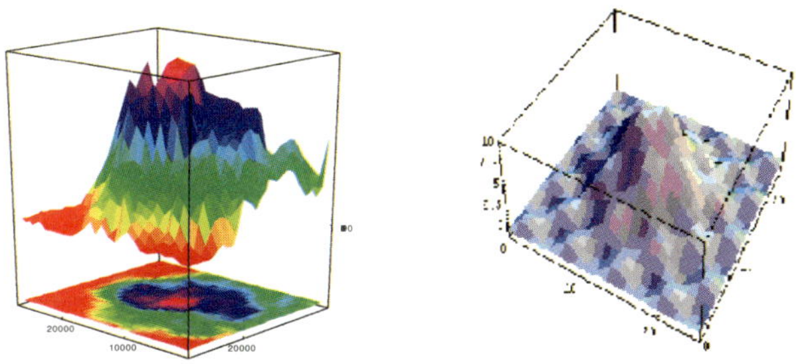

(a) (b)

Fig 4 (a) Three dimensional plot of MOT cloud and (b) Simulation of the cloud with the function A sinx siny +B $G(x-x_0)$ $G(y-y_0)$

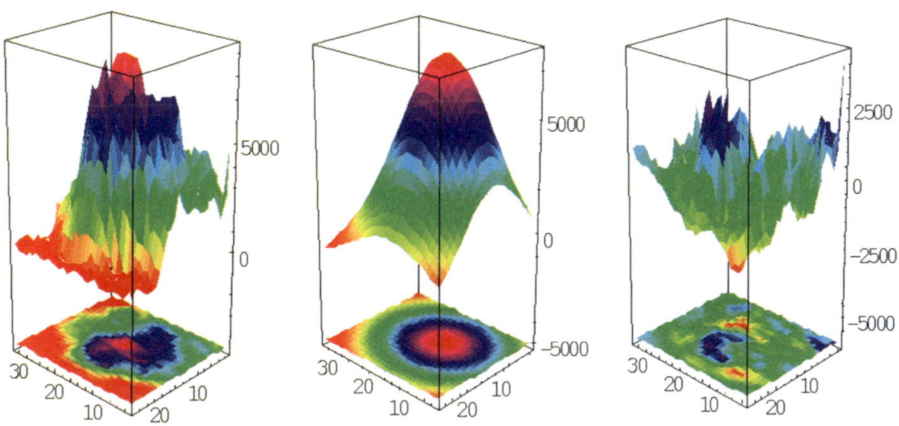

Observed Simulated Residual

Fig 5 The observed and simulated Gaussian with the residual showing the presence of fringes.

Conclusion

A magneto-optic trap based on a glass cell is set up and Rb atoms are cooled to a temperature of the order 100 μK. An analysis of the cloud with Mathmatica shows fringes on the fluorescence of the atom cloud. Further investigations are in progress.

Acknowledgements

PNG thanks the BRNS-DAE, New Delhi for the award of a research project. This project is also supported by the FIST programme of DST, New Delhi. PNG also thanks Professor H Helm of Freiburg University, Germany for useful discussions. We are thankful to the members of the Physics workshop of Calcutta University for fabrication of several components used in our setup. We are grateful to the workers of the glass blowing section, BARC for the fabrication of the glass cell (MOT chamber). The authors thank D. Bhattacharya for assistance during the progress of the work.

REFERENCES

1. C. Wieman, et. al., "Inexpensive laser cooling and trapping experiment for undergraduate laboratories' Am. J. Phys., **63 (4)**, p.317-330

2. E. L. Raab, et. al., "Trapping of neutral Sodium atoms with radiation pressure", Phy. Rev. Lett., **59 (23)** p.2631-2634

Pair-correlation in Bose-Einstein Condensate and Fermi Superfluid of Atomic Gases

Bimalendu Deb
Physical Research Laboratory, Navrangpura, Ahmedabad 380 009, India

Abstract. We describe pair-correlation inherent in the structure of many-particle ground state of quantum gases, namely, Bose Einstein condensate and Cooper-paired Fermi superfluid of atomic gases. We make a comparative study on the pair-correlation properties of these two systems. We discuss how to probe this pair-correlation by stimulated light scattering. This intrinsic pair-correlation may serve as a resource for many-particle entanglement.

PACS numbers: 03.75.Ss,74.20.-z,32.80.Lg

1. Introduction

The realization of Bose-Einstein condensation [1, 2] in dilute atomic gases [3] a decade ago marked a breakthrough revitalizing many areas of physics, particularly, atomic and molecular physics. Bose-Einstein condensate (BEC) is a special state of matter in which quantum properties, in particular matter-wave property, are manifested conspicuously on a macroscopic scale. Predicted about eighty years ago by Einstein [2] based on Bose's quantum statistics [1] of indistinguishable particles, BEC of gaseous systems had long been thought to be an object of mere theoretical interest beyond experimental reach until its first realization in 1995. This experimental feat has been possible because of tremendous technological advancement in cooling and trapping of neutral atoms in 80's and 90's.

One of the most significant advantages of experimentation with cold atoms is the ability to tune atom-atom interaction over a wide range by a magnetic field Feshbach resonance. This provides an unique opportunity to explore physics of interacting many-particle systems in a new parameter regime. In this context, cold atoms obeying Fermi-Dirac statistics have currently attracted enormous research interest. Fermions are the basic constituents of matter, therefore research with trapped Fermi atoms [4, 5, 6, 7] under controllable physical conditions has important implications in materials science. In particular, it has significant relevance in the field of superconductivity.

The first achievement of quantum degeneracy in Fermi gas of ^{40}K atoms by Colorado group [4] in 1999 marked a turning point in the research with cold atoms. Since then, cold Fermi atoms have been of prime research interest in physics today. In a series of experiments, several groups [5, 6, 8, 9, 10, 11] have demonstrated many new features of degenerate atomic Fermi gases. In a recent experiment, Ketterle's group [7] has realized quantized vortices as a signature of Fermi superfluidity in a trapped atomic gas. Two groups [12, 13] have independently reported the measurement of pairing gap in Fermi atoms. Collective oscillations [14, 15] which are indicative of the occurrence of Fermi superfluidity [16] have been previously observed. The crossover [17, 18, 19] between BCS state of atoms and BEC of molecules formed from Fermi atoms has become a key issue of tremendous research interest. Several groups have achieved BEC [20] of molecules formed from Fermi atoms. There have been several other experimental [21] and theoretical investigations [22] on various aspects of interacting Fermi atoms.

Both atomic BEC and superfluid atomic Fermi gas have some common quantum features: (a) both are macroscopic quantum objects (b) the thermal de-Broglie wave-length greatly exceeds the interparticle separation; (c) both have off-diagonal long range order (ODLRO) or coherence; (d) the ground state of both the systems has a structure whose constituents include pair-correlated states; (e) both have ground state of broken symmetry; (f) both must possess long wave-length phonon modes for restoration of symmetry that is broken by their respective ground state. Our focus here would be the common feature (d) to investigate how this pair-correlation can be probed.

In the next section, we make a comparative study between BEC and BCS ground states. Our objective is to show that a nontrivial pair-correlation naturally arises in BEC [23] and BCS matter, and possibly it is a generic feature of all macroscopic quantum objects. In section 3, we discuss briefly some relevant features of trapped Fermi gas. In subsequent sections, we describe stimulated light scattering as a means of probing Cooper-pairing. We find that using stimulated scattering of circularly polarized light, it is possible to scatter selectively either partner atom of a Cooper-pair [24]. In the low momentum transfer regime, this may be useful in exciting Anderson-Bogoliubov phonon mode of broken symmetry.

2. A comparison between BEC and BCS states

Bose-Einstein condensate (BEC) of a weakly interacting Bose gas and Bardeen-Cooper-Schrieffer (BCS) state of an interacting Fermi gas are important in studies of macroscopic quantum physics. Both refer to special states of matter in which conspicuous quantum effects appear on a macroscopic scale. Both are quantum degenerate matter. Quantum degeneracy refers to a physical situation in which thermal de-Broglie wavelength of matter wave exceeds inter particle separation. As a result, matter wave properties play a crucial role in determining not only the microscopic nature but also the bulk properties of matter. Particle-particle interaction in degenerate Bose and Fermi gas leads respectively to Bose and Fermi superfluidity.

Let us now discuss some striking similarities as well as differences in BEC and BCS ground states of interacting systems. Let us begin by writing the ground states of uniform interacting systems in momentum space

$$\text{BEC}: \quad \Psi_0^{\text{BEC}} = \prod_{k\neq 0} \phi_k = \prod_{k\neq 0} \frac{1}{u_k} \sum_{n=0}^{\infty} \left(-\frac{v_k}{u_k}\right)^n \mid n_k, n_{-k}\rangle \tag{1}$$

$$\text{BCS}: \quad \Psi_0^{\text{BCS}} = \prod_{k} \psi_k = \prod_{k} \left(u_k \mid \mathbf{0}\rangle + v_k \mid \mathbf{1}_{k\uparrow}, \mathbf{1}_{-k\downarrow}\rangle\right) \tag{2}$$

where u_k and v_k are amplitude of corresponding Bose or Fermi quasiparticle associated with celebrated transformation that bears Bogoliubov's name. BEC ground state as expressed in Eq. (1) is a product of all possible nonzero momentum states ϕ_k which is a coherent superposition of two mutually opposite momentum states $\mathbf{k}$ and $-\mathbf{k}$ occupied by equal number of particles n ranging from zero to infinity. In other words, ϕ_k is a superposition of all possible pair states $\mid n_k, n_{-k}\rangle$. All the nonzero momentum states compose the non-condensate part of BEC, while zero-momentum state is the condensate part. Clearly, nonzero momentum states form the structure in the ground state. At zero temperature, non-condensate part consisting of nonzero momentum states arises because of particle-particle interaction. Therefore, we can infer that interaction leads to nontrivial pairing correlation which may be used as a resource for generation of continuous variable entanglement. How to extract this correlation by light scattering and thereby to entangle two spatially separated BECs in number and phase variables by a pair of common laser beams passing through both the condensates has been discussed

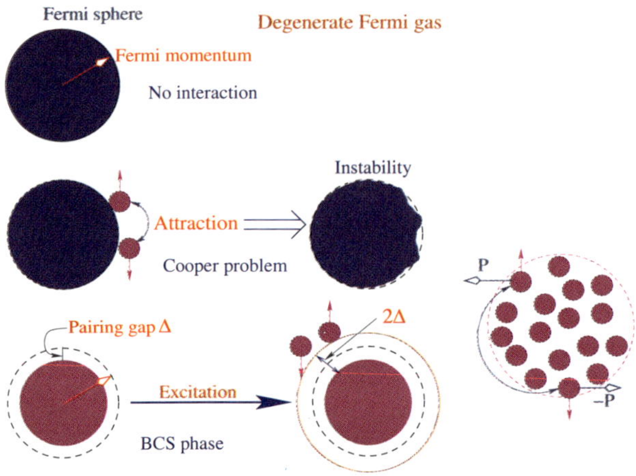

Figure 1. A naive pictorial illustration of Cooper-pair formation and its consequence. For a noninteracting (ideal) Fermi gas, the ground state is simply the Fermi sphere which is completely filled up to Fermi surface and completely empty above the surface. As shown first by Cooper, an attractive inter-fermion interaction, even if it is very weak, leads to formation of an exotic pair-bound state (Cooper-pair) which in turn leads to instability in the Fermi surface. Note that this pairing is basically a many-body effect, since for this effect to occur, quantum degeneracy or near degeneracy is essential. Bardeen, Cooper and Shrieffer then demonstrated that the ground state of a Fermi system with an attractive inter-particle interaction has a gap Δ which is known as pairing gap. Naively speaking, this ground state forms a sphere in momentum space with a radius which is less than Fermi energy by an amount equal to Δ. To break Coopr-pairs and thereby to excite single-particle excitations, a minimum of 2Δ energy is required to be imparted on the system. However, various collective modes among which Bogoliubov-Anderson mode is most significant one can be excited below the gap energy. The sphere at the extreme right is to be considered in real space and drawn to illustrate the fact that the two particles whose distance may exceed enormously the average inter-particle separation can form the pairing state.

elsewhere [25]. Similar experimental configuration has been recently used to produce and subsequently measure phase difference between two spatially separated BECs [26].

Now let us turn our attention to Eq. (2) which expresses the ground state of an attractively interacting spin-half Fermi system. Figure 1 shows pictorially and very naively what happens to the ground state of noninteracting Fermi system when interaction is switched on. Like BEC ground state, it has a structure that is based on particle-particle pairing (Cooper-pairing) in mutually opposite momentum, albeit in opposite spin up ($\uparrow$) and down ($\downarrow$) states. The structure of BCS ground state differs from that of BEC because of Pauli's exclusion principle which forbids more than one fermion to occupy a single quantum state. Hence in a uniform Fermi system, there is only one particle having momentum $\mathbf{k}$ and spin up, if it has to form pairing with another particle with opposite momentum and down spin, it will find only one such partner particle. Since pairing occurs in opposite momentum states, the center-of-mass (COM) momentum of a Cooper-pair is zero. Furthermore, the pairing state is in spin-singlet and hence antisymmetric with respect to spin degrees of freedom. Therefore its spatial part must be symmetric. This means pairing must occur in even number of relative angular momentum l. In low temperature weak-coupling superconductor, Cooper-pairing occurs in s-wave ($l = 0$) state. Although, a Cooper-pair is a kind of two-particle bound state, it is fundamentally different from familiar bound states like diatomic molecule. Cooper-pairing is basically a many-body phenomenon. It occurs only when fermions attract one another under quantum degenerate condition. In contrast a diatomic molecule can be formed by three body interaction. A single molecule can exist in isolation. In contrast, any attempt to isolate a single Cooper-pair from many-body degenerate environment will result in its breaking up into individual fermions. When molecule formation takes place, only nearest neighbor particles form molecular bonding. Cooper-paring can occur between two fermions lying far apart, their distance can greatly exceed average inter-fermion separation. Cooper-pairs can condense into zero (COM) momentum. In fact, a crossover from BCS state of atoms to BEC state of molecules formed from atoms due to a magnetic field Feshbach resonance is an important object of current research interest.

3. BCS state of trapped Fermi gas of atoms

To illustrate the main idea, we specifically consider trapped ^{6}Li Fermi atoms in their two lowest hyperfine spin states $\mid g_1\rangle = \mid 2S_{1/2}, F = 1/2, m_F = 1/2\rangle$ and $\mid g_2\rangle = \mid 2S_{1/2}, F = 1/2, m_F = -1/2\rangle$. For s-wave pairing to occur, the atom number difference δN of the two components should be restricted by $\frac{\delta N}{N} \leq T_c/\epsilon_F$ where T_c is the critical temperature for superfluid transition and ϵ_F is the Fermi energy at the trap center. Unequal densities of the two components result in interior gap (IG) superfluidity [27, 28]. We have suggested in Ref. [28] that it is possible to experimentally realize IG state in two-component Fermi gas of ^{6}Li atoms by making density mistmatch between the two spin-components. In two remarkable recent experiments [29, 30] using two-component ^{6}Li gas, some results which indicate the occurrence of IG state have been obtained. We here consider only the case $N_{1/2} = N_{-1/2}$ which is the optimum condition for s-wave Cooper pairing.

Let us consider a cylindrical harmonic trap characterized by the radial (axial) length scale $a_{\perp(z)} = \sqrt{\hbar/(m\omega_{\perp(z)}}$. One can define a geometric mean frequency $\omega_{ho} = (\omega_\perp^2\omega_z)^{1/3}$ and a mean length scale by $a_{ho} = \sqrt{\hbar/(m\omega_{ho})}$. In Thomas-Fermi local density approximation (LDA) [31], the state of the system is governed by $\epsilon_F(\mathbf{r}) + V_{ho}(\mathbf{r}) + U(\mathbf{r}) = \mu$, where $\epsilon_F(\mathbf{r}) = \hbar^2 k_F(\mathbf{r})^2/(2m)$ is the local Fermi energy, $k_F(\mathbf{r})$ denotes the local Fermi momentum which is related to the local number density by $n(\mathbf{r}) = k_F(\mathbf{r})^3/(6\pi^2)$. Here U represents the mean-field interaction energy and μ is the chemical potential. At low energy, the mean-field interaction energy depends on the two-body s-wave scattering amplitude $f_0(k) = -a_s/(1 + ia_s k)$, where a_s represents s-wave scattering length and k denotes the relative wave number of two colliding particles. In the dilute gas limit ($|a_s|k << 1$), U becomes proportional to a_s in the form $U(\mathbf{r}) = \frac{4\pi\hbar^2 a_s}{2m} n(\mathbf{r})$. In the unitarity limit $|a_s|k \to \infty$, the scattering amplitude $f_0 \sim i/k$ and hence U becomes independent of a_s. It then follows from a simple dimensional analysis that in this limit, U should be proportional to the Fermi energy: $U(\mathbf{r}) = \beta\epsilon_F(\mathbf{r})$ where β is the constant. In this limit, the pairing gap also becomes proportional to the Fermi energy.

Under LDA, the density profile of a trapped Fermi gas is given by

$$n(\mathbf{r}) = n(\mathbf{0})(1 - r_\perp^2/R_\perp^2 - r_z^2/R_z^2)^{3/2}, \tag{3}$$

where $n(\mathbf{0}) = 1/(6\pi^2\hbar^3)[2m\mu/(1+\beta)]^{3/2}$ is the density of the atoms at the trap center. Here $R_{\perp(z)}^2 = 2\mu/(m\omega_{\perp(z)}^2)$ is the radial(axial) Thomas-Fermi radius. The normalization condition on eq. (3) gives an expression for $\mu = (1+\beta)^{1/2}(6N_\sigma)^{1/3}\hbar\omega_0$ where N_σ is the total number of atoms in the hyperfine spin σ. The Fermi momentum $k_F = [3\pi^2 n(\mathbf{0})]^{1/3} = (1+\beta)^{-1/4}k_F^0$ where $k_F^0 = (48N_\sigma)^{1/6}/a_{ho}$ is the Fermi momentum of the noninteracting trapped gas.

4. Stimulated light scattering in Cooper-paired Fermi atoms

To unravel the nature of Fermi superfluid of atomic gases, it is important to analyze the possible response of this quantum gas due to an external perturbation. A method has been suggested to use resonant light [32] to excite one of the spin components into an excited electronic state and thereby making an interface between normal and superfluid atoms as in superconductive tunneling. This has a threshold equal to the gap energy Δ. This method has been applied in recent experiments [12, 33] with the use of rf field for estimating gap energy. There have been a number of proposals [32, 34] for probing pairing gap.

We calculate response function of superfluid Fermi gas due to stimulated light scattering that does not cause any electronic excitation in the atoms. We particularly emphasize the role of light polarization in single-particle excitations which have a threshold 2Δ. We present a scheme by which it is possible to have single-particle excitation in only one partner atom (of a particular hyperfine spin state) of a Cooper-pair using proper light polarizations in the presence of a magnetic field. This may lead to better precision in spin-selective time-of-flight detection of scattered atoms. Furthermore, spin-selective light scattering allows for unequal energy and momentum transfer into the two partner atoms of a Cooper-pair. This may be useful in exciting Bogoliubov-Anderson (BA) phonon mode of symmetry breaking by making small difference in momentum transfers received by the two partner atoms from the photon fields. Recently, a number of authors [35, 36, 37] have studied Bogoliubov-Anderson (BA) mode [38, 39, 40] in fermionic atoms as a signature of superfluidity. BA mode is associated with long wave Cooper-pair density fluctuations. In electronic superconductor, this mode is suppressed due to long wave Coulomb interaction. In neutral superfluid Fermi system such as trapped atomic Fermi gas, this mode is well defined and should in principle be experimentally observable. However, its experimental detection poses a challenging problem since it is a near zero-energy zero-momentum mode.

Figure 2 shows the schematic level diagram for stimulated light scattering by two-component ^{6}Li atoms in the presence of an applied magnetic field which is tuned near the Feshbach resonance ($\sim$ 834 Gauss) results in strong inter-component s-wave interaction. At such high magnetic fields, the splitting between the two ground hyperfine states is $\sim$ 75 MHz while the corresponding splitting between the excited states $\mid e_1\rangle = \mid 2P_{3/2}, F = 3/2, m_F = -1/2\rangle$ and $\mid e_2\rangle = \mid 2P_{3/2}, F = 3/2, m_F = -3/2\rangle$ is $\sim$ 994 MHz. Two off-resonant laser beams with a small frequency difference are impinged on atoms, the scattering of one laser photon is stimulated by the other photon. In this process, one laser photon is annihilated and reappeared as a scattered photon propagating along the other laser beam. The magnitude of momentum transfer is $q \simeq 2k_L \sin(\theta/2)$, where θ is the angle between the two beams and k_L is the momentum of a laser photon. Let both the laser beams be σ_- polarized and tuned near the transition $\mid g_2\rangle \rightarrow \mid e_2\rangle$. Then the transition between the states $\mid g_1\rangle$ and $\mid e_2\rangle$ would be forbidden while the transition $\mid g_1\rangle \rightarrow \mid e_1\rangle$ will be suppressed due to the large detuning $\sim$ 900 MHz. This leads to a situation where the Bragg-scattered atoms remain in the same initial internal state $\mid e_2\rangle$. Similarly, atoms in state $\mid g_1\rangle$ only would undergo Bragg scattering when two σ_+ polarized lasers are tuned near the transition $\mid g_1\rangle \rightarrow \mid 2P_{3/2}, F = 3/2, m_F = 3/2\rangle$. Thus, we infer that in the presence of a high magnetic field, it is possible to scatter atoms selectively of either spin components only by using circularly polarized Bragg lasers. We assume that both the laser beams are σ_- polarized and tuned near the transition $\mid g_2\rangle \rightarrow \mid e_2\rangle$. Under such conditions, considering a uniform gas of atoms, the effective laser-atom interaction Hamiltonian [24] in electric-dipole approximation is $H_I \propto \sum_{\mathbf{k},\sigma=1,2} \gamma_{\sigma\sigma} \hat{c}^\dagger_\sigma(\mathbf{k}+\mathbf{q})\hat{c}_\sigma(\mathbf{k})$, where $\hat{c}_\sigma(k)$ represents annihilation operator of an atom with momentum $\mathbf{k}$ in the internal state σ. The subscript $\sigma = 1(2)$ refers to the state $\mid g_1\rangle$ ($\mid g_2\rangle$). The bare vertex $\gamma_{\sigma\sigma}$ is given by the Kramers-Heisenberg formula [41]

$$\gamma_{\sigma\sigma} = \frac{e^2\mathcal{E}_1\mathcal{E}_2}{m_e\sqrt{n}\hbar\omega_1\omega_2} \sum_{i=1,2} \frac{(\mathbf{d}_{\sigma\sigma}.\hat{\mathcal{E}}_2)(\mathbf{d}_{\sigma\sigma}.\hat{\mathcal{E}}_1)}{\hbar^2(\omega_{\sigma\sigma}-\omega_i)} \tag{4}$$

where d_{ii} is the dipole matrix element between $\mid g\rangle_i$ and $\mid e\rangle_i$ and n is the incident photon number which is assumed to be equal for both the laser beams. Here m_e and e are the mass and charge, respectively, of the valence electron; $\hat{\mathcal{E}}_i$ and ω_i represent the electric field and frequency, respectively, of i-th laser beam and $\omega_{\sigma\sigma}$ is the atomic frequency between the states $\mid g\rangle_\sigma$ and $\mid e\rangle_\sigma$. For the particular case of σ_- polarization in the presence of magnetic field as discussed above, one finds $\gamma_{22} >> \gamma_{11}$. On the other hand, in the absence of magnetic field, one has $\gamma_{11} \simeq \gamma_{22}$.

4.1. The response function

We assume that, except the center-of-mass momentum, the spin or any other internal degrees of atom does not change due to light scattering.

Now, one can define the density operators by $\rho_q^{(0)} = \sum_{\sigma,\mathbf{k}} a^\dagger_{\sigma,\mathbf{k}+\mathbf{q}} a_{\sigma,\mathbf{k}}$ and

$$\rho_q^{(\gamma)} = \sum_{k,\sigma} \gamma_{\sigma\sigma} a^\dagger_{\sigma,\mathbf{k}+\mathbf{q}} a_{\sigma,\mathbf{k}} \tag{5}$$

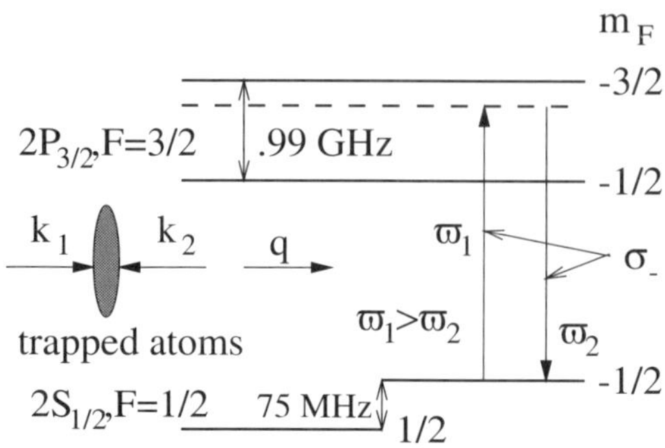

Figure 2. A schematic level diagram for polarization-selective light scattering in two-component Fermi gas of ^{6}Li atoms

One can identify the operator $\rho_q^{(0)}$ as the Fourier transform of the density operator in real space. The scattering probability is related to the susceptibility

$$\chi(\mathbf{q}, \tau - \tau') = -\langle T_\tau[\rho_q^{(\gamma)}(\tau)\rho_{-q}^{(\gamma)}(\tau')]\rangle. \tag{6}$$

where T_τ is the complex time τ ordering operator and $\langle \cdots \rangle$ means thermal averaging. The dynamic structure factor is related to χ by $\chi(\mathbf{q}, \omega_n)$ as

$$S(\mathbf{q}, \omega) = -\frac{1}{\pi}[1 + n_B(\omega)]\mathrm{Im}[\chi(\mathbf{q}, \mathrm{z} = \omega + \mathrm{i}0^+)]. \tag{7}$$

This follows from generalized fluctuation-dissipation theorem. In order to treat collective excitations, it is essential to go beyond Hartree approximation and apply either a kinetic equation or a time-dependent Hartree-Fock equation or a random phase approximation [40]. The essential idea is to take into account the residual terms which are neglected in the BCS approximation and thereby treat the off-diagonal matrix elements (vertex functions) of single-particle operators in a more accurate way [42, 40].

To study light scattering in Cooper-paired fermionic atoms, we apply Nambu-Gor'kov formalism [43, 44] of superconductivity [42]. Using the familiar Pauli matrices, the susceptibility can be expressed as

$$\chi(\mathbf{q}, \omega) = \int \frac{d^4k}{(2\pi)^4 i}\mathrm{Tr}[\tilde{\gamma}_{\mathrm{k}}\mathbf{G}(\mathrm{k}_+)\Gamma(\mathrm{k}_+, \mathrm{k}_-)\mathbf{G}(\mathrm{k}_-)] \tag{8}$$

where the Green function has a matrix form as

$$G(k) = \frac{k_0\tau_0 + \xi_k\tau_3 + \Delta_k\tau_1}{k_0^2 - E_k^2 + i\delta}, \tag{9}$$

where $E_k = \sqrt{\xi_k^2 + \Delta_k^2}$ and $\xi_k = \epsilon_k - \mu$ with $\epsilon_k = \hbar^2k^2/(2m)$. Here τ_0 is a 2×2 unit matrix. The vertex equation is

$$\Gamma(k_+, k_-) = \tilde{\gamma} + i\int \frac{d^4k'}{(2\pi)^4}\tau_3\mathbf{G}(k'_+)\Gamma(k'_+, k'_-)\mathbf{G}(k'_-)\tau_3 V(\mathbf{k}, \mathbf{k}'), \tag{10}$$

where $k_\pm = k \pm q/2$ and $k = (\mathbf{k}, k_0)$ is the energy-momentum 4-vector whose components are $k_3 = \xi_k$ and $k_4 = ik_0$. The bare vertex is a diagonal matrix: $\tilde{\gamma} = \mathrm{Diag.}[\gamma_{11}, -\gamma_{22}]$.

Using Pauli matrices τ_0 and τ_3, this can be rewritten as $\tilde{\gamma} = \gamma_0(k)\tau_0 + \gamma_3(k)\tau_3$, where $\gamma_0 = [\gamma_{11} - \gamma_{22}]/2$ and $\gamma_3 = [\gamma_{11} + \gamma_{22}]/2$. In the specific case of σ_- polarization as discussed above, we have $\gamma_{11} \simeq 0$ and $\gamma_{22} \neq 0$ implying $\gamma_3 \simeq -\gamma_0 \simeq \gamma_{22}/2$. For our numerical analysis, we use $\gamma_{22} \to 1$ meaning $\gamma_3^2 = \gamma_0^2 \to 1/4$.

If the potential $V(\mathbf{k}, \mathbf{k}')$ is separable in $\mathbf{k}$ and $\mathbf{k}'$, Eq. (10) admits an analytical solution, albeit under certain approximations such as particle-hole scattering being negligible. We replace $V(\mathbf{k}, \mathbf{k}')$ by the potential $V = 4\pi\hbar^2 a_s/(2\tilde{m})$, ($\tilde{m} = m/2$ being the reduced mass) expressed in terms of s-wave scattering length a_s. The four-dimensional integrals of Eq. (10) can be performed following the method of relativistic quantum electrodynamics as applied for studying collective excitations in a superconductor [45]. The detailed method of solution is discussed elsewhere [24]. We here present the final result

$$\chi(\mathbf{q}, \omega) = 2N(0)\gamma_0^2\langle B\rangle + 2N(0)\left[\langle A\rangle + \frac{\omega^2\langle f\rangle^2}{4\Delta^2\langle\beta^2 f\rangle}\right]\gamma_3^2 \tag{11}$$

where $N(0)$ is the density of states at the Fermi surface and

$$A = \frac{(\mathbf{v}_k.\mathbf{p}_q)^2 - \omega^2 f}{\omega^2 - (\mathbf{v}_k.\mathbf{p}_q)^2}, \qquad B = \frac{(\mathbf{v}_k.\mathbf{p}_q)^2(1-f)}{\omega^2 - (\mathbf{v}_k.\mathbf{p}_q)^2}. \tag{12}$$

Here $\mathbf{p}_q = \hbar\mathbf{q}$ and $\mathbf{v}_k$ is the velocity of the atoms with momentum $\mathbf{k}$, $f(q) = \sin^{-1}(\beta)/[\beta(1-\beta^2)^{1/2}]$ and $\beta^2 = [\omega^2 - (\mathbf{v}_k.\mathbf{p}_q)^2]/(4\Delta^2)$. The symbol $\langle X\rangle$ implies averaging of a function X over the chemical potential surface: $\langle X\rangle = [N(0)]^{-1}\int d^3\mathbf{k}\delta(\epsilon_k)X$. As $\omega \to \omega + i0^+$, $\beta \to \beta + i0^+$, we have the following analytic property of $f(\beta)$:

$$f(\beta) = -\frac{\text{arcsinh}\sqrt{\beta^2-1}}{\beta\sqrt{\beta^2-1}} + \frac{i\pi/2}{\beta\sqrt{\beta^2-1}}, \quad \beta > 1 \tag{13}$$

4.2. gap equation

The chemical potential μ and the gap Δ_k are to be obtained by solving the gap equation

$$\frac{m}{4\pi\hbar^2 a_s} = \frac{1}{V}\sum_{\mathbf{k}}\left(\frac{1}{2\epsilon_k} - \frac{1}{2E_k}\right) \tag{14}$$

along with the equation

$$n = \frac{1}{6\pi^2}k_F^3 = \frac{1}{V}\sum_{\mathbf{k}}\left(1 - \frac{\xi_k}{E_k}\right). \tag{15}$$

of the density of single component. Note that the eq. (14) is obtained by regularizing the zero-temperature BCS gap equation with a mean-field parameterized by two-body scattering length as done in Ref. [18]. This approach fails to account for pairing fluctuation effects which are particularly significant near T_c in strong-coupling regime. However, far below T_c, the correction due to the pairing fluctuation is very small [18]. Based on this regularized mean-field approach and local density approximation (LDA), the zero-temperature density profiles [46], momentum distribution [47] and the finite temperature effects [48] of superfluid trapped Fermi atoms have been recently studied

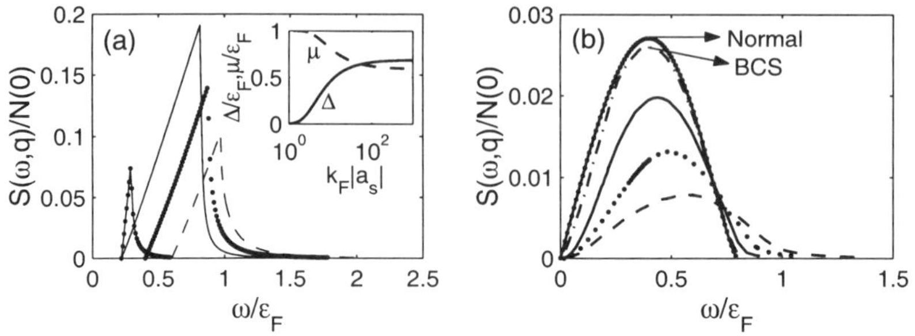

Figure 3. (a) Dimensionless DSF $S(\omega, \mathbf{q})/N(0)$ for single-particle excitations of a uniform superfluid Fermi gas is plotted as a function of dimensionless energy transfer ω/ϵ_F for different values of the scattering length $|a_s| = 2.76k_F^{-1}$ (solid), $|a_s| = 3.89k_F^{-1}$ (dotted), $|a_s| = 5.47k_F^{-1}$ (dashed) for a fixed momentum transfer $q = 0.8k_F$. The dash-dotted curve is plotted for $|a_s| = 2.76k_F^{-1}$ and $q = 0.4k_F$. The inset to Fig. (a) shows the variation of the gap Δ and the chemical potential μ as a function of $|a_s|$. For large a_s, μ and Δ saturate at $0.59\epsilon_F$ and $0.68\epsilon_F$, respectively. (b) Same as in Fig. (a) but for a trapped superfluid Fermi gas for a fixed momentum transfer $q = 0.8k_F$ (k_F refers to the Fermi momentum at the trap center). Also shown are the DSF for small $\Delta = 0.05$ (BCS) and $\Delta = 0$ (normal).

The two coupled eqs. (14) and (15) admit analytical solutions [49]. In the unitarity limit the solutions yield $\Delta \simeq 1.16\mu$, $\mu = (1 + \beta)\epsilon_F$, where $\beta = -0.41$ is a constant.

5. Results and discussions

Figure 3 shows $S(\omega, \mathbf{q})$ as a function of ω for a uniform and trapped gas for different values of a_s. In the case of trapped gas, we use LDA with local chemical potential $\mu(\mathbf{r})$ determined from equation of state of interacting Fermi atoms in a harmonic trap. When a_s is large, the behavior of $S(\delta, \mathbf{q})$ is quite different from that of normal as well as weak-coupling BCS superfluid. This can be attributed to the occurrence of large gap for large a_s. In contrast to the case of a uniform superfluid, $S(\delta, \mathbf{q})$ for a superfluid trapped Fermi gas has a structure below $2\Delta(0)$, where $\Delta(0)$ is the gap at the trap center. As the energy transfer decreases below $2\Delta(0)$, the slope of $S(\delta, \mathbf{q})$ gradually reduces. Particularly distinguishing feature of $S(\delta, \mathbf{q})$ of a superfluid compared to normal fluid is gradual shift of the peak as a_s or Δ increases. The quasiparticle excitations occur only when $2\Delta(\mathbf{x}) < \omega$. This implies that, when ω is less than $2\Delta(0)$, the atoms at the central region of the trap can not contribute to quasiparticle response.

6. Conclusion and outlook

In conclusion, we have studied long-ranged pair-correlation inherent in both the BEC and BCS states. We have also investigated polarization-selective light scattering in

Cooper-paired Fermi atoms as a means of estimating the gap energy. Our results suggest that it is possible to detect the pairing gap by large-angle (i.e., large q) Bragg scattering. Small angle polarization-selective stimulated light scattering may be useful in exciting BA mode. The pair-correlation which may be a generic feature of all macroscopic quantum systems with long-range order may serve as a potential resource for may-particle robust entanglement that is central to quantum information science.

References

[1] Bose S. N. 1924 *Z. Phys.* **26** 178.

[2] Einstein A. 1924 *Sitzber. Kgl. Preuss. Akad. Wiss.* 261 1925 *ibid* 3

[3] Anderson M., Ensher J. R., Matthews M. R., Wieman C. E., and Cornell E. A. 1995 *Science* **269** 198; Bradley C. C., Sackett C. A., Tollett J. J. , and Hulet R. G. 1995 *Phys. Rev. Lett.* **75** 1687 ; Davis K. B. , Mewes M. O., Andrews M. R., van Druten N. J., Durfee D. S., Kurn D. M. , and Ketterle W. 1995 *Phys. Rev. Lett.* **75** 3969

[4] B. DeMacro and D. S. Jin 1999 *Science* **285** 1703

[5] A. G. Truscott, K. E. Strecker, W. I. McAlexander, G.B. Patridge, and R. G. Hulet 2001 *Science* **291** 2570

[6] S. R. Granade, M. E. Gehm, K. M. O'Hara, and J. E. Thomas 2002 Phys.Rev.Lett. **88**, 120405; O'Hara *et al.* 2002 *Science* **298** 2179

[7] Zwierlein M. W., Abo-Shaeer J. R., Schirotzek A., Schunck C. H., Ketterle W. 2005 *Nature* **435** 1047

[8] F. Schreck *et al.* 2001 *Phys. Rev. Lett.* **87** 080403;T. Bourdel *et al.* 2003 *ibid.* **91** 020402

[9] Z. Hadzibabic *et al.*, *Phys. Rev. Lett.* **88**, 160401 (2002).

[10] G. Roati, F. Riboli, G. Modungo, and M. Inguscio, *Phys. Rev. Lett.* **89**, 150403 (2002).

[11] Jochim S. *et al.* 2003 *Science* **302** 2101

[12] Cin C. *et al.* 2004 *Science* **305** 1128

[13] Greiner M., Regal C. A., and Jin D. S. 2005 *Phys. Rev. Lett.* **94** 070403

[14] Kinast, J. *et al.* 2004 *Phys. Rev. Lett.* **92** 150402

[15] Bartenstein M. *et al.* 2004 *Phys. Rev. Lett.* **92** 203201

[16] Stringari S. 2004 *Europhys. Lett.* **65** 749

[17] Noziéres P. and Schmitt-Rink S. 1985 *J. Low. Temp. Phys.* **59** 195

[18] Sa de Melo C.A.R., Randeria M. and Engelbrecht J.R. 1993 *Phys. Rev. Lett.* **71** 3202; Engelbrecht J.R., Randeria M. and Sa de Melo C.A.R. 1997 *Phys. Rev. B* **55** 15153

[19] Holland M., Kokkelmans S. J. J. M. F., Chiofalo M. L. and Wasler R. 2001 *Phys. Rev. Lett.* **87** 120406; Timmermans E. *et al.* 2001 *Phys. Lett A* **285** 228; Ohashi Y. and Griffin A. 2002 *Phys. Rev. Lett.* **89** 130402; Hofstetter W. *et al.* 2002 *Phys. Rev. Lett.* **89** 220407

[20] Greiner M., Regal C. A. and Jin D. S. 2003 *Nature* **426** 537; Jochim S. *et al.* 2003 *Science* **302** 2101; Zwierlein M. W. *et al.* 2003 *Phys. Rev. Lett.* **91** 250401

[21] Modugno G. *et al.* 2002 *Science* **297** 2240; Strecker K. E. *et al.* 2003 *Phys. Rev. Lett.* **91** 080406; Cubizolles J. *et al.* 2003 *Phys. Rev. Lett.* **91** 240401

[22] Falco G. M. and Stoof H. T. C. 2004 *Phys. Rev. Lett.* **92** 130401; Carr L. D., Shlyapnikov G. V. and Castin Y. 2004 *Phys. Rev. Lett.* **92** 150404; Heiselberg H. 2003 *Phys. Rev. A* **68** 053616, Perali A., Pieri P. and Strinati G. C. 2003 *Phys. Rev. A* **68** 031601; Perali A., Pieri P., Pisani L. and Strinati G. C. *lanl e-print cond-mat/0311309.*

[23] Deb B. and Agarwal G. S., 2002 *Phys.Rev.A* **65** 063618.

[24] Deb B., 2006 *J. Phys. B: At. Mol. & Opt. Phys.* **39**, 529.

[25] Deb B. and Agarwal G. S., 2003 *Phys.Rev.A* **67** 023603.

[26] Saba M., Pasquini T. A., Sanner C., Shin Y., Ketterle W., Pritchard D. E., 2005 *science* **307** 1945.

[27] W. V. Liu and F. Wilczek 2003 *Phys.Rev.Lett.* **90** 047002

[28] Deb B., Mishra A., Mishra H. and Panigrahi P. K. 2004 *Phys. Rev. A* **70** 011604

[29] Zwierlein M. W., Schirotzek A., Schunck C. H., and Ketterle W, 2006 *Science* **311** 492

[30] Partridge G. B., Lui W., Kamar R. I., Liao Y., and Hulet R. G., 2006 *Science* **311** 503

[31] Houbiers M. *et al.* 1997 *Phys. Rev. A* **56** 4864; Vichi L. and Stringari S. 1999 *Phys. Rev. A* **60** 4734

[32] Törmä P. and Zoller P. 2000 *Phys. Rev. Lett.* **85** 487; Bruun G. M. *et al.* 2001 *Phys. Rev. A* **64** 033609; Kinnunen J., Rodriguez M., and Törmä P 2004 Phys. Rev. Lett. **92** 230403; Bruun G. M. and Baym G. 2004 *Phys. Rev. Lett.* **93** 150403; Büchler H. P., Zoller P., Zwerger W. 2004 *Phys. Rev. Lett.*, **93** 080401

[33] Kinnunen J., Rodriguez M., and Törmä P 2004 *Science* **305** 1131

[34] Zhang W., Sackett C. A. and Hulet R. G. 1999 *Phys. Rev. A* **60** 504; Ruostekoski J. 1999 *Phys. Rev. A* **60** 1775; Rodriguez M. and Törmä P. 2002 *Phys. Rev. A* **66** 033601.

[35] Bruun G. M. and Mottelson B. R. 2001 *Phys. Rev. Lett.* **87** 270403

[36] Ohashi Y. and Griffin A. 2003 *Phys. Rev. A* **67** 063612; Ohashi Y. and Griffin A. *lanl archive cond-mat/0503641*

[37] Minguzzi A., Ferrari G. and Castin Y. 2001 *Eur. Phys. J. D.* **17** 49

[38] Bogoliubov N. N. 1958 *Nuovo Cimento* **7** 6; Bogoliubov N. N., Tolmachev V. V., and Shirkov D. V. 1959 *A New Method in the Theory of Superconductivity* (Consultants Bureau, NY).

[39] Anderson P. W. 1958 *Phys. Rev.* **112** 1900

[40] Martin P. C., in 1969 *Superconductivity*, Vol.1, edited by Parks R. D. (Dekker, NY).

[41] Sakurai J. J. 1967 *Advanced Quantum Mechanics* (Pearson Education, Inc.)

[42] Schrieffer J. R. 1964 *Theory of Superconductivity* (W. A. Benjamin)

[43] Nambu Y. 1960 *Phys. Rev.* **117** 648

[44] Abrikosov A. A., Gorkov L. P., and Dzyaloshinski 1963 *Methods of Quantum Field Theory in Statistical Physics* (Dover, NY).

[45] Vaks V. G., Galitskii, and Larkin A. I. 1962 *Soviet Physics JETP* **14** 1177

[46] Perali A. Pieri P. and Strinati G. C. 2003 *Phys. Rev. A* **68** 031601

[47] Viverti L., Giorgini S., Pitaevskii L. P. and Stringari S. 2004 *Phys. Rev. A* **69** 013607

[48] Perali A. Pieri P., Pisani L. and Strinati G. C. 2004 *Phys. Rev. Lett* **92** 110401

[49] Marini M., Pistolesi F. and Strinati G. C. 1998 *Eur. Phys. J. B* **1** 151

Properties of Trapped Bose gas in the large-gas-parameter regime

Arup Banerjee
Laser Physics Application Section, Raja Ramanna Centre for Advanced Technology
Indore 452013, India
**E-mail: banerjee@cat.ernet.in*
www.cat.ernet.in

1. Introduction

Successful achievement of Bose-Einstein condensation (BEC) in atomic gases has provided a good opportunity to test the applicability of the predictions of theoretical methods developed in the context of superfluid helium. The mean-field Gross-Pitaevskii (GP) theory[1] has been quite successful in explaining both static and dynamic properties[2–6] of the Bose-Einstein condensates produced in alkali-metal atoms confined in magnetic or optical traps. It is well known that the GP theory is valid for condensates satisfying the dilute gas condition $na^3 << 1$ (where n is the atomic density, a is the s-wave scattering length of the interatomic potential, and the parameter $x = na^3$ is called the gas parameter). Typically, the values of the gas parameter in most of the experiments were in the range of $10^{-4} - 10^{-5}$ and therefore the GP theory works very well in predicting the properties of these trapped condensates. However, in a recent experiment, condensates with peak gas parameter of the order of 10^{-2} have been achieved by enhancing the scattering length a with the help of Fesbach resonance.[7] It is important to note here that the diffusion Monte Carlo (DMC) simulation of uniform Bose gas has revealed that for the values of gas parameter beyond 10^{-3}, the results of GP theory deviate substantially from the results of the simulation.[8] Therefore, it becomes quite natural to test the validity of the mean-field GP theory with increasing value of the gas parameter, especially for condensates with the values of the gas parameter more than 10^{-3} . We go beyond GP theory by including the effect of depletion on the condensates, which is caused by the increased value of the interatomic interaction strength in the large-gas-parameter regime. The inclusion of depletion leads to the modification of the GP equation; it is referred to as the modified GP (MGP) equation. The main aim of this paper is to study the properties of the trapped condensates in the large-gas-parameter regime by employing both MGP and GP equations. The basic idea behind such study is to estimate the size of the corrections, introduced by the MGP equation over the GP equation, in various physical properties of the condensates in the large-gas-parameter regime.

In this paper, we consider two different states of the trapped condensates, namely, the ground and the excited or the vortex states, to investigate the effect of depletion on the condensates. The first part of the paper concentrates on the calculation of static and dynamic properties of the trapped condensates in their ground states. The ground state density profile plays an important role in characterizing the occurence of the BEC in atomic gases. The density profiles characterize the static property of the condensates. On the other hand, the collective oscillations of the trapped condensates characterizing the dynamic properties, are important phenomena associated with the many-body systems. The collective oscillations in the atomic condensates have been extensively studied both theoretically and experimentally. It is important to note that the mean-field GP equation has been quite successful in accurate evaluation of the frequencies of the collective oscillations - the agreement between the experimental and the theoretical results is excellent at zero temperature. It is then natural to study the effect of depletion on the collective oscillations in the large-gas-parameter regime. To this end, we calculate the monopole and the quadrupole modes of the collective oscillations of the trapped condensates in their groud state by employing both MGP and GP equations.

In the second part of this paper, we consider the properties of the condensates carrrying a quantized vortex. The vortex states of the condensates correspond to the excited states of the GP or the MGP equations. Consequently, the vortex states are also termed as the excited states of the condensates. These states play an important role in establishing the superfluid properties of BEC.[2,3] In the light of the above discussion, it is quite natural to ask how the properties of vortex states are modified in the large-gas-parameter regime. Moreover, it is well known that a quantized vortex state not only affects the static properties of the condensate, but it also modifies the dynamic properties. For example, the presence of a quantized vortex state leads to splitting of the two modes of the quadrupole oscillations with opposite values of the third component of angular momentum, which are degenerate in the absence of a quantized vortex state.[9,10] The vortex state breaks the time reversal symmetry, which in turn results in the removal of the degeneracy of the two modes of oscillations carrying opposite values of angular momentum. Therefore, splitting of the two quadrupole modes of collective oscillations can be employed to detect the presence of a quantized vortex state in BEC,[11] as the measurement of frequencies of the collective oscillations can be carried out with high precision. This has motivated us to study the effect of depletion on the static and the dynamic properties of the condensates carrying a single quantized vortex.

The paper is organized in the following manner. In section II, we describe the theoretical methods employed in this paper and briefly describe the MGP theory followed by the variational method employed to obtain the wave function of the vortex state and other physical observables mentioned above and the sum-rule approach for the calculation of frequencies of collective state oscillations. The section III is devoted to a discussion of the results. The paper is concluded in section IV.

2. Theory

The ground-state energy functional associated with a condensate of N bosons each with mass m confined in a trap potential $V_t(\mathbf{r})$ can be written as

$$E[\Psi] = \int d\mathbf{r} \left[\frac{\hbar^2}{2m} |\nabla\Psi|^2 + V_t(\mathbf{r})|\Psi|^2 + \epsilon(n)|\Psi|^2 \right]. \tag{1}$$

where $\Psi(\mathbf{r})$ is the the condensate wave function (order parameter) and $n(\mathbf{r})$ represents the corresponding density and it is given by $n(\mathbf{r}) = |\Psi(\mathbf{r})|^2$. The condensate wave function $\Psi(\mathbf{r})$ can be determined by minimizing the above energy functional. In the above equation, the first, second, and the third terms represent the kinetic energy of bosons, the energy due to the trapping potential, and the interatomic interaction energy within the local density approximation (LDA), respectively. To go beyond the GP theory, we make use of the perturbative expansion for $\epsilon(n)$ in terms of the gas parameter na^3,

$$\epsilon(n) = \frac{2\pi\hbar^2 an}{m} \left[1 + \frac{128}{15\sqrt{\pi}} \left(na^3\right)^{\frac{1}{2}} + 8\left(\frac{4\pi}{3} - \sqrt{3}\right) \left(na^3\right) \ln\left(na^3\right) + \mathcal{O}\left(na^3\right) \right]. \tag{2}$$

The first term in the above expansion, which corresponds to the energy of the homogenous Bose gas within the mean-field theory as considered in the GP theory, was calculated by Bogoliubov.[12] The second term was obtained by Lee, Huang, and Yang (LHY),[13] while the third term was first calculated by Wu[14] using the hard-sphere model for the interatomic potential. The DMC simulation of Ref.[8] has clearly demonstrated that the above expansion up to LHY term gives accurate results even for the gas parameter of the order of 10^{-2}. On the other hand, inclusion of logarithmic term leads to severe mismatch with DMC simulation results. Consequently, we do not consider the logarithmic term in the expansion (2) for all the calculations in this study. The trapping potential $V_t(\mathbf{r})$ is taken to be axially symmetric characterized by two angular frequencies $\omega_\perp^0$ and ω_z^0 ($\omega_x^0 = \omega_y^0 = \omega_\perp^0 \neq \omega_z^0$). It is given by

$$V_t(\mathbf{r}) = \frac{m{\omega_\perp^0}^2}{2}\left(x^2 + y^2 + \lambda_0^2 z^2\right), \tag{3}$$

where $\lambda_0 = \omega_z^0/\omega_\perp^0$ is the anisotropy parameter of the trapping potential ($\lambda_0 = 1$ corresponds to a spherically symmetric trap).

As mentioned above, the minimization of above energy functional with respect to $\Psi(\mathbf{r})$ with the constraint

$$\int |\Psi(\mathbf{r})|^2 d\mathbf{r} = N, \tag{4}$$

leads to the MGP equation for the condensate wave function

$$\left[-\frac{\hbar^2}{2m}\nabla^2 + V_t(\mathbf{r}) + \frac{4\pi\hbar^2 a}{m}|\Psi|^2\left(1 + \frac{32a^{3/2}}{3\pi^{1/2}}|\Psi|\right)\right]\Psi(\mathbf{r}) = \mu\Psi(\mathbf{r}), \tag{5}$$

where μ is the chemical potential arising from the constraint condition given by Eq. (4). The GP equation for the condensate wave function can be obtained from Eq. (5) by neglecting the interaction energy term proportional to $|\psi|^3$ from the lefthand side of this equation.

The ground state energy functional given by Eq. (1) can be easily generalized to include the vortex states by writing the condensate wave function as

$$\Psi(\mathbf{r}) = \psi(\mathbf{r})e^{i\kappa\phi}, \tag{6}$$

where ϕ is the angle around the z-axis and κ is an integer denoting the quantum circulation, and the total angular momentum along the z-axis is given by $N\hbar\kappa$. Substituting the above complex wave function of Eq.(6) in Eq. (1), we get the MGP energy functional for the condensates with a single vortex state as[4]

$$E[\psi] = \int d\mathbf{r}\left[\frac{\hbar^2}{2m}|\nabla\psi|^2 + \frac{\hbar^2\kappa^2}{2mr_\perp^2}|\psi|^2 + v_{ext}(\mathbf{r})|\psi|^2 + \epsilon(n)|\psi|^2\right]. \tag{7}$$

In the above equation $r_\perp = \sqrt{x^2+y^2}$ and the density $n(\mathbf{r}) = |\psi(\mathbf{r})|^2$. The presence of a centrifugal term due to the vortex state makes the above functional different from Eq. (1).

Various numerical techniques have been developed to solve the nonlinear MGP equation for studying properties of trapped Bose-Einstein condensates.[15] In our calculations, we employ a variational method to obtain the condensate wave function.[16,17,20,21] The main advantage of this method is that with a suitable choice for the form of the wave function, one can get quite accurate results with significantly less computational effort. We have chosen following variational forms for the wave functions describing ground and the vortex states of the trapped condensates

$$\Psi(\mathbf{r}_1) = Ae^{-\frac{1}{2}\left(\frac{\omega_\perp}{\omega_\perp^0}\right)^p (r_{1\perp}^2+\lambda z_1^2)^p}, \tag{8}$$

and

$$\psi(\mathbf{r}_1) = Ar_{1\perp}^q e^{-\frac{1}{2}\left(\frac{\omega_\perp}{\omega_\perp^0}\right)^p (r_{1\perp}^2+\lambda z_1^2)^p}, \tag{9}$$

respectively with q, λ, $\omega_\perp$, and p as the variational parameters and A is the normalization constant. The variational parameters are obtained by minimizing the energy functionals, given by (1) for the ground-state, and (7) for the vortex state, with respect to these parameters. In the above equation, we use the scaled co-ordinates $\mathbf{r}_1 = \mathbf{r}/a_{ho}$ with $a_{ho} = \left(\hbar/m\omega_\perp^0\right)^{\frac{1}{2}}$. We note that $n(\mathbf{r_1})$ is normalized to unity and from Eq. (4) it can be seen that

$$n(\mathbf{r}) = \frac{N}{a_{h0}^3} n(\mathbf{r_1}). \tag{10}$$

It is important to note here that the above variational forms render analytical expressions for the energy components and it has been demonstrated that they describe the ground and the excited (vortex) states of Bose-Einstein condensates quite accurately for a wide range of particle numbers. In the next section, we present the results of our calculation for the equilibrium and the dynamic properties of the condensates with the values of the gas parameter falling in the large-gas-parameter regime.

3. Results and Discussion

Before, we proceed with the discussion of results in the large-gas-parameter regime, it is necessary to establish the accuracy of variational method adopted in this paper. For this purpose, we calculate the total energy, chemical potential, and the column density profile of a harmonically trapped ^{87}Rb condensates both for the ground and the vortex states (with a single quantized vortex, $\kappa = 1$). The column density profile is generally measured in the laboratory to characterize the density profile of the trapped atomic cloud. Theoretically, the column density is calculated by using the expression

$$n_c(r_{1\perp}) = \int dzn(r_{1\perp}, z). \tag{11}$$

The z-direction along which the above integration is performed, coincides with that of laser beam used to image the atomic cloud. Using the above-mentioned variational forms for the wave functions, physical observables can be expressed analytically in terms of the four variational parameters. For example, the total energy $E_1 = E/\hbar\omega_\perp^o$ and the chemical potential $\mu_1 = \mu/\hbar\omega_\perp^o$ can be written as

$$\frac{E_1}{N} = T + U + E_{rot} + E_{int}^1 + E_{int}^2, \tag{12}$$

and

$$\mu_1 = T + U + E_{rot} + 2E_{int}^1 + \frac{5E_{int}^2}{2}, \tag{13}$$

respectively. Here T, U, and E_{rot} (for the ground-state $E_{rot} = 0$) denote the average kinetic, trapping potential, and the rotational energies per particle, respectively. The term E_{int}^1 gives the interaction energy per particle in the mean-field approximation as considered in the GP theory, while E_{int}^2 gives the correction due to the LHY term in the expansion (2) resulting in the MGP functional. The analytical expressions for these energy components are

$$T = \frac{\omega_\perp(1+2q)\left[(1+2p)(1+\lambda/2) + q(2p+2q+\lambda)\right]\Gamma((1+2q)/2p)}{\omega_\perp^0 4(3+2q)\Gamma((3+2q)/2p)}, \tag{14}$$

$$U = \frac{\omega_\perp^0\left(1+q+\lambda_0^2/2\lambda\right)\Gamma((5+2q)/2p)}{\omega_\perp(3+2q)\Gamma((3+2q)/2p)}, \tag{15}$$

$$E_{rot} = \frac{\kappa^2\omega_\perp(1+2q)\Gamma((1+2q)/2p)}{4\omega_\perp^0 q\Gamma((3+2q)/2p)}, \tag{16}$$

$$E_{int}^1 = 2N\tilde{a}\left(\frac{\omega_\perp}{\omega_\perp^0}\right)^{3/2} \frac{p\sqrt{\lambda}\Gamma^2(q+3/2)\Gamma(2q+1)\Gamma((4q+3)/2p)}{2^{(4q+3)/2p}\sqrt{\pi}\Gamma^2((2q+3)/2p)\Gamma^2(q+1)\Gamma(2q+3/2)}, \tag{17}$$

and

$$E^2_{int} = \frac{128}{15\sqrt{2}\pi} \frac{2\sqrt{2}}{\pi^{9/4}} N^{3/2} \tilde{a}^{5/2} \left(\frac{\omega_\perp}{\omega^0_\perp}\right)^{9/4} \left(\frac{2}{5}\right)^{(5q+3/2p)} p^{3/2} \lambda^{3/4} \times \frac{\Gamma^{5/2}(q+3/2)\Gamma(5q/2+1)\Gamma((5q+3)/2p)}{\Gamma^{5/2}((2q+3)/2p)\Gamma^{5/2}(q+1)\Gamma((5q+3)/2)}, \tag{18}$$

where $\tilde{a} = a/a_{ho}$ and $\Gamma(m)$ is the gamma function. The analytical expressions for the different energy components corresponding to the ground state of the condensates can be obtained by taking the limit κ and $q \longrightarrow 0$ of the Eqs. (14) - (18).[20,21] For particular values of N and $\tilde{a}$, the parameters $\omega_\perp$, q, λ, and p are obtained by minimizing the energy. Having described the variational method for obtaing the the wave function, we now proceed

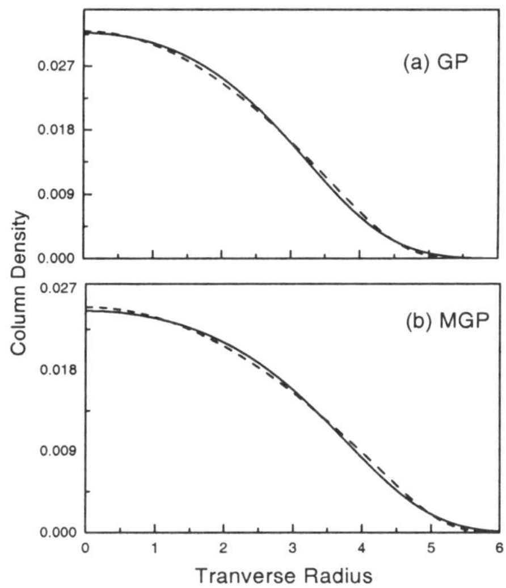

Fig. 1. Comparison of (a) GP and (b) MGP ground-state column density profiles as a function of the radial distance $r_\perp$, obtained by our variational formalism (solid lines) with the ones obtained by the method of steepest-descent in Ref.[22] (dashed lines). The calculations are performed for a condensate of $N = 500$ ^{85}Rb atoms trapped in an anisotropic trap with $\lambda_0 = \sqrt{8}$ and $\omega^0_z = 2\pi \times 220Hz$. The scattering length is $\tilde{a} = a/a_{ho} = 0.15155$.

with the presentation of the results for the equlibrium properties. In Figs. 1 and 2, we show the column density profiles of the ground and the vortex states respectively, obtained by employing both GP and MGP equations. It can be clearly seen that the ground state density profiles obtained by the variational method are almost similar to the ones obtained numerically. On the other hand, for vortex states the variational results vary slightly

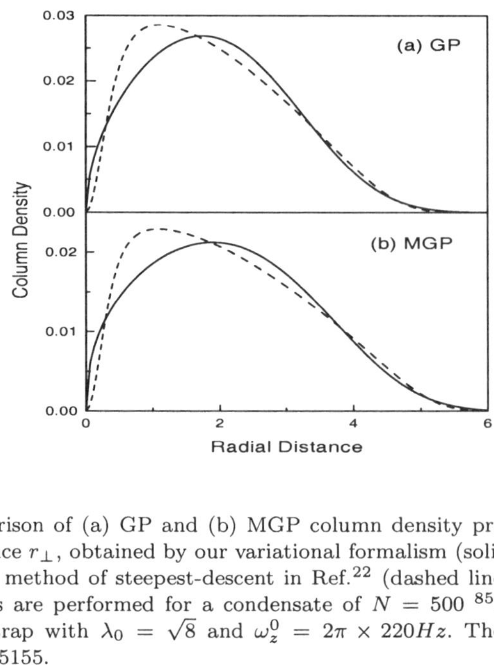

Fig. 2. Comparison of (a) GP and (b) MGP column density profiles as a function of the radial distance $r_\perp$, obtained by our variational formalism (solid lines) with the ones obtained by the method of steepest-descent in Ref.[22] (dashed lines) for a vortex state. The calculations are performed for a condensate of $N = 500$ ^{85}Rb atoms trapped in an anisotropic trap with $\lambda_0 = \sqrt{8}$ and $\omega_z^0 = 2\pi \times 220Hz$. The scattering length is $\tilde{a} = a/a_{ho} = 0.15155$.

in comparison to the steepest-descent results. There are some differences in the core regions. The densities attain their maximum more rapidly for the steepest-descent method described in Ref.[22] As a result of this, the peak value and peak position of the density profiles obtained by the two methods are somewhat different. On the other hand, the tail regions of the density profiles obtained by the two methods match quite well. In Table I, we present the results for the total energy and the chemical potential for the ground and the vortex states of the condensate. It can be seen from Table I that both for the ground and the vortex states the results obtained by the variational method are slighly higher than the steepest-descent numbers.

The results of Table I and those of Figs. 1 and 2 clearly demonstrate that the variational approach yields quite accurate results for the equilibrium properties of the condensates both for the ground and the vortex states in the large-gas-parameter regime. At this point we note that in addition

Table 1. Results for the chemical potential and the energies in units of $\hbar\omega_\perp^0$ obtained from the GP and the MGP calculations for the condensate of of $N = 500$ ^{85}Rb atoms trapped in an anisotropic trap with $\lambda_0 = \sqrt{8}$ and $\omega_z^0 = 2\pi \times 220Hz$. The scattering length is $\tilde{a} = a/a_{ho} = 0.15155$. Numbers in parenthesis are results of Ref.[22]

	Ground state		Vortex state	
	μ_1	E_1/N	μ_1	E_1/N
GP	13.008	9.519	13.334	9.889
	(12.980)	(9.497)	(13.187)	(9.7836)
MGP	15.490	11.088	15.81	11.443
	(15.453)	(11.061)	(15.623)	(11.305)

to the comparison with the ab-initio results, the virial relation among the different energy components also provides a way of checking the correctness and the accuracy of the variational solutions. The general virial relations for the ground and the vortex states of the condensates are given by

$$2T - 2U + 3E_{int}^1 + \frac{9}{2}E_{int}^2 = 0, \tag{19}$$

and

$$2T - 2U + 2E_{rot} + 3E_{int}^1 + \frac{9}{2}E_{int}^2 = 0, \tag{20}$$

respectively. These expressions have been derived by using the variational nature of the energies and the scaling transformation $n(\mathbf{r}_\nu) \longrightarrow \nu^3 n(\nu\mathbf{r})$. In all our calculations, the respective virial relations given above are satisfied up to the sixth decimal place or better. This further indicates that the variational method employed in this paper yields quite accurate results for the equlibrium properties of trapped condensates in the large-gas-parameter regime.

Now we focus our attention on the study on the equilibrium and dynamic properties of the condensates in the large-gas-parameter regime by employing MGP theory. For the purpose of calculations, the parameters of the trap and the condensate are chosen in accordance with the experiment of Cornish et al.[7] Following the experiment, we choose $N = 10^4$ ^{85}Rb atoms confined in an anisotropic trap with the trap frequencies $\omega_\perp^0 = 2\pi \times 17.5$ Hz and $\omega_z^0 = 2\pi \times 6.9$ Hz and the scattering length is varied from $a = 1400a_0$ to $a = 10000a_0$, where a_0 is the Bohr radius of hydrogen atom. We note here that this range of values of the scattering length falls in the large-gas-parameter regime as the maximum value of a corresponds to the peak gas parameter $x_{peak} \approx 10^{-2}$.

Table 2. Results for the chemical potential and the total energy per unit number of bosons in units of $\hbar\omega_\perp^0$ obtained from the GP and the MGP calculations for the condensate of $N = 10^4$ ^{85}Rb atoms trapped in an anisotropic trap with $\lambda_0 = 0.39$ and $\omega_\perp^0 = 2\pi \times 17.5Hz$.

a/a_0	GP		MGP	
	μ_1	E_1/N	μ_1	E_1/N
1400	9.845	7.107	10.257	7.357
3000	13.300	9.557	14.553	10.349
8000	19.637	14.067	24.486	17.184
10000	21.463	15.367	27.918	19.535

Table 3. Results for the chemical potential and the total energy per unit number of bosons in units of $\hbar\omega_\perp^0$ obtained from the GP and the MGP calculations for the condensate carrying a single vortex ($\kappa = 1$)of $N = 10^4$ ^{85}Rb atoms trapped in an anisotropic trap with $\lambda_0 = 0.39$ and $\omega_\perp^0 = 2\pi \times 17.5Hz$.

a/a_0	GP		MGP	
	μ_1	E_1/N	μ_1	E_1/N
1400	10.223	7.545	10.621	7.790
3000	13.649	9.955	14.898	10.739
8000	19.955	14.422	24.798	17.528
10000	21.774	15.715	28.223	19.871

In Tables II and III, we show the MGP results for the total energy and the chemical potential for the ground and the vortex states respectively, and compare them with the corresponding GP numbers. These results clearly show that the difference between the MGP and the GP results increases with the increase in the scattering length. For $a/a_0 = 1400$, the differences in the total energy and the chemical potential both for the ground and the vortex states are approximately 4%. On the other hand, for the maximum value of $a/a_0 = 10,000$, the MGP results for the total energy and

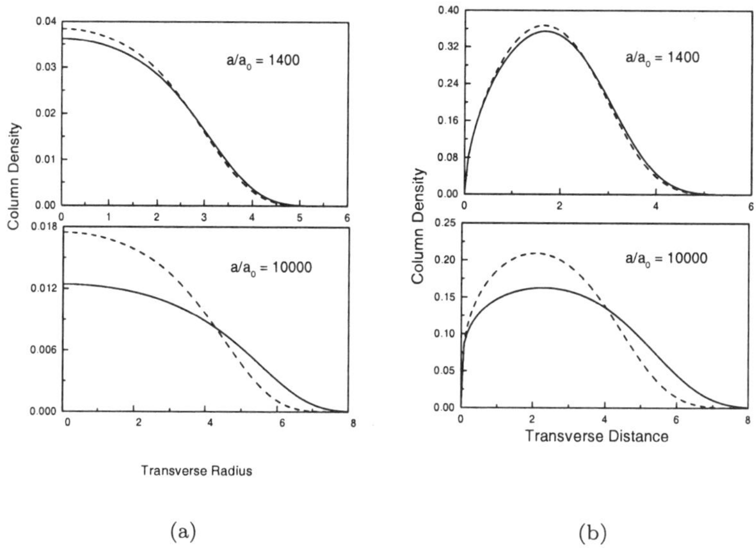

Fig. 3. Comparison of the GP (dashed line) and the MGP (solid line) column density profiles of a ground state (a) and the vortex state (b) as a function of the radial distance $r_\perp$ (in units of a_{ho}) from the z-axis for two different values of scattering length: $a/a_0 = 1400$ and $a/a_0 = 10000$. The condensate consists of $N = 10^4$ ^{85}Rb atoms trapped in an anisotropic trap with $\lambda_0 = 0.39$ and $\omega_\perp^0 = 2\pi \times 17.5Hz$.

the chemical potential are higher by around 30% over the corresponding GP numbers. Next the effect of large gas parameter on the density profiles of the ground and the vortex states are displayed in Fig. 3. The column density, which is measured to characterize the density profile of the condensates are shown for the two extreme values of a/a_0, namely, $a/a_0 = 1400$ and $10,000$. It can easily be seen from Fig. 3 that for $a/a_0 = 1400$, the differences between the column densities obtained by the GP and the MGP equations are quite small, both for the ground and the vortex states. In contrast to this, for $a/a_0 = 10000$, the two results for the density profiles show substantial differences. From these results, we conclude that in the large-gas-parameter regime the MGP results for the equlibrium properties of a condensate in the ground state and a condensate with a single quantized vortex are substantially altered as compared to the corresponding GP results. The next section is devoted to the discussion of the results for the collective oscillations of the trapped condensates.

3.1. *Collective Oscillations*

Having described the equilibrium properties, we now focus our attention on the calculation of collective oscillation frequencies of the trapped condensates. The main aim of this section is to study how the collective oscillation frequencies get modified in the large-gas-parameter regime. To calculate the frequencies of collective oscillations and also splitting of the two modes due to the presence of a quantized vortex, we employ sum-rule approach of many-body response theory. In the following, we briefly describe the sum-rule approach and present some results which are relevant for the present paper. For detail of the sum-rule approach, we refer to Ref.[23,24] One of the basic result of the sum-rule approach, which has been extensively exploited to calculate the frequencies of collective oscillations, is that the upper bound of the lowest excitation energy is given by

$$\hbar\Omega_{ex} = \sqrt{\frac{m_3}{m_1}}, \tag{21}$$

where

$$m_r = \sum_n |\langle 0|F|n\rangle|^2 \left(\hbar\omega_{n0}\right)^r, \tag{22}$$

is the r-th order moment of the excitation energy associated with the excitation operator F and Ω_{ex} is the frequency of the excitation. Here, $\hbar\omega_{n0} = E_n - E_0$ is the excitation energy of the eigenstate $|n\rangle$ of the Hamiltonian H. The upper bound given by Eq. (21) is close to the exact lowest excited state when this state is highly collective, that is, when the oscillator strength is almost exhausted by a single mode. This approximation is satisfied by trapped bosons in most of the cases. An important property of these moments is that for a given r some of the moments can be expressed as expectation values of the commutators between F and H in the ground state $|0\rangle$. For example, m_1 and m_3 can be expressed as

$$\begin{aligned} m_1 &= \frac{1}{2}\langle 0| \left[F^\dagger, [H, F]\right] |0\rangle, \\ m_3 &= \frac{1}{2}\langle 0| \left[\left[F^\dagger, H\right], [[H, [H, F]]\right] |0\rangle. \end{aligned} \tag{23}$$

The main advantage of the sum-rule approach is that it allows us to calculate the dynamic properties like excitation frequencies of many-body systems with the knowledge of the ground state $|0\rangle$ (or the ground state density) only. Thus, with the knowledge of a reasonably accurate ground state wave function (density), obtained by the variational method as described in the previous section, we calculate the expectation values given in Eq.(23)

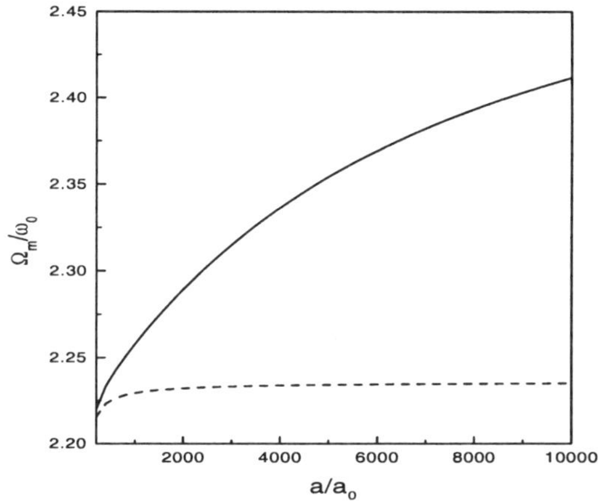

Fig. 4. The frequencies (in units of trap frequency $\omega_0 = 2\pi \times 12.83Hz$) of the monopole mode of 10^4 ^{85}Rb atoms confined in a spherically symmetric trap as a function of scattering length a/a_0. The solid and the dashed lines represent the MGP and the GP results respectively.

to determine the frequencies of collective oscillations. In this paper, we consider the monopole and the quadrupole modes of the collective oscillations. To perform the calculations, the general form of the excitation operator is chosen as

$$F = \sum_i \left(x_i^2 + y_i^2 - \alpha z_i^2 \right) . \tag{24}$$

For spherically symmetric trap, $\alpha = -1$ and $\alpha = 2$ denote monopole ($l = 0, m = 0$) and quadrupole ($l = 2, m = 0$) modes, respectively. On the other hand, for axially symmetric trap ($\lambda_0 \neq 1$) two modes get coupled, and α is determined by minimization of excitation energy.[20] These two modes constitute the upper and the lower branches of the breathing mode of the collective oscillations.

First, we present the results for the spherically symmetric trap. For this case, the frequencies of the monopole and the quadrupole modes are given by

$$\frac{\Omega_m}{\omega_0} = \left(5 - \frac{T}{U} + \frac{27}{8} \frac{E_{int}^2}{U} \right)^{1/2} , \tag{25}$$

and

$$\frac{\Omega_q}{\omega_0} = 2\left(1 + \frac{T}{U}\right)^{1/2}, \tag{26}$$

respectively, where ω_0 is the frequency of the spherically symmetric trap. The frequency of the quadrupole mode does not depend on the interaction energy, and the same expression is obtained for the GP case also. On the other hand, the frequency of the monopole mode explicitly depends on the interaction energy arising due to the LHY term. Therefore, in the large-gas-parameter regime, the MGP results for the frequency of the monopole mode will deviate significantly from that of GP numbers. We show this in Fig. 4 by plotting Ω_m as a function of a/a_0, both for the GP and the MGP cases. The GP result for the monopole frequency remains almost constant with the increase in the value of the scattering length in the large-gas-parameter regime. In contrast to this, the frequency of the monopole mode obtained by the MGP equation shows an increasing trend as a function of the interaction strength. Next we focus our attention on the anisotropic case ($\lambda_0 \neq 0$). Unlike spherically symmetric case, it is not possible to obtain analytic expressions for the frequencies for the anisotropic traps. In Figs. 5, we show the upper (Ω_u) and the lower (Ω_l) frequencies as a function of the dimensionless parameter a/a_0. Notice that the maximum value of the parameter a/a_0 is chosen to be 10^4, in accordance with the experimental number,[7] and the corresponding peak gas parameter is $\approx 10^{-2}$. For comparison, we also show in these figures, the corresponding results obtained within the mean-field GP theory.

It can be clearly seen from Fig. 5 that for small values of the gas parameter (proportional to a) the numbers obtained by employing the MGP and the GP equations are quite close. The difference between these two results, however, increases with the increase in the values of a. For example, for $a/a_0 = 8000$, the MGP number for Ω_u is 6% higher than the corresponding GP result. Thus we conclude that the introduction of LHY term in the interatomic interaction energy results in significant correction to the upper frequency of $m = 0$ mode of the collective oscillations. Considering the accuracy of current experiments in measuring the frequencies of the collective excitations, it should be possible to observe these corrections experimentally. It is also important to note that the frequency Ω_u obtained within the GP theory, almost remains constant with the increase in the parameter a/a_0. In contrast to this, for the MGP case Ω_u grows monotonically with a/a_0.

Next we present the results for the lower frequency of the breathing

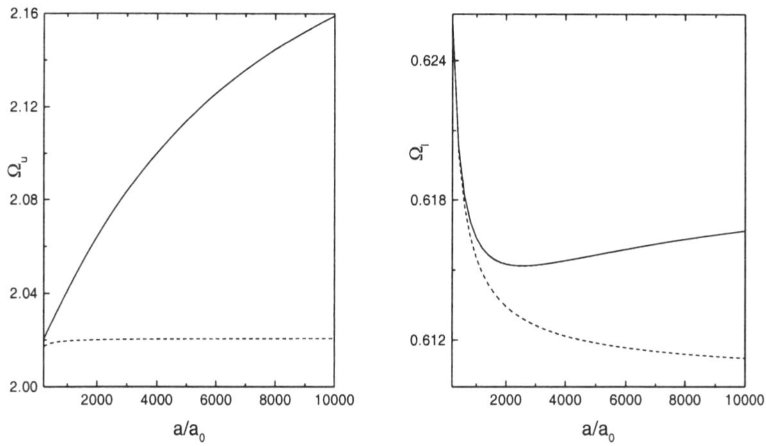

(a) (b)

Fig. 5. The frequencies (in units of trap frequency $\omega_\perp^0$) of the upper (a) and the lower (b) braches of the mode $m = 0$ of 10^4 ^{85}Rb atoms confined in an axially symmetric trap as a function of scattering length a/a_0. The trap parameters are $\lambda_0 = 0.39$ and $\omega_\perp^0 = 2\pi \times 17.5Hz$. The solid and the dashed lines represent the MGP and the GP results respectively.

mode which is also shown in Fig. 5. For small values of the gas parameter, the frequency of the lower branch obtained by the GP equation is very close to the corresponding MGP results. In the same regime of the gas parameter, the frequency Ω_l decreases with the increase in the gas parameter both for the GP and the MGP cases. However, in contrast to the GP case frequency of lower branch for the MGP case starts increasing, albeit slowly, after a critical value of $a/a_0 \approx 2200$. It is also noteworthy that the change in the frequency Ω_l obtained by employimg the MGP equation is considerably lower than that in the upper frequency Ω_u. For example, for $N = 10^4$ and $a/a_0 = 8000$, the change in the lower frequency is just 0.8% while that in the upper frequency is 6%. Finally, we consider the effect of quantized vortex state on the collective oscillations of the trapped condensates. In the presence of a quantized vortex state two degenerate quadrupolar modes $(l = 2, m \pm 2)$ undergo a splitting due to the violation of time reversal symmetry. This splitting of the two modes can be employed to detect the presence of a vortex in the condensate, which is otherwise difficult to observe from the static measurements. The two quadrupolar modes considered in

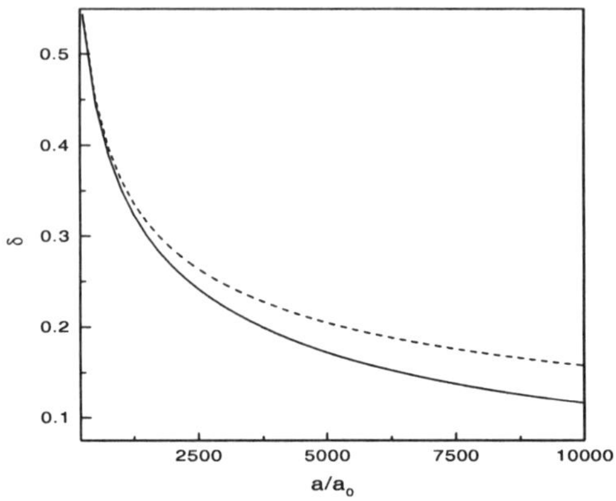

Fig. 6. The frequency shift (in unit of $\omega_\perp^0$) of the quadrupole oscillations of a condensate due to the presence of a single quantized vortex ($\kappa = 1$) as a function of the scattering length a/a_0. The solid line corresponds to the MGP result and the dashed line is obtained with the GP calculation. The trap parameters are $\lambda_0 = 0.39$ and $\omega_\perp^0 = 2\pi \times 17.5 Hz$.

this paper are $m = 2$ and $m = -2$, and they are excited by the operators F_+ and F_- respectively. These operators are given by

$$F_\pm = (x \pm iy)^2. \tag{27}$$

By applying the sum-rule approach, we obtain following expression for the splitting between the two quadrupolar modes

$$\hbar\delta = \hbar\left(\omega_+ - \omega_-\right) = \frac{m_2^-}{m_1^+}, \tag{28}$$

where

$$m_1^+ = \langle 0| \left[F^\dagger, [H, F]\right] |0\rangle,$$
$$m_2^- = \langle 0| \left[\left[F^\dagger, H\right], [H, F]\right] |0\rangle.$$

The above expectation values of the commutators can be calculated, and they yield following expression for the splitting between the two frequencies under single mode approximation[21]

$$\delta = \frac{2}{m}\frac{N\hbar\kappa}{\langle r_\perp^2\rangle}, \tag{29}$$

where $\langle r_\perp^2 \rangle$ is the transverse size of the condensate determined by the average

$$\langle r_\perp^2 \rangle = \int \left(x^2 + y^2\right) n(\mathbf{r}) d\mathbf{r}. \tag{30}$$

It is important to note that although the frequency shift does not explicitly depend on the boson-boson coupling paprameter, implicit dependence on the two-body interaction enters through the wave function or the density of the condensate, which crucially depends on the nature of boson-boson interaction. The MGP and the GP results for the splitting will vary significantly in the large-gas-parameter regime, as the tranverse sizes of the condensates obtained by employing the two equations differ signifiacntly in this regime. We show the results for the MGP and the GP frequency shifts as a function of the scattering length in Fig. 6. The frequency shift between the two quadrupole modes decreases with the increase in the scattering length due to the fact that higher values of the scattering length correspond to greater repulsive interaction between the bosons, resulting in condensates with larger values of the transverse size. For up to around $a/a_0 = 1000$, the GP and the MGP frequency shifts are nearly identical, and as the scattering length is increased beyond this value, the difference between the two results starts growing. For example, at the maximum value of the scattering length ($a/a_0 = 10000$), we find that the GP frequency shift is around 26% higher in comparison to the MGP result.

From the above results, we conclude that in the large-gas-parameter regime, the MGP and the GP results for the frequencies of the collective oscillations and the shift of the two quadrupole modes are substantially different. The range of scattering length considered in this paper have already been achieved by tuning the Fesbach resonance and these changes are quite large in magnitude to be easily observed in the measurements performed with the condensates as achieved in Ref.[7]

4. Conclusion

In this paper we have studied the equlibrium and the dynamic properties of the trapped condensates in the large-gas-parameter regime. We have considered both ground and the vortex states of the condensates. For this purpose, we have employed the MGP theory which has been obtained by including the second term (LHY term), arising from the depletion of the condensate, in the perturbative expansion of the interatomic interaction energy per particle of uniform Bose gas within the local density approximation. We have used the variational approach to solve the MGP equation by employing a suitable ansatz for the wave function representing the ground and vortex states of the condensates. The wave function obtained by the variational approach is then, employed to calculate the equlibrium properties like the total energy, chemical potential, and the column density profile for a wide range of scattering length lying well within the large-gas-parameter regime. The correctness and accuracies of our solutions are checked by verifying the generalized virial relation and also comparing them with the solutions obtained by the steepest-descent method. By using, the variational wave function and the sum-rule approach, we have calculated the frquencies of monopole and quadrupole modes, and the frequency shift between the two quadrupole modes of the collective oscillations. To test the accuracy of GP theory, we have made a detailed comparison of the results of the GP and the MGP calculations for all the observables mentioned above. We have found that the MGP calculations introduce sizable corrections in all the properties of the condensates lying in the large-gas-parameter regime. These changes are quite large in magnitude, and therefore, can easily be observed in measurements involving condensates with large gas parameter as achieved in Ref.[7] The comparison of the results obtained in this paper with the experimental results will also test the validity of the GP theory for the description of condensates with large values of the gas parameter.

5. Acknowledgment

I wish to thank Dr. M. P. Singh for the collaborative work presented in this paper. I also wish to thank Prof. M. Hjorth-Jensen and Dr. J. K. Nilsen for providing us the data for the density profile reported in Ref.[22]

References

1. L. P. Pitaevskii, Sov. Phys. JETP **13**, 451 (1961); E. P. Gross, Nuovo Cimento **20**, 454 (1961); J. Math. Phys. **4**, 195 (1963).
2. C. J. Pethik and H. Smith, *Bose-Einstein Condensation in Dilute Gases* (Cambridge University Press, Cambridge, 2002).
3. L. Pitaevskii and S. Stringari, *Bose-Einstein Condensation*, (Clarendon Press, Oxford, 2003).
4. F. Dalfovo, S. Giorgini, L. Pitaevskii and S. Stringari, Rev. Mod. Phys. **71** 463 (1999).
5. A. L. Fetter, *Proc. International School of Physics Enrico Fermi*, Course **CXL**, Ed. M. Inguscio, S. Stringari and C. E. Wieman (IOS Press, Amsterdam, 1999).

6. Y. Castin in *Coherent Matter Waves, Les Houches* ***LXXII***, Eds. R. Kaiser, C. Westbrook and F. David (Springer, Berlin, 1999).
7. S. L. Cornish, N. R. Claussen, J. L. Roberts, E. A. Cornell, and C. E. Wieman,Phys.Rev. Lett **85**, 1795 (2000).
8. S. Giorgini, J. Boronat, and J. Casulleras, Phys. Rev. A **60**, 5129 (1999).
9. F. Zambelli and S. Stringari, Phys. Rev. Lett. **81**, 1754 (1998).
10. A. Svidzinsky and A. L. Fetter, Phys. Rev. A **58**, 3168 (1998).
11. P. C. Haljan, I. Coddington, P. Engels, and E. A. Cornell, Phys. Rev. Lett. **87**, 210403 (2001).
12. N. N. Bogoliubov, J. Phys. (Moscow) **11**, 23 (1947).
13. T. D. Lee, K. Huang and C. N. Yang, Phys. Rev. A **106**, 1135 (1957).
14. T. T. Wu, Phys Rev. **115**, 1390 (1959).
15. A. Minguzzi, S. Succi, F. Toschi, M. P. Tosi, and P. Vignolo, Phys. Rep. **395**, 223 (2004); W. Bao and W. Tang, J. Comp. Phys. **187**, 230 (2003) and references their in.
16. M. P. Singh and A. L. Satheesha, Eur. Phys. J. D **7**, 391 (1998).
17. A. Banerjee and M. P. Singh, Phys. Rev. A. **64**, 063604 (2001).
18. A. Fabrocini and A. Polls, Phys. Rev. A **64**, 063610-1 (2001).
19. L. Pitaevskii amd S. Stringari, Phys. Rev. Lett. **81**, 4541 (1998).
20. A. Banerjee and M. P. Singh, Phys. Rev. A. **66**, 043609 (2002).
21. A. Banerjee and M. P. Singh, Phys. Rev. A. To be published.
22. J. K. Nilsen, J. Petit, M. Guilleumas, M. H-Jensen and A. Polls, Phys Rev. A **71**, (2005).
23. O. Bohigas, A. M. Lane and J. Martorell, Phys. Rep. **51**, 267 (1971).
24. E. Lipparini and S. Stringari, Phys. Rep. **175**, 103 (1989).

A Feynman-Kac path integral study of Rb gas

S. Datta

S. N. Bose National Centre for Basic Sciences, India

We study the ground and excited states of a weakly interacing Rb gas (with positive length) in connection with Bose Einstein condensation to test the validity of using the mean field theory and Born approximation at $T = 0$ by path integral technique. We also study thermodynamical properties Rb gas. Within numerical limitations, this method is exact in priciple and turns out to be a better alternative to GP as all the ground and excited states properties can be calculated in a much simpler way.

Keywords: Generalized Feynman Kac method(GFK), Exciation Spectrum, Condensation fraction

1. Introduction

With the experimental realization of Bose Einstein Condensation in alkali gases[1], the study of many boson systems has become an area of active research interst. Previous numerical procedures are based on mean field theory like Gross Pitaevski [2] etc. They seem to work well for ground state properties but turn out to be approximate as they fail to include correlations in the many body theory. Investigations of effects beyond mean field theory is an important task and makes the many boson systems interesting even from the many body perspective[3]. Moreover earlier calculations with δ function potential do not solve the many body problem exacxtly but only within a perturbation theory as in a system of Bose gas with δ function potential, particles do not collide [4]. As a result, speculations (particularly excitation frequencies etc) based on these methods differ drastically for different experimental modes. So an alternative to GP was necessary which can describe the effect of interaction in a more reliable way and predict the excitation frequencies and other properties more accurately. Eventhough Monte Carlo techniques are slow, computationally expensive and faces sign problem for fermionic systems, these are the only numerical techniques available for these kind of many body systems, which are exact and can include corrlations in a reliable way. We also test the validity of

Born approximation at low energy and temperature.

Thermodynamics of Bose gases was studied before at a higher temperature ($k_BT >> \hbar\omega$) by a semiclassical treatment[5]. Since effects of interactions become more pronounced at low temperatures we restrict our discussions at low but finite temperature ($k_BT < \mu < T_c$). At low temperature the de Broglie wavelength of the atoms become appreciable, the study of thermodynamic behaviour at low temperatures (of the order of harmonic oscillator temperature) requires a quantum description of a lowlying elementary modes. As Quantum Monte Carlo technque and many body theory are closely connected, in this write up we present a quantum monte Carlo method namely Generalized Feynman-Kac method (GFK)[6,7] to study the thermodynamic properties of a Bose gas. From the equivalence of the imaginary time propagator and temperature dependent density matrix, finite temperature results can be obtained from the same zero temperature code by running it for finite time. We calculate temperature variation of condensation fraction, total energy, release energy, frequency shift, chemical potential for system of 100 Rb^{87} atoms.

2. Theory

2.1. *Path integral Theory at T=0*

2.1.1. *Feynman-Kac Path integretion*

For the Hamiltonian $H = -\Delta/2 + V(x)$ consider the initial value problem

$$i\frac{\partial u}{\partial t} = (-\frac{\Delta}{2} + V)u(x,t)$$
$$u(0,x) = f(x) \tag{1}$$

with $x \in R^d$ and $u(0,x) = 1$. The solution of the above equation can be written in Feynman-Kac representation as

$$u(t,x) = E_x exp\{-\int_0^t V(X(s))ds\} \tag{2}$$

where X(t) is a Brownian motion trajectory and E is the average value of the exponential term with respect to these trajectories. The lowest energy eigenvalue for a given symmetry can be obtained from the large deviation priniciple of Donsker and Varadhan [8],

$$\lambda = -\lim_{t\to\infty}\frac{1}{t}lnE_x exp\{-\int_0^t V(X(s))ds\} \tag{3}$$

Now to speed up the convergence we use Generalized Feynman-Kac (GFK) method.

2.1.2. *Generalized Feynman Kac path integretion*

To formulate the (GFK) method, we first rewrite the Hamiltonian as $H = H_0 + V_p$, where $H_0 = -\Delta/2 + \lambda_T + \Delta\psi_T/2\psi_T$ and $V_p = V - (\lambda_T + \Delta\psi_T/2\psi_T)$. Here ψ_T is a twice differentiable nonnegative reference function and $H\psi_T = \lambda_T\psi_T$. The expression for the energy can now be written as

$$\lambda = \lambda_T - \lim_{t\to\infty} \frac{1}{t} lnE_x exp\{-\int_0^t V_p(Y(s))ds\} \tag{4}$$

where Y(t) is the diffusion process which solves the stochastic differential equation

$$dY(t) = \frac{\Delta\psi_T(Y(t))}{\psi_T(Y(t))}dt + dX(t) \tag{5}$$

The presence of both drift and diffusion terms in this expression enables the trajectory Y(t) to be highly localized. As a result, the important regions of the potential are frequently sampled and Eq (3) converges rapidly.

2.2. *Fundamentals of BEC*

Even though the phase of Rb vapors at T=0 is certainly solid, Bose condensates are preferred in the gasous form over the liquids and solids because at those higher densities interactions are complicated and hard to deal with on an elementary level. They are kept metastable by maintaining a very low density. So keeping the density low only two body collisions are allowed as a result of which dilute gas approximation [9] still holds for condensates which tantamounts to saying $na^3 << 1$ (a is the scattering length of s wave). Now defining $n = N/V = r_{av}^{-3}$ as a mean distance between the atoms (definition valid for any temperature), the dilute gas condition reads as $a << r_a v$ and zero point energy dominates (dilute limit). In the dense limit, for $a \approx r_{av}$ on the other hand the interatomic potential dominates .The gas phase is accomplished by reducing the material density through evaporative cooling.

2.3. *Schroedinger formalism for condensate dynamics*

At low energy the motion of condensate can be represented as

$$[-\Delta/2 + V_{int} + \frac{1}{2}\sum_{i=1}^{N}[{x_i}^2 + {y_i}^2 + \lambda {z_i}^2]\psi(\vec{r}) = E\psi(\vec{r}) \tag{6}$$

where $\frac{1}{2}\sum_{i=1}^{N}[x_i{}^2 + y_i{}^2 + \lambda z_i{}^2]\psi(\vec{r})$ is the anisotropic potential with anisotropy factor $\lambda = \frac{\omega_z}{\omega_x}$. Now

$$V_{int} = V_{Morse} = \sum_{i,j} V(r_{ij}) = \sum_{i<j} D[e^{-\alpha(r-r_0)}(e^{-\alpha(r-r_0)} - 2)] \quad (7)$$

The above Hamiltonian is not separable in spherical polar coordinates because of the anisotropy. In cylindrical coordinates the noninteracting part behaves as a system of noninteracting harmonic oscillators and can be writtem as follows :

$$\begin{aligned} &[-\frac{1}{2\rho}\frac{\partial}{\partial\rho}(\rho\frac{\partial}{\partial\rho}) - \frac{1}{\rho^2}\frac{\partial^2}{\partial\phi^2} - \frac{1}{2}\frac{\partial^2}{\partial z^2} \\ &+\frac{1}{2}(\rho^2 + \lambda^2 z^2)]\psi(\rho, z) \\ &= E\psi(\rho, z) \end{aligned} \quad (8)$$

The energy 'E' of the above equation can be calculated exactly which is

$$E_{n_\rho n_z m} = (2n_\rho + |m| + 1) + (n_z + 1/2)\lambda \quad (9)$$

In our guided random walk we use the solution of Schoroedinger equation for harmonic oscillator in d-dimension as the trial function as follows [10]:

$$\psi_{n_\rho n_z m}(\vec{r}) \simeq \exp^{\frac{-z^2}{2}} H_{n_z}(z) \times e^{im\phi}\rho^m e^{-\rho^2/2} L_{n_\rho}{}^{(m)}(\rho^2) \quad (10)$$

2.4. *Positive scattering length: Rb*

As the potential does not sustain any many body bound state (which is ensured by suitably choosing the value of the parameters α and r_0) and the scattering length is positive the system behaves as a gas or as a metastable state which can be long-lived at very low densities[3]. In the table below, we explicitly show the expectation values of trap potential, interatomic potential and kinetic energy as three components of total energy for different number of particles and it is observed that virial theorem is satisfied in each case[11].

$$2E_{kin} - 2E_{HO} + 3E_{pot} = 0 \quad (11)$$

From Fig 1, 2 and 3, we see that energy/particle rises with increase in number of atoms in the trap for different symmetry states.

Table 1. Results for ground state of Rb with $\lambda = \sqrt{8}$ Chemical potential and energy are in units of $\hbar\omega_\perp$ and length is in units of $a_\perp$. Numbers in the brackets correspond to the reference[20]

N	μ	E/N	E/N_{kin}	E/N_{ho}	E/N_{pot}	$\sqrt{<x^2>/<z^2>}$
1	2.414213	2.414213	1.207409	1.206803	0.0	1.679563
						(1.68333)
10	2.448952	2.431595(5)	1.202488	1.211725	.017369	1.684900
40	2.564350	2.489287(1)	1.196455	1.217758	0.075068	1.68732
70	2.678893	2.546602(6)	1.188591	1.225621	0.132339	1.688079
100	2.792482	2.603549(3)	1.180656	1.233556	0.189134	1.688960
	(2.88)	(2.66)				(1.79545)
200	3.149535	2.79075(7)	1.061734	1.35247	0.367660	1.690056
	(3.21)	(2.86)				(1.88888)

Next we show the variation of Energy/particle with number of atoms for different symmetry states.

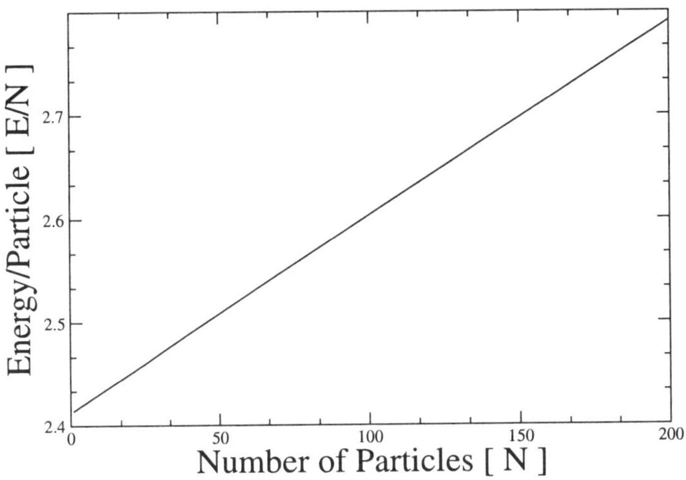

Fig. 1. A plot for the Condensate Energy/Particle versus Number of atoms in trap for 200 particles for the ground state : $n_z = n_\rho = m = 0$; this work

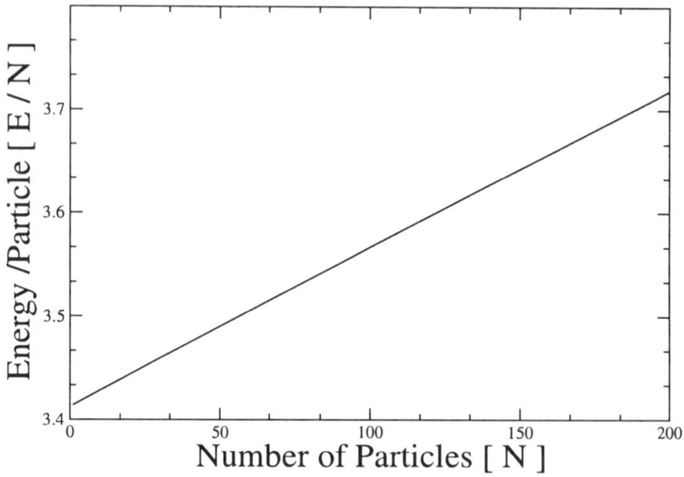

Fig. 2. A plot for the Condensate Energy/Particle versus Number of atoms in trap for 200 particles for the 1st excited state : $n_z = n_\rho = 0, m = 1$;this work

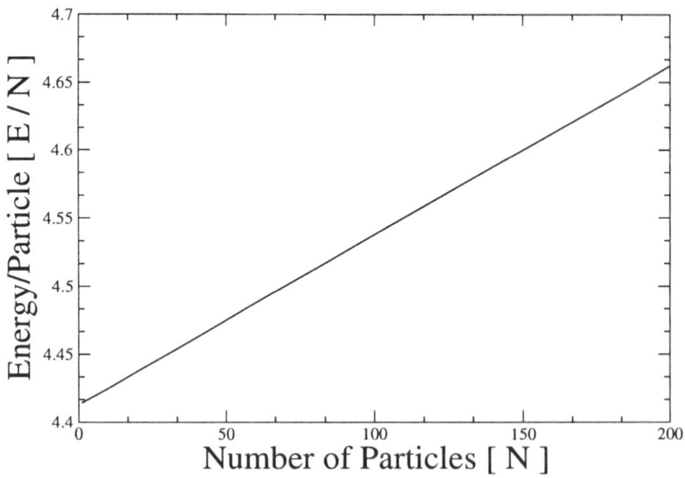

Fig. 3. A plot for the Condensate Energy/Particle versus Number of atoms in trap for 200 particles for the 2nd excited state: $n_z = n_\rho = 0, m = 2$; this work

In the following table we show the frequencies for different symmetries.

Table 2. **frequency ω for lowest lying modes**

N	quantum numbers	ω
100	$n_z = 0 = n_\rho = m = 0$	0
100	$n_z = 0 = n_\rho, m = 2$	1.994966

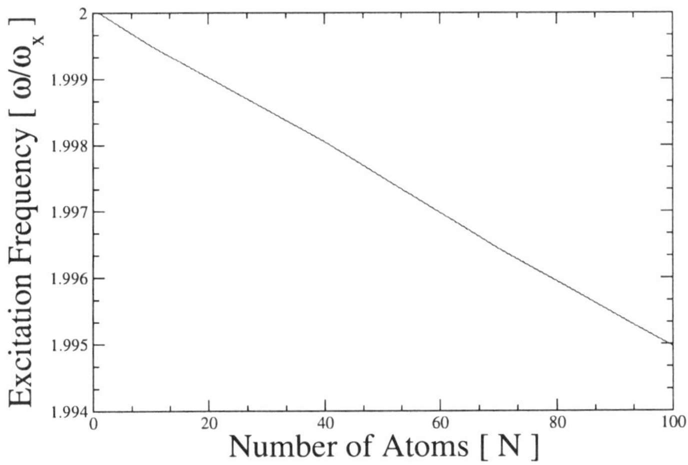

Fig. 4. A plot of Excitation Frequency vs Number of Atoms for lowest lying m=2;this work

2.5. *Effects of temperature on the frequency shifts; comparison with other experiments and theories*

Underneath from JILA data[12], one observes a large temperature dependent frequency shift for both m=0 and m=2 modes. For m=2 mode, starting from Stringari limit it decreases all the way up to $0.9T_c$ whereas for m=0 mode it shows a rising trend with rise in temperature. Our data [Fig. 6] agrees with JILA TOP data and theoretical data in Fig. 7 Ref [13,14,15] all the way to $0.9T_c$. Temperature variation of m=0 mode from our data [Fig 8] also agrees with JILA data when dynamics of thermal cloud is considered.

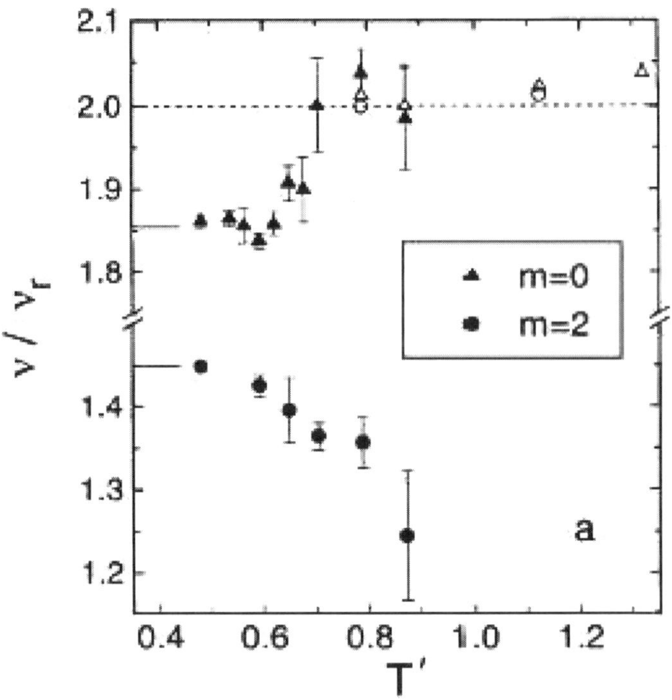

Fig. 5. Effects of temperature on m=2 mode; JILA data; The data marked by triangles represent the m=0 mode and solid circles represent m=2 mode

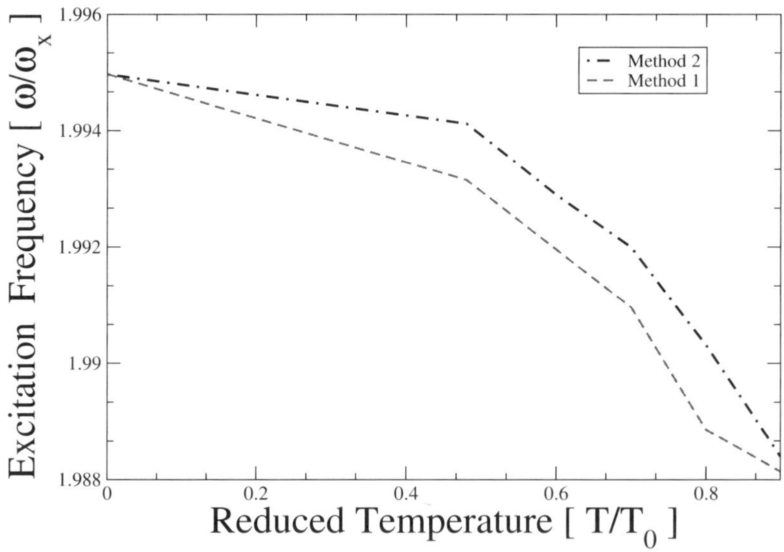

Fig. 6. Effects of temperature on m=2 mode; this work. The top cuve from equivalent T=0 system[method 2], the bottom curve by putting temperature directly[method 1]. Both show agreement with experimental data all the way up to $0.9T_c$

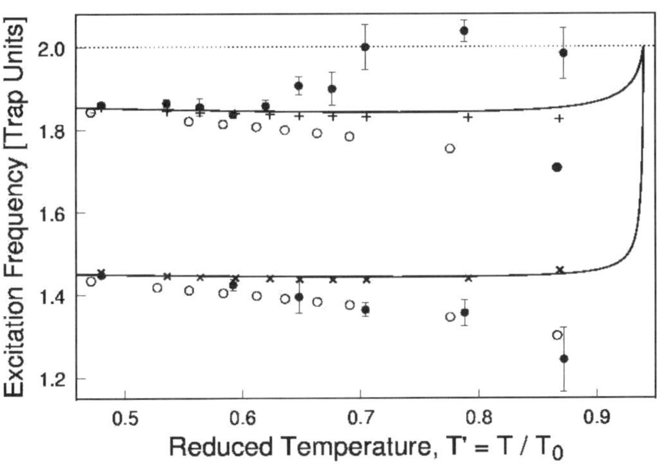

Fig. 7. Effects of temperature on m=2 mode; Ref 5 which agrees with JILA for m=2 mode but shows opposite trend for m=0 mode

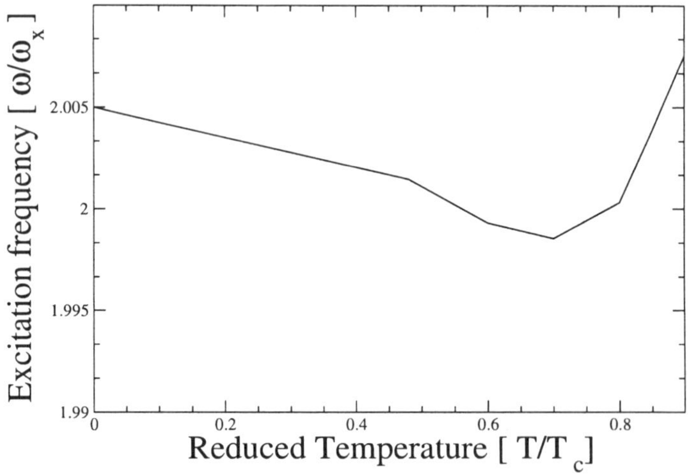

Fig. 8. Effects of temperature on m=0 mode from GFK considering noncondensate dynamics[this work], shows resemblence with JILA and Ref 15

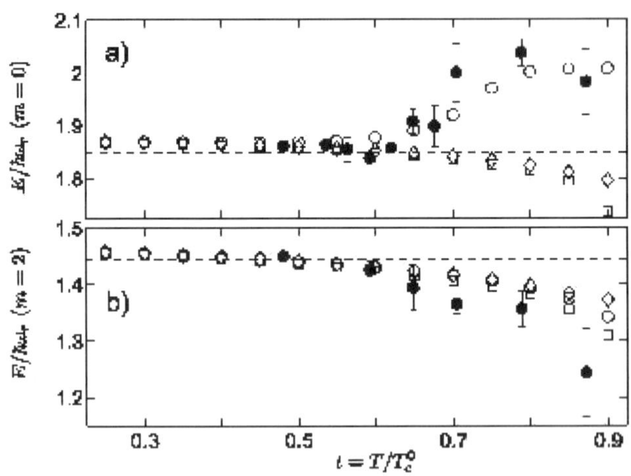

Fig. 9. Effects of temperature on m=0 mode; Morgan data

2.6. *Effects of temperature on condensation fraction*

Density of condensate atoms decreases in the trap as temperature increases. This lowers the interaction energy of the condensate atoms resulting in a shift in the critical temperature. As a matter of fact in the interacting case, the critical temperature decreases. This is a very unique feature of trapped gas. In the case of uniform gas it is just the other way around.

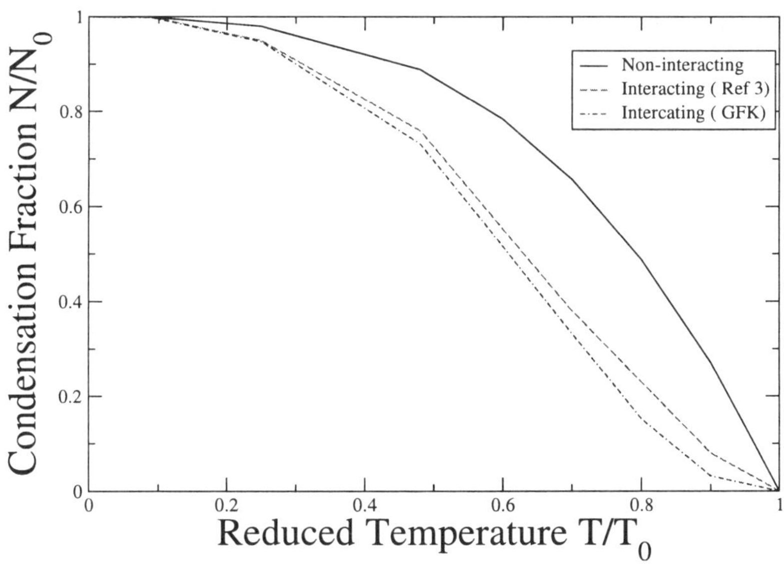

Fig. 10. Condensation fraction vs Reduced Temperature ; this work. The inner curve corresponds to the 100 interacting atoms and the outer one corresponds to the noninteracting case. The number of condensed particles decreases with the interaction

2.7. *The effect of temperature on condensate density*

Underneath we plot the axial density due to condensate along x axis.

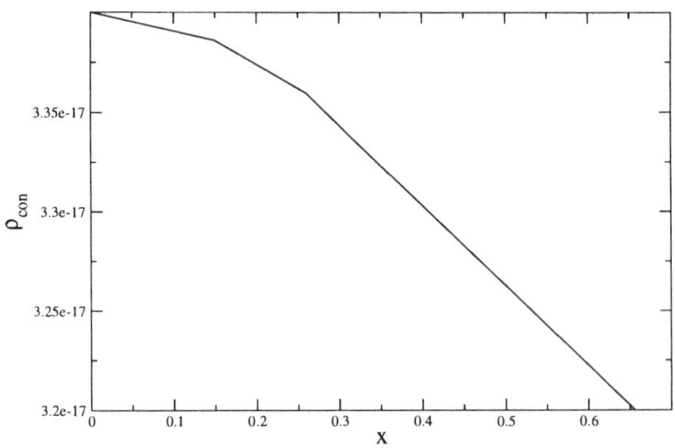

Fig. 11. Axial density profile due to Condensate at temperature T=0.48

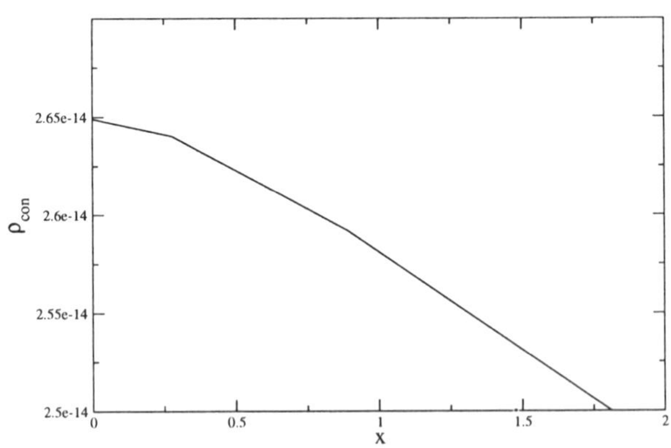

Fig. 12. Axial density profile due to condensate at temperature T=0.6

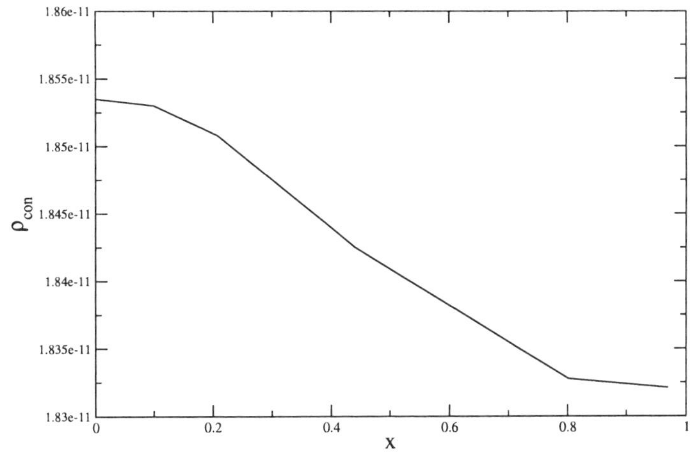

Fig. 13. Axial density profile due to condensate at temperature T=0.7

2.8. *Effect of temperature on the total energy of the Bose gas*

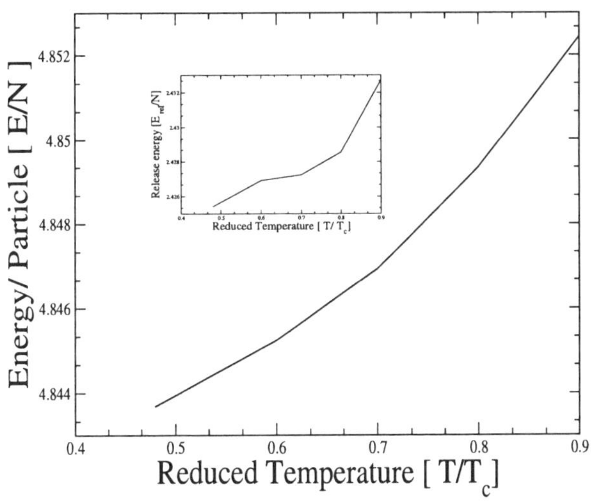

Fig. 14. Total energy/particle as a function of reduced temperature. Inset: Release energy as a function of temperature

2.9. *Effects of temperature on Chemical potential*

The chemical potential μ can be written in terms of different contributions to the energy, namely E_{kin}, E_{ho} and E_{int} as follows[4].

$$\mu = \frac{1}{N}(E_{kin} + E_{ho} + 2E_{int}) \tag{12}$$

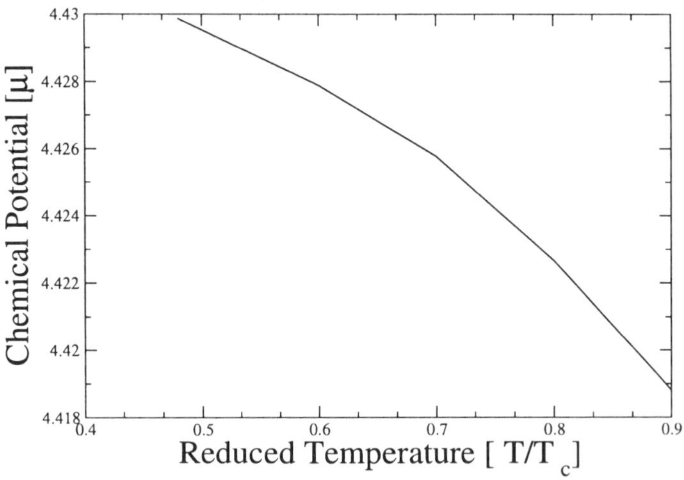

Fig. 15. Chemical Potential vs Reduced Temperature

Acknowledgements:
Financial help from DST (under Young Scientist Scheme (award no. SR/FTP/-76/2001)) is gratefully acknowledged. The author would like to thank Prof J. K . Bhattacharjee, Indian Association for the Cultivation of Science, India for suggesting the problem and many stimulating discussions and also Prof C. W. Clark of NIST, USA for suggesting very usuful references.

References

1. M. H. Anderson, J.R. Ensher, M.R. Matthews, C. E. Wieman E. A. Cornell, Science **269**,198 (1995)
2. V. L. Ginzburg and L.P. Pitaevski, Zh. Eksp. Teor Fiz, **34** 1240(1958)[Sov. Phys. JETP 7, 858 (1958)], E.P. Gross, J. Math Phys.**4**, 195(1963)
3. S. Giorgini, J. Boronat and J. Casulleras, Phys. Rev A **60** 5129 (1999).
4. B. D. Esry and C. H. Green, Phys. Rev A **60** 1451, (1999)
5. S. Giorgini, L. p. Pitaevskii and S. Stringari, Phys. Rev Lett. **78** 3987 (1997), arXiv cond-mat/9704014 (1997).
6. M.Cafferel and P. Claverie, J. Chem Phys. **88** , 1088 (1988), **88**, 1100 (1988)
7. S. Datta, J. L Fry, N. G. Fazleev, S. A. Alexander and R. L. Coldwell, Phys Rev A **61** (2000) R030502, Ph. D dissertation, University of Texas at Arlington, 1996
8. M. D. Donsker and M. Kac, J. Res. Natl. Bur. Stand, **44**
9. J. L. duBois, Ph D dissertation, University of Delaware,(2003).
10. R. J. Dodd, J Res. Natl Inst. Stand. Technol **101**,545(1996)
11. F. Dalfovo and S. Stringari, Phys. Rev A **53**, 2477(1996)
12. D. S. Jin, M. R. Mathews, J. R. Ensher, C. E. Wieman and E. A. Cornell Phys Rev Lett **78** 764 (1997)
13. D. A. Hutchinson, R. J. Dodd ans K Burnett, Phys. Rev. Lett **81**, 2198 (1998)
14. S. A. Morgan, J.Phys. B **33**,3847-3893, 2000
15. S. A. Morgan, M.Rusch, D. A. W. Huchinson, K. Burnett, Phys. Rev Lett.,**91**, 250403, 2003

Mean Field Theory for Interacting Spin-1 Bosons on a Lattice

Ramesh V. Pai

Department of Physics, Goa University,
Taleigao Plateau, Goa 403 206, India
E-mail: rvpai@unigoa.ac.in

K. Sheshadri

686, BEL Layout, 3rd Block,
Vidyaranyapura, Bangalore 560 097, India
E-mail: kshesh@gmail.com

Rahul Pandit

Centre for Condensed Matter Theory,
Department of Physics,
Indian Institute of Sciences,
Bangalore 560 012, India
E-mail: rahul@physics.iisc.ernet.in

We develop a simple mean-field theory for studying spin-1 Bose-Hubbard model for cold atoms in an optical lattice model and obtain its zero temperature phase diagram. The superfluid phase can be polar (ferro) for antiferromagnetic (ferromagnetic) interaction. The superfluid to Mott insulator transition is continuous for ferromagnetic interaction. However, the polar superfluid to Mott insulator phase is a first order transition for even densities and continuous for odd densities.

Keywords: Optical cooling of atoms, spinor condensates, superfluid to Mott insulator transition.

1. Introduction

There has been a resurgence of interest in quantum phase transitions;[1] and in this context, interacting bosons have attracted considerable attention[2,3] with a fruitful interplay between theory,[2–13] numerical simulations,[14–17] and experiments. Experimental systems include liquid ^{4}He in porous media like vycor or aerogel,[18] microfabricated Josephson-junction arrays,[19,20] the disorder-driven superconductor-insulator transition in thin films of superconducting materials like bismuth,[21] and flux lines in type-II superconductors pinned by columnar defects aligned with an external magnetic field.[22] But by far the cleanest realization of such bosonic systems are ultracold atoms[23] in optical lattices created by using the interference pattern of intersecting laser beams. A convincing experimental realization is the system of spin-polarized ^{87}Rb where a superfluid (SF) to Mott insulator (MI) transition has been demonstrated[24] by changing the strength of the onsite potential.[7] Jaksch et.al.[25] have shown that the behaviour of the atoms in an optical lattice can be described by Bose-Hubbard models: atoms can be confined on separate lattice sites and their interactions controlled. The absence of disorder in optical lattices makes them easier to model than their condensed-matter counter parts.

Alkalis with nuclear spin $I = 3/2$ such as ^{23}Na, ^{39}K and ^{87}Rb have hyperfine spin $F = 1$. In conventional magnetic traps, usually employed to confine the condensate, these spins cannot be regarded as a degree of freedom, so the atoms can be treated as spinless bosons. By contrast, a recently realized optical trap[26] confines atoms independently of their spin orientations offering the possibility of studying spinor condensates.[27]

In this paper we give a brief review of theoretical studies of the zero temperature phase diagram[28,29] of spin-1 Bose-Hubbard model. These studies use a mean-field approximation that generalize the technique used to study the phase transition of the spinless Bose-Hubbard model.[7] Two other groups have carried out mean-field studies for the spin-1 Bose-Hubbard model.[28,29] Our version of this mean-field theory is described below; it can be generalized to finite temperature as we will show elsewhere.[30] We introduce the Bose-Hubbard model for spin-1 bosons and study the effects of bosonic spin degrees on the SF-MI transitions. We discuss various mean field approaches used in the study of the Bose-Hubbard model in the next section. Section 3 gives a brief outline of our results for the spin-1 Bose-Hubbard model at zero temperature. Our finite-temperature results will be discussed elsewhere.[30]

2. Model and Mean-Field Approximation

The Bose-Hubbard model for spin-1 bosons is given by

$$\mathcal{H} = -t \sum_{<i,j>,\sigma} (a^{\dagger}_{i,\sigma} a_{j,\sigma} + h.c) + \frac{U_0}{2} \sum_i \hat{n}_i (\hat{n}_i - 1) + \frac{U_2}{2} \sum_i (\vec{F}_i^2 - 2\hat{n}_i) - \sum_i \mu_i \hat{n}_i. \tag{1}$$

The first term in Eq. (1) represents the usual kinetic energy associated with the hopping of bosons from site i to its nearest-neighbor site j with amplitude t; $a^{\dagger}_{i,\sigma}$ ($a_{i,\sigma}$) is the boson creation (annihilation) operator at site i with the spin component σ taking three possible values $\{1, 0, -1\}$. $\hat{n}_i = \sum_{\sigma} a^{\dagger}_{i,\sigma} a_{i,\sigma}$ is the total number of atoms and $\vec{F}_i = \sum_{\sigma,\sigma'} a^{\dagger}_{i,\sigma} \vec{F}_{\sigma,\sigma'} a_{i,\sigma'}$ is the total spin on the site i. $\vec{F}_{\sigma,\sigma'}$ are the usual spin-1 matrices. The second term in Eq. (1) is the on-site Hubbard repulsion between atoms; and the third term penalizes non-zero spin configurations on individual lattice sites. The spin-dependent term in Eq. (1) follows from the difference between the scattering lengths a_0 and a_2, respectively, for the $S = 0$ and $S = 2$ channels.[31] The interactions U_0 and U_2 are related to the scattering lengths a_0 and a_2 by $U_0 = 4\pi\hbar^2(a_0 + 2a_2)/3M$ and $U_2 = 4\pi\hbar^2(a_2 - a_0)/3M$, where M is the mass of the atom.[27] For the scattering of ^{23}Na, $a_2 = 54.7a_B$ and $a_0 = 49.4a_B$, where a_B is the Bohr radius, so $U_2 > 0$, whereas for ^{87}Rb, with $a_2 = (107 \pm 4)a_B$ and $a_0 = (110 \pm 4)a_B$, U_2 can be negative. The parabolic trapping potential with strength V_T is represented by the last term with $\mu_i = \mu - V_T|R_i|^2$, where R_i is the distance from the center of the trap potential. However, in this study we have neglected the trap potential ($V_T = 0$) to focus on the effects of the spin degrees of freedom on the SF-MI transition.

We study model (1) by using a simple mean-field theory in which we decouple the hopping term as was done in Ref.:[7]

$$\begin{aligned} a^{\dagger}_{i,\sigma} a_{j,\sigma} \approx &< a^{\dagger}_{i,\sigma} > a_{j,\sigma} + a^{\dagger}_{i,\sigma} < a_{j,\sigma} > \\ &- < a^{\dagger}_{i,\sigma} >< a_{j,\sigma} > . \end{aligned} \tag{2}$$

The resulting mean-field Hamiltonian approximating Eq. (1) can be written as a sum of single-site terms, i.e.,

$$\mathcal{H} = \sum_i \mathcal{H}_i^{MF}, \tag{3}$$

where

$$\begin{aligned} \mathcal{H}_i^{MF} &= \frac{U_0}{2}\hat{n}_i(\hat{n}_i - 1) + \frac{U_2}{2}(\vec{F}_i^2 - 2\hat{n}_i) - \mu\hat{n}_i \\ &- \sum_{\sigma} \psi_{\sigma}(a^{\dagger}_{i,\sigma} + a_{i,\sigma}) + \sum_{\sigma} |\psi_{\sigma}|^2, \end{aligned} \tag{4}$$

where $\psi_{\sigma} =< a^{\dagger}_{i,\sigma} >=< a_{i,\sigma} >$ (that we have assumed to be real[7]) is the superfluid order parameter for spin component σ. We set the energy scale by setting $zt = 1$, where z is the number of nearest-neighbors. The superfluid order parameter ψ_{σ} is non-zero in a superfluid phase. The magnetic properties of the superfluid state are obtained from[27]

$$< \vec{F} >= \frac{\sum_{\sigma,\sigma'} \psi_\sigma \vec{F}_{\sigma,\sigma'} \psi_{\sigma'}}{\sum_\sigma |\psi_\sigma|^2}, \tag{5}$$

where $\vec{F}_{\sigma,\sigma'}$ are the usual spin matrices for spin-1 particles; the explicit forms of these matrices yield

$$\begin{aligned} < \vec{F} > &= \sqrt{2}\frac{(\psi_1\psi_0 + \psi_{-1}\psi_0)}{\sum_\sigma |\psi_\sigma|^2}\hat{x} + \frac{(\psi_1^2 - \psi_{-1}^2)}{\sum_\sigma |\psi_\sigma|^2}\hat{z}; \\ < \vec{F} >^2 &= 2\frac{(\psi_1\psi_0 + \psi_{-1}\psi_0)^2}{\sum_\sigma |\psi_\sigma|^4} + \frac{(\psi_1^2 - \psi_{-1}^2)^2}{\sum_\sigma |\psi_\sigma|^4}. \end{aligned} \tag{6}$$

The superfluid states with $< \vec{F} >= 0$ and $< \vec{F} >^2 = 1$ are referred to as polar and ferromagnetic, respectively.

In order to calculate the superfluid order parameter ψ_σ for the ground state, we first obtain the matrix elements of the mean-field Hamiltonian $\mathcal{H}^{MF}$ in the onsite, occupation number (Fock) basis $\{|n_{-1}, n_0, n_1 >\}$ truncated at a finite value, say n_{max}, of total number of bosons per site $n = \sum_\sigma n_\sigma$. If $n_{max} = 4$, then the mean-field Hamiltonian is a 36×36 matrix. We then diagonalize the mean-field Hamiltonian matrix to obtain its eigenvalues and eigenfunctions and identify the lowest eigenvalue as the energy $E_g(\mu, U_0, U_2; \psi)$; this is a function of the three components of the superfluid order parameter $\psi = \{\psi_{-1}, \psi_0, \psi_1\}$. We then minimize E_g with respect to these three superfluid order parameters to obtain the ground-state energy as a function of μ, U_0 and U_2 and thence the zero-temperature phase diagram. The order parameters are obtained by analyzing the ground-state wavefunction. The density of bosons ρ of this ground state can be obtained from

$$\rho = -\frac{\partial E_g}{\partial \mu} = \sum_\sigma < \{\psi_\sigma\}|\hat{n}_\sigma|\{\psi_\sigma\} > . \tag{7}$$

The thermodynamic phase of the model for any point in the parameter space $\{\mu, U_0, U_2\}$ is determined as follows: When the superfluid density $\rho_s = \sum_\sigma |\psi_\sigma|^2 > 0$, we have a superfluid; further, whenever $\langle\vec{F}\rangle^2 = 0$, we have a polar superfluid (PSF); and if $\langle\vec{F}\rangle^2 = 1$ we have ferro superfluid(FSF). When $\rho_s = 0$, the phase is a Mott insulator (MI). In the next Section, we discuss the results obtained based on this mean-field calculation.

The mean-field theory discussed here is equivalent to the mean-field theory of Fisher, et al[2] and Gutzwiller approximation.[5,6] These methods have been applied to the spinless case. The generalization of these methods to the spin-1 case is straightforward[28] as discussed below. By defining

$$\phi_{i,\sigma} = \sum_j < a_{j,\sigma} >, \tag{8}$$

the mean-field Hamiltonian can be written as $H^{MF} = H_0 + H_1^{MF}$, where H_0 is the Hamiltonian when $t = 0$,

$$H_1^{MF} = -\sum_{i,\sigma}(\phi^*_{i,\sigma} a_{i,\sigma} + h.c) + \sum_{i,j,\sigma} t^{-1}_{ij}\phi^*_{i,\sigma}\phi_{j,\sigma}, \tag{9}$$

and t_{ij} is the hopping matrix element between the sites i and j. Here we allow for infinite-range hopping. The partition function of the full Hamiltonian is

$$Z = Z_0 \int D\phi^*(\tau) D\phi(\tau) e^{-\Delta F[\phi]}, \tag{10}$$

where the free energy

$$\begin{aligned} \Delta F[\phi] &= \sum_{ij,\sigma} t^{-1}_{ij} \int_0^\beta d\tau \phi^*_{i,\sigma}\phi_{j,\sigma} \\ &- \sum_{i,\sigma} \ln \left\langle T_\tau e^{\left[\int_0^\beta d\tau \phi^*_{i,\sigma}(\tau) a_{i,\sigma}(\tau) + h.c\right]} \right\rangle. \end{aligned} \tag{11}$$

The mean-field theory of Fisher et al[2] is the saddle-point value of $\Delta F[\phi]$ and it is obtained by assuming that $\phi_{i,\sigma}(\tau)$ is independent of τ and the same for all the sites. This saddle-point solution is exact for the case of infinite-range hopping t_{ij}. Thus the solution for $\phi_{i,\sigma}$ is obtained from the saddle-point condition $\frac{\partial \Delta F[\phi]}{\partial \phi_{i,\sigma}} = 0$ which is same as (8). If we substitute this solution in $\Delta F[\phi]$ we get Eq. (4). Hence our mean-field theory is equivalent to the mean-field theory of Fisher et al.[2]

It is easy to see that the Gutzwiller approximation[5,6] is equivalent to our mean-field theory. The single-site, Gutzwiller wave function is equivalent to writing the many-body state as a product of identical, onsite states $|\Psi_i>$; our mean-field factorization Eq. (2) is exact for such a state. When the order parameters ψ_σ are nonzero, the state $|\Psi_i>$ can in general be written as $|\Psi_i> = \sum_{n_i} f(n_i)|n_i>$, where $f(n_i)$ has the same meaning as in Refs.[5,6] We determine the coefficients $f(n_i)$ by evaluating the matrix elements of the Hamiltonian Eq. (4) in the occupation-number basis $\{|n_{-1}, n_0, n_1>\}$ and then diagonalizing it. The $f(n_i)$'s thus determined depend on U_0, U_2 and ψ_σ. Our minimization of E_g with respect to ψ_σ is equivalent to its minimization with respect to the general functions $f(n_i)$'s. In this method the minimization is accomplished via a particular parametrization of $f(n_i)$. In the Gutzwiller approximation, the filling is fixed via a delta-function in the wave function, which works in the canonical ensembles. However, our mean-field theory works in the grand canonical ensemble. We achieve fixed density via a chemical potential acting as a Lagrange multiplier.

Though our mean-field method is equivalent to other similar methods, there are many advantages in our method like the inclusion of finite temperature effects, disorder, and the determination of the excitation spectrum.[7] The effects of finite temperature will be reported elsewhere.[30] We restrict ourselves to zero temperature below.

3. Results and Discussions

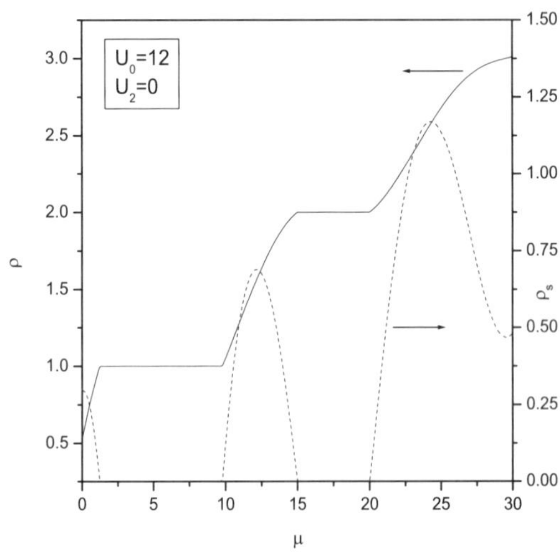

Fig. 1. The density of bosons ρ (solid line) and the superfluid density ρ_s (dotted line) plotted against the chemical potential μ for $U_0 = 12$, $U_2 = 0$. The plateau regions with $\rho = 1$ and $\rho = 2$ show the Mott insulator phase which has $\rho_s = 0$.

In order to discuss various aspects of the model (1), we consider three different parameter regions: (i) $U_2 = 0$, (ii) $U_2 > 0$, ($U_2/U_0 = 0.03$), and (iii) $U_2 < 0$ ($U_2/U_0 = -0.03$). We begin our discussion with the case $U_2 = 0$. This limit is attained experimentally when the scattering lengths are equal i.e., $a_0 = a_2$. In this limit, bosons with different spin components do not mix with each other, the interaction between bosons with various spin component is symmetric, and thus the boson numbers $n_1 = n_0 = n_{-1}$. Therefore, model (1) maps onto a spinless Bose-Hubbard model.[7] For example, the superfluid density ρ_s and the density of the bosons ρ are given for $U_0 = 12$ in Fig. (1). The plateau regions at $\rho = 1$ and $\rho = 2$ have vanishing superfluid density ρ_s. These regions, thus represents the Mott-Insulator phases, respectively, for densities $\rho = 1$ and $\rho = 2$. All the other regions represent the superfluid phase with $\psi_1 = \psi_0 = \psi_{-1}$. The critical interaction U_{0c} is equal to 5.83 and 9.9, respectively, for $\rho = 1$ and $\rho = 2$. The SF-MI transition is continuous because the density ρ does not have any discontinuity at the transition (see Fig. 1). The phase diagram for the case $U_2 = 0$ in the (μ, U_0) plane is given in Fig. (2).

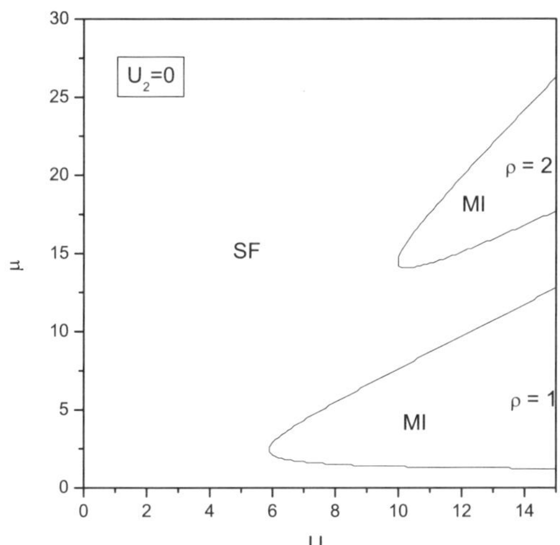

Fig. 2. Mean-field phase diagram in the (μ, U_0) plane for $U_2 = 0$ showing the superfluid(SF) phase with $\psi_1 = \psi_0 = \psi_{-1}$. The lobes marked MI are the Mott- Insulator phases. Only the first two lobes are shown.

When the scattering length $a_2 > a_0$, U_2 becomes repulsive. For ^{23}Na, from the estimated value[27] of a_0 and a_2, the ratio $U_2/U_0 \simeq 0.03$. For a finite and positive U_2, we can easily see from Eq. (4) that the ground-state

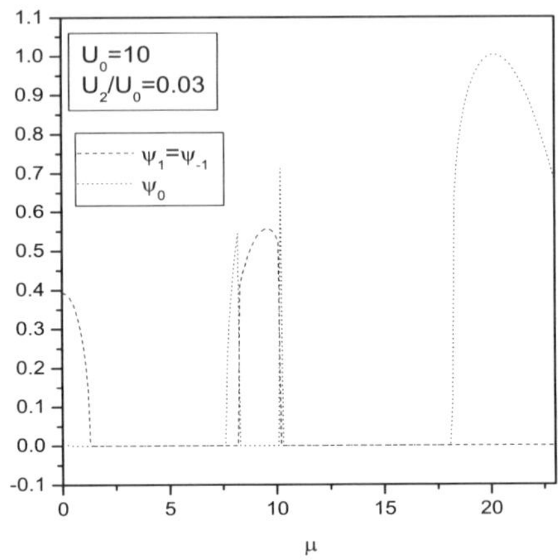

Fig. 3. The mean-field values of the superfluid order parameters $\psi_1 = \psi_{-1}$ and ψ_0 plotted as functions of μ for $U_0 = 10$, $U_2/U_0 = 0.03$. See the text for details.

energy is minimum when $<\vec{F}>= 0$, which means that the superfluid phase should be polar. The polar state has a symmetry group $U(1) \times S^2$, where $U(1)$ denotes the phase angle θ and S^2 the surface of a unit sphere [all orientations (α, β) of the spin-quantization axis]. Thus the superfluid order parameter can be written as

$$\begin{pmatrix} \psi_1 \\ \psi_0 \\ \psi_{-1} \end{pmatrix} = \sqrt{\rho_s}\ e^{\imath\theta} \begin{pmatrix} -\frac{1}{\sqrt{2}} e^{-\imath\alpha} \sin\beta \\ \cos\beta \\ \frac{1}{\sqrt{2}} e^{\imath\alpha} \sin\beta \end{pmatrix}. \tag{12}$$

Since we have assumed the superfluid order parameters ψ_σ to be real in our calculation, we restrict ourselves to some particular values of (θ, α, β). This lead to $\psi_1 = \psi_{-1}$ in our calculation. Thus from Eq. (6), $<\vec{F}>= 0$ is satisfied if either $\psi_1 = \psi_{-1} = 0$ or $\psi_0 = 0$. Our mean-field results indeed satisfy this relation in the entire region of the superfluid phase, thus confirming the polar nature of the superfluid phase for $U_2 > 0$. For a representative case, the calculated mean-field values of ψ_σ are given in Fig. (3) for $U_0 = 10$ and $U_2/U_0 = 0.03$. The superfluid density ρ_s and the density of bosons ρ are plotted as function of μ in Fig. (4) for $U_0 = 10$ and $U_2/U_0 = 0.03$. As in the case of $U_2 = 0$, the plateau regions represent the MI phases with $\rho_s = 0$. The nature of the SF-MI transition for $\rho = 1$ is the same as that for the case $U_2 = 0$ as discussed above. The transition is continuous: U_{0c}, for $U_2/U_0 = 0.03$, is not significantly different from the U_{0c} for $U_2 = 0$. For odd integer density, the spin degrees of freedom do not have any significant effect on the SF-MI transition.

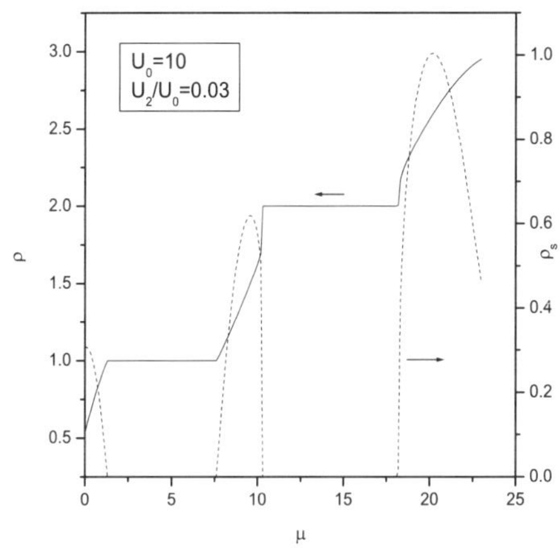

Fig. 4. The density of bosons ρ and the superfluid density ρ_s plotted against the chemical potential μ for $U_0 = 10$, $\frac{U_2}{U_0} = 0.03$. The plateau regions with $\rho = 1$ and $\rho = 2$ show the insulator phases with $\rho_s = 0$.

However, for $\rho = 2$, the SF-MI transition is first order: the density of bosons ρ shows a discontinuity at the SF-MI transition as shown in Fig. (4) and Fig. (5). This was missed by the earliest theoretical studies but has been obtained also in the mean-field studies of Refs.[28,29] At the transition the superfluid density goes to zero discontinuously. This first-order nature of the transition can be easily understood in the following way. When there are two bosons per site, the total spin at every site can be $F = 0$ or 2. The finite value of U_2 causes an energy difference between $F = 0$ and $F = 2$ states with the singlet state having a lower energy. Thus for MI-SF transition, this singlet state has to be broken and the energy required for breaking this state is of the order of U_2. This is the zero-temperature analogue of the *latent heat* for this first-order transition. The critical value of U_{0c} for finite U_2 for $\rho = 2$ SF-MI transition is significantly lower than that for the case $U_2 = 0$. This shows that the MI phase is strongly stabilized when $\rho = 2$. This was predicted earlier by a a variational study that used Gutzwiller approximation discussed earlier .[28,29] Though the superfluid order parameter vanishes discontinuously at the transition,

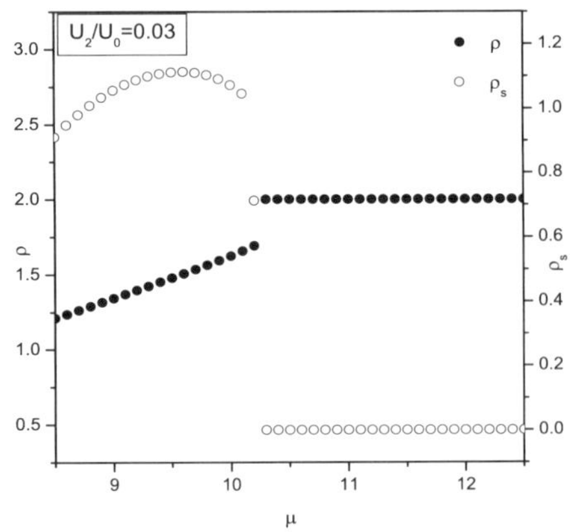

Fig. 5. The same plot as Fig. 4 but with the region around $\rho = 2$ enlarged. ρ_s goes to zero discontinuously at the PSF-MI transition; and ρ shows a discontinuity signaling a first-order PSF-MI transition. Here PSF stands for polar superfluid.

the gap Δ vanishes continuously for both $\rho = 1$ and $\rho = 2$ as shown in Fig. (7). The gap, however, is smaller for $\rho = 2$ than for $\rho = 1$.

When the scattering length $a_0 > a_2$, as in the case of ^{87}Rb, U_2 become attractive.[27,32] In this limit, the mean-field energy Eq. (4) has a minimum when $< \vec{F} >^2 = 1$ and the corresponding superfluid phase should be ferromagnetic.[27] The symmetry group of this phase is $SO(3)$. Our mean-field values of ψ_σ are consistent with this ferromagnetic nature of the superfluid phase. The superfluid order parameters are plotted as functions of μ for $U_0 = 12$, $U_2/U_0 = -0.03$ for a representative case in Fig. (8). The SF-MI phase transition, when $U_2 < 0$, is similar to that for the $U_2 = 0$ case with the critical U_{0c} being slightly increased. For example, the superfluid density ρ_s and the density of bosons ρ are given for $U_0 = 12$, $U_2/U_0 = -0.03$, in Fig. (9). ρ is continuous at the SF-MI transition. The phase diagram in the (μ, U_0) plane is given in Fig. (10). Thus the phase diagram and the nature of the SF-MI transition for attractive U_2 are similar to those in the case of

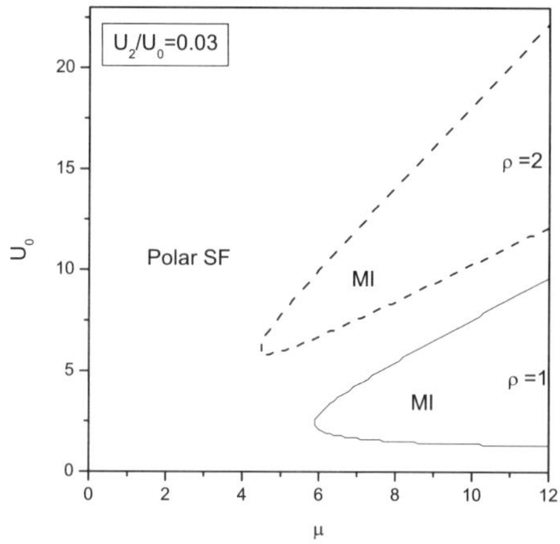

Fig. 6. Mean-field phase diagram in the (μ, U_0) plane for $U_2/U_0 = 0.03$. The lobes marked MI are the Mott-Insulator phases. The rest of the phase diagram contains a polar superfluid with $< F >= 0$. The dashed line represents the first-order boundary of the PSF-MI transition for $\rho = 2$. Only the first two lobes are shown.

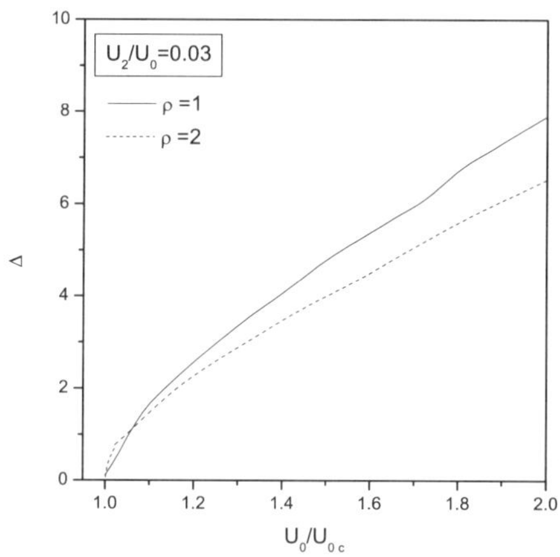

Fig. 7. The gap Δ in the MI phase versus U_0/U_{0c} for $\rho = 1$ and $\rho = 2$ for $U_2/U_0 = 0.03$.

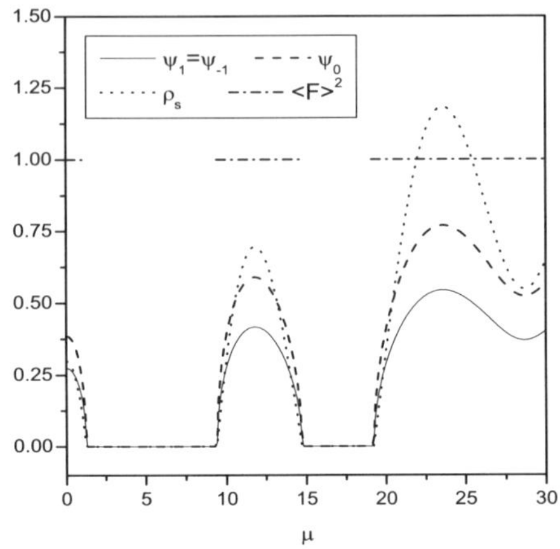

Fig. 8. The superfluid order parameters ψ_σ plotted against the chemical potential μ for $U_0 = 12$, $U_2/U_0 = -0.03$. The quantity $< \vec{F} >^2$ is always equal to 1 in the ferromagnetic superfluid phase.

$U_2 = 0$ but with the SF phase being ferromagnetic. For $U_2/U_0 = -0.03$, the critical values U_{0c} are 6.0 and 10.2, respectively, for $\rho = 1$ and $\rho = 2$.

In summary, we have developed a simple mean-field theory for the spin-1 Bose-Hubbard model and obtained its phase diagram at zero temperature which comprises superfluid and Mott insulator phases. We show that the superfluid phase is either polar or ferromagnetic depending upon the values of the scattering lengths in different spin channels. The polar-superfluid to Mott-insulator transition is first-order if there are an even number of bosons per site because of the formation of singlets (this is quite different from the spin-0 case). However, all other superfluid to Mott-insulator transitions are continuous as in the spin-0 case. Finally we note that while this paper being written two similar studies, one using Gutzwiller approximation[28] and second using a method similar to ours[29] appeared. Analyzing the behaviour of ground state energy, these works also predicted a first-order transition for the PSF-MI transition when density of bosons equal to an even integers. In addition to these results, we have also obtained the finite jump in the superfluid density at the first-order polar-superfluid to the Mott-Insulator transition. The gap at this transition, however, does not show any discontinuity. Our finite-temperature results for the spin-1 Bose-Hubbard model

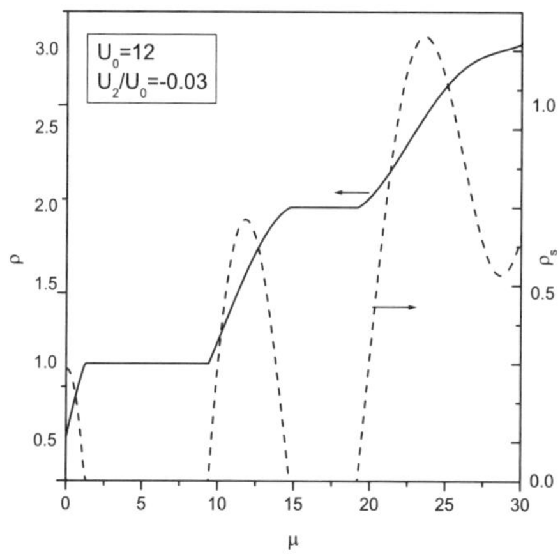

Fig. 9. The density of bosons ρ and the superfluid density ρ_s plotted against the chemical potential μ for $U_0 = 12$, $\frac{U_2}{U_0} = -0.03$. The plateau regions with $\rho = 1$ and $\rho = 2$ show the Mott-Insulator phases with $\rho_s = 0$.

will be discussed elsewhere.[30]

Spin-1 bosons can exhibit different magnetic orderings in the Mott-Insulator phases because of the spin degrees of freedom. These phases can be studied by mapping the Bose-Hubbard model (1) onto a spin model in the limit $U_0 \to \infty$. Imambekov *et. al.*[33] has predicted that, for an odd number of bosons per site, the resultant spin model always exhibits a nematic phase.[34] However, for an even number of bosons per site, they predict a first-order transition from a nematic to a spin-singlet phase. These prediction cannot be checked in our mean-field calculation. Our mean-field approximation leads to a single-site Hamiltonian and thus neglects spin interactions between different sites in the insulating phase. So we do not see the nematic phase in our approximation. However, we are able to predict the spin-singlet phase accurately because it is dominated by the on-site spin interactions which have been taken into account exactly in our model. In order to obtain the spin-spin interaction within our model, we need to go beyond the mean-field level, which is outside the scope of our present work.

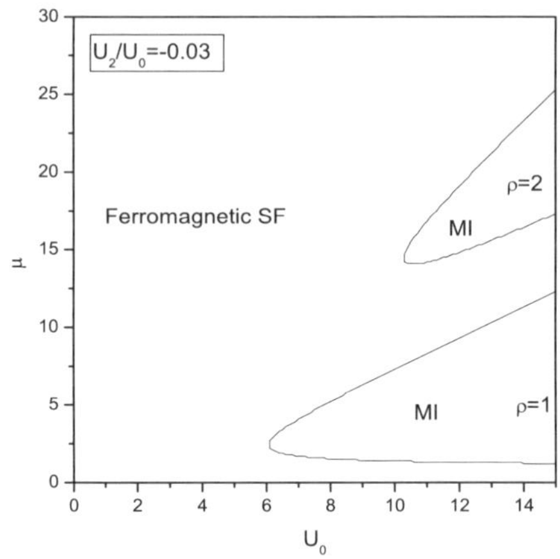

Fig. 10. Mean-field phase diagram in the (μ, U_0) plane for $U_2/U_0 = -0.03$. The lobes marked MI are the Mott-Insulator phases. The rest of the phase diagram consists of a ferromagnetic superfluid.

Acknowledgments

One of us (RVP) thanks the Jawaharlal Nehru Centre for Advanced Scientific Research and the Department of Physics, Indian Institute of Science, Bangalore for hospitality during the time when a part of this paper was written. This work was supported by DST, India (Grants No. SP/S2/M-60/98 and SP/I2/PF-01/2000) and UGC, India.

References

1. S. Sachdev, Quantum Phase Transitions, Cambridge University Press (1999).
2. M. P. A. Fisher, P. B. Weichman, G. Grinstein and D. S. Fisher, Phys. Rev. B **40** 546 (1989).
3. T. V. Ramakrishnan, Ordering Disorder: Prospect and Retrospect in Condensed Matter Physics, eds. V. Srivastava, A. K. Bhatnagar and D. G. Naugle AIP Conference Proceedings **286** 38 (1994) and references therein.
4. M. Ma, B. I. Halperin, and P. A. Lee, Phys. Rev. B **34** 3136 (1989); P. Nisamaneephong, L. Zhang, and M. Ma, Phys. Rev. Lett. **71** 3830 (1993).
5. W. Krauth, M. Caffarel, and J. P. Bouchaud, Phys. Rev. B **45** 3137 (1992).
6. D. S. Rokhsar and B. G. Kotliar, Phys. Rev. B **44** 10328 (1991).
7. K. Sheshadri, H. R. Krishnamurthy, R. Pandit, and T. V. Ramakrishnan, Europhys. Lett. **22** 257 (1993); Phys. Rev. Lett. **75** 4075 (1995).
8. L. Amico and V. Penna, Phys. Rev. Lett. **80** 2189 (1998).

9. R. V. Pai, R. Pandit, H. R. Krishnamurthy and S. Ramasesha, Phys. Rev. Lett. **76** 2937 (1996).
10. R. V. Pai and R. Pandit, Phys. Rev. B **71** 104508 (2005), R. V. Pai and R. Pandit, Proc. Indian Academy of Sciences (Chemical Sciences) **115** 721 (2003).
11. R. Baltin and K. H. Wagenblast, Europhys. Lett. **39** 7 (1997).
12. T. D. Kühner and H. Monien, Phys. Rev. B **58** R14741 (1998), T. D. Kühner, S. R. White and H. Monien H, Phys. Rev. B **61** 12474 (2000).
13. V. A. Kashurnikov and B. V. Svistunov, Phys. Rev. B. **53** 11776 (1996).
14. G. G. Batrouni, R. T. Scalettar, G. T. Zimanyi and A. P. Kampf, Phys. Rev. Lett. **74** 2527 (1995); P. Niyaz, R. T. Scalettar, C. Y. Fong and G. G. Batrouni, Phys. Rev. B. **44** 7143 (1991).
15. W. Krauth, N. Trivedi, and D. Ceperley, Phys. Rev. Lett. **67** 2307 (1991); N. Trivedi and M. Makivic, Phys. Rev. Lett. **74** 1039 (1995).
16. M. Wallin, E. S. Sorensen, S. M. Girvin, and A. P. Young, Phys. Rev. B **49** 12115 (1994).
17. S. Zhang, N. Kawashima, J. Carlson, and J. E. Gubernatis, Phys. Rev. Lett. **74** 1500 (1995).
18. M. H. W. Chan, K. I. Blum, S. Q. Murphy, G. K. S. Wong, and J. D. Reppy, Phys. Rev. Lett. **61** 1950 (1988).
19. E. Chow, P. Delsing and D. B. Haviland, Phys. Rev. Lett **81** 204 (1998).
20. A. van Oudenaarden, and J. E. Mooij, Phys. Rev. Lett. **76** 4947 (1996); A. van Oudenaarden, B. van Leeuwen, M. P. M. Robbens and J. E. Mooij, Phys. Rev. B **57** 11684 (1998).
21. D. B. Haviland, Y. Liu, and A. M. Goldman, Phys. Rev. Lett. **62** 2180 (1989).
22. D. R. Nelson and V. M. Vinokur, Phys. Rev. B **48** 13060 (1993).
23. F. Dalfovo, S. Giorgini, L. P. Pitaevskii and S. Stringari, Rev. Mod. Phys. **71** 463 (1999).
24. M. Greiner, O. Mandel, T. Esslinger, T. W. Hänsch and L. Bloch Nature (London) **415** 39 (2002).
25. D. Jaksch, C. Bruden, J. I. Cirac, C. W. Gardiner and P. Zoller Phys. Rev. Lett. **81** 3108 (1998).
26. D. M. Stamper-Kurn *et al* Phys. Rev. Lett. **80** 2027 (1998).
27. T. L. Ho, Phys. Rev. Lett. **81** 742 (1998).
28. S. Tsuchiya, S. Kurihara and T. Kimura Phys. Rev. Lett., **94** 110403 (2005).
29. K. V. Krutitsky and R. Graham, Phys. Rev. A, **70** 063610 (2004)
30. R. V. Pai, K Sheshadri and R. Pandit, to be published.
31. C. K. Law, H. Pu and N. P. Bigelow Phys. Rev. Lett. **81** 5257 (1998).
32. H. Schmaljohann *et al.*, Rev. Lett. **92** 040402 (2004).
33. A. Imambekov, M Lukin and E. Demler, Phys. Rev. A **63** 063602 (2003).
34. G. Fáth and J. Sólyom, Phys. Rev. B **51** 3620 (1995).

MIXED INTERNAL-EXTERNAL STATE APPROACH FOR QUANTUM COMPUTATION WITH NEUTRAL ATOMS ON ATOM CHIPS

E. CHARRON

Laboratoire de Photophysique Moléculaire du CNRS, Bâtiment 210,
Université Paris-Sud 11, 91405 Orsay cedex, France

M. A. CIRONE

ECT, Strada delle Tabarelle 286, I-38050 Villazzano, Trento, Italy*
Dipartimento di Fisica, Università di Trento and BEC-CNR-INFM, I-38050 Povo, Italy

A. NEGRETTI

ECT, Strada delle Tabarelle 286, I-38050 Villazzano, Trento, Italy*
Dipartimento di Fisica, Università di Trento and BEC-CNR-INFM, I-38050 Povo, Italy
Institut für Physik Universität Potsdam Am Neuen Palais 10, 14469 Potsdam, Germany

J. SCHMIEDMAYER

Physikalisches Institut, Universtät Heidelberg,
Am Neuen Palais 10, 14469 Potsdam Germany

T. CALARCO

ECT, Strada delle Tabarelle 286, I-38050 Villazzano, Trento, Italy*
Dipartimento di Fisica, Università di Trento and BEC-CNR-INFM, I-38050 Povo, Italy
ITAMP and Department of Physics, Harvard University, Cambridge, MA 02138, USA

We present a realistic proposal for the storage and processing of quantum information with cold ^{87}Rb atoms on atom chips. The qubit states are stored in hyperfine atomic levels with long coherence time, and two-qubit quantum phase gates are realized using the motional states of the atoms. Two-photon Raman transitions are used to transfer the qubit information from the internal to the external degree of freedom. The quantum phase gate is realized in a double-well potential created by slowly varying dc currents in the atom chip wires. Using realistic values for all experimental parameters (currents, magnetic fields, ...) we obtain high gate fidelities (above 99.9%) in short operation times (~ 10 ms).

1. Introduction

Beyond their fundamental interest in physics, coherence and entanglement of quantum states are the building blocks of quantum information[1]. Performing very simple operations on a limited number of qubits is a real experimental challenge since quantum information, stored in the amplitude and phase of two-state quantum systems, is usually very sensitive to experimental noise or unwanted interactions. Trapped ions, cavity QED, nuclear spins (NMR), and cold neutral atoms have long coherence times, and are thus well known candidates for the implementation of qubits and multiple-qubit gates. The achievement of a quantum computer acting on a limited number of quantum registers would already lead to an intrinsic speed-up of calculation that is not possible with a classical computer[2,3]. This requires the physical implementation of a universal set of single-qubit and two-qubit operations[4]. Since the design of single-qubit operations is usually relatively straightforward, we concentrate our investigations on the implementation of a two-qubit π conditional phase gate with cold atoms trapped on an atom chip. A conditional phase gate $P(\varphi)$, defined by the transformation

$$\begin{array}{lcr} |00\rangle & \to & |00\rangle \\ |01\rangle & \to & |01\rangle \\ |10\rangle & \to & |10\rangle \\ |11\rangle & \to & e^{i\varphi}\,|11\rangle \end{array}, \tag{1}$$

induces some degree of entanglement between two qubits by selectively changing the state $|11\rangle$ into $e^{i\varphi}\,|11\rangle$, while leaving other states unchanged. In practice, it is often simpler to implement a phase gate which changes the different qubit states according to

$$\begin{array}{lcr} |00\rangle & \to & e^{i\varphi_{00}}\,|00\rangle \\ |01\rangle & \to & e^{i\varphi_{01}}\,|01\rangle \\ |10\rangle & \to & e^{i\varphi_{10}}\,|10\rangle \\ |11\rangle & \to & e^{i\varphi_{11}}\,|11\rangle \end{array}. \tag{2}$$

This transformation can be reduced to the conditional phase gate $P(\varphi)$ described previously, with $\varphi = \varphi_{00} + \varphi_{11} - \varphi_{10} - \varphi_{01}$, by using additional single-qubit operations. The case $\varphi = \pi$ is of interest since this particular operation can be used to transform a separable two-qubit state into a maximally entangled state. As a consequence, a judicious combination of $P(\pi)$ together with two one-qubit Hadamard gates can replace a controlled-NOT.

In this study we propose the implementation of such a conditional π phase gate with neutral atoms trapped on atom chips. An atom chip consists in a series of nano-fabricated wires on a surface[5]. By varying the currents flowing through these wires, neutral atoms are subjected to varying magnetic fields and can be manipulated above the surface. We concentrate here on a particular scheme where the qubits, initially stored in the hyperfine levels of the atoms, will be first transferred to the motional states of the atoms and then entangled as a result of a selective interaction taking place between some of these motional states. This entanglement will take place in a double-well potential created by homogeneous bias magnetic fields and by magnetic fields created by dc (but time-dependent) currents in the atom chip. The spin of the slow, cold atoms stays constantly aligned with the magnetic field and the trapping magnetic potential is expressed in the weak field approximation by

$$V(\mathsf{r}) = g_F\,\mu_B\,m_F\,B(\mathsf{r})\,, \tag{3}$$

where μ_B is the Bohr magneton, g_F is the Landé factor, m_F is the azimuthal quantum number, and $B(\mathsf{r})$ is the magnetic field.

Different approaches have been proposed and/or implemented for the manipulation of cold atoms in a double-well potential[6]. A simple configuration of wires which can create such a potential is shown in Figure 1. A longitudinal wire along x (hereafter, quadrupole wire) carrying a dc current $I_0(t)$ and a uniform bias magnetic field B_{0y} perpendicular to the wire create a quadrupole potential, with a zero magnetic field along a line parallel to the quadrupole wire. Along this line the magnetic field is minimum, however a vanishing field cannot trap the atoms, so the minimum is shifted to a non-zero value with the addition of a second uniform bias magnetic field B_{0x}, orthogonal to the first one and parallel to the chip surface. Two more wires (hereafter, left and right wire, respectively), perpendicular to the quadrupole wire, carry a dc current $I_1(t) = I_2(t) = \alpha(t)I_0(t)$, whose magnetic fields give rise to a modulation of the trapping potential.

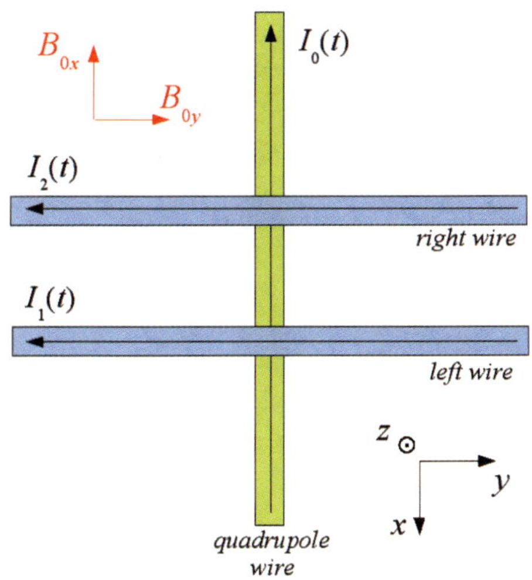

Figure 1. Schematic view of the atom chip configuration. The two wires along the y-axis lie on the chip surface and are separated (direction z) from the quadrupole wire by 200 nm. The quadrupole wire is therefore located below the surface. The left and right wires are separated (direction x) by 1.6 μm. The section of all wires is 700 nm × 200 nm.

For the calculation of the trapping potential of Equation (1), we have assumed infinitely long wires of finite section 700 nm × 200 nm. A trapping potential with two well separated minima, as shown in Figure 2 (a), is created for $I_0 = 40.89$ mA, $\alpha = 70.25 \times 10^{-3}$, $B_{0x} = -9.90$ G, and $B_{0y} = 50.0$ G. The centers of the left and right wires are 1.60 μm apart. The center of the quadrupole wire is at a distance $z_Q = 400$ nm under the chip surface, whereas the left and right wires lie on the chip. We stress that the values for the currents, bias fields, size and distances of the wires are within current laboratory conditions.

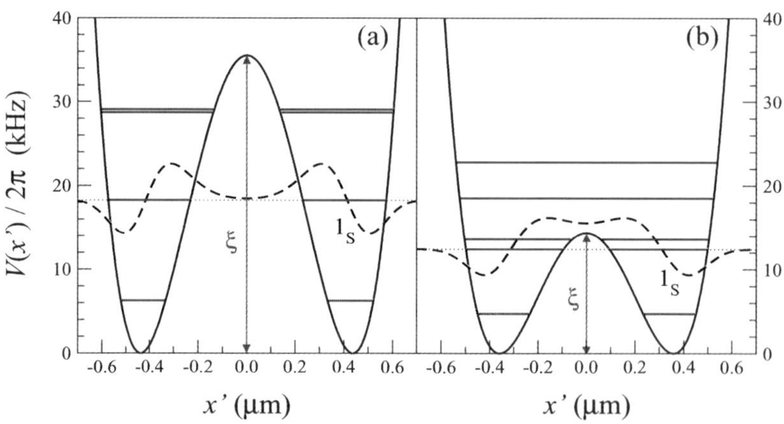

Figure 2. Double well potentials created by the atom chip configuration shown in Figure 1. The energies of the first six eigenstates are shown as red horizontal lines. The blue dashed line represents the wavefunction of the third eigenstate labeled as 1_S because it originates from the symmetric combination of the $v = 1$ trapped levels, also labeled as $|e\rangle$ in the text. (a) Highest barrier $\xi/2\pi = 35.4$ kHz obtained with $I_0 = 40.89$ mA and $\alpha = 70.25 \times 10^{-3}$. (b) Lowest barrier $\xi/2\pi = 14.4$ kHz obtained with $I_0 = 42.01$ mA and $\alpha = 69.70 \times 10^{-3}$. In both cases the bias magnetic fields are equal to $B_{0x} = -9.90$ G and $B_{0y} = 50.0$ G.

The two potential minima are at a distance of 1.19 μm from the surface, and the line joining them is slightly tilted by an angle $\beta \simeq 14.8°$ from the x axis. This angle defines the new axis x' along which the dynamics will take place (see M. A. Cirone *et al* [7] for details). We also define a new axis y' parallel to the chip surface and perpendicular to x'. The z axis remains unchanged. The trapping frequencies at the two minima verify $\nu_{x'} \ll \nu_{y'} \simeq \nu_z$, a clear indication that the trapped atoms will experience a quasi one-dimensional (1D) dynamics[8]. This necessary transverse confinement has already been approached even in the early atom chip experiments with nano fabricated chips[9]. The required potential smoothness[10] and wire performance has been demonstrated with semi-conductor substrate based atom chip fabrication[11].

The value of the magnetic field at the two minima is $B_{\min} \simeq 3.23$ G. This value minimizes the decoherence induced by fluctuations of the dc currents for the hyperfine states $|F=2, m_F=1\rangle$ and $|F=1, m_F=-1\rangle$ of the $5\,\mathrm{S}_{1/2}$ ground state of ^{87}Rb [12,13]. These clock states will therefore be used to store the qubit information at the end of the gate operation. For a detailed description of the two-photon Raman process involved for the transfer of the qubit information from the internal to the external degree of freedom, we refer the reader to References 7 and 14.

2. Two-qubit π conditional phase gate

As it can be noticed in Figure 2 (a), when the barrier is high the translational wavefunctions of the atoms do not overlap in the inter-well region. In this type of environment, the atoms do not interact. On the other hand, when the barrier is lowered, as in Figure 2 (b), tunneling takes place and the probability of finding the atoms in the classically forbidden region is not negligible any more. As a consequence, the energy splitting between the symmetric and anti-symmetric state combinations increases quickly when the barrier height ξ is lowered. This effect is clearly selective in the sense that it affects differently the ground $|g\rangle$ and excited $|e\rangle$ translational states. It therefore constitutes an interesting candidate for the implementation of a conditional logical gate.

In the present scheme, the barrier height $\xi(t)$ is controlled by varying simultaneously the intensities $I_0(t)$ and $I_1(t) = I_2(t) = \alpha(t) I_0(t)$ in the quadrupole and in the perpendicular wires. In a first and simple implementation of the phase gate, we impose a linear variation of the barrier height $\xi(t)$ with time. The phase gate is decomposed in three steps:

- When $0 \leqslant t \leqslant T_0$ the barrier is lowered and the double-well potential changes from the one of Figure 2 (a) to the one of Figure 2 (b).
- When $T_0 \leqslant t \leqslant T_0 + T_1$ the inter-well barrier is fixed at its lowest value, such that a large inter-atomic interaction takes place.
- Finally, when $T_0 + T_1 \leqslant t \leqslant 2T_0 + T_1$ the inter-well barrier is raised again until the initial condition is recovered.

The linear variation of $\xi(t)$ is obtained by changing $I_0(t)$ and $\alpha(t)$ only, as depicted by the solid lines shown in Figure 3 (a). We have verified that this simultaneous variation of the dc currents does not modify the direction x' along which the dynamics is taking place. The value of the magnetic field at the potential minima also remains equal to 3.23 G during the whole gate operation. A quasi-adiabatic dynamics is therefore likely if $T_0 \gg 1/\nu_{x'} \simeq 77\,\mu$s.

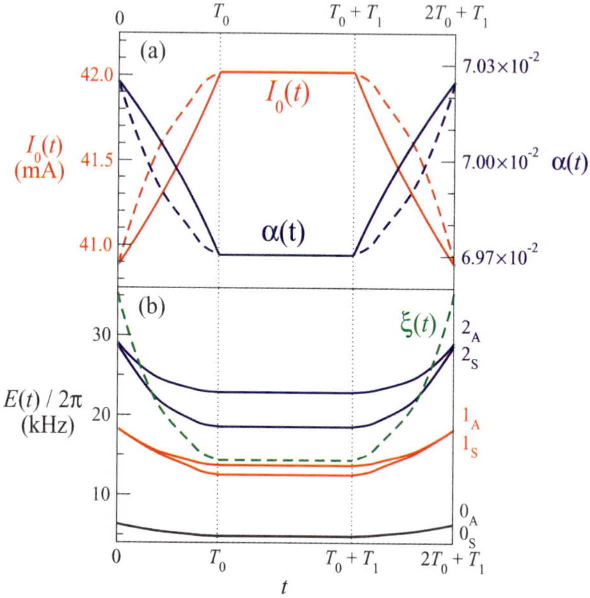

Figure 3. The times T_0, $(T_0 + T_1)$ and $(2T_0 + T_1)$ delimit the three steps which constitute the conditional phase gate. (a) Variation of the dc current $I_0(t)$ (red lines) in the quadrupole wire and of $\alpha(t) = I_1(t)/I_0(t) = I_2(t)/I_0(t)$ (blue lines) during the gate operation. The solid lines correspond to a simple linear gate and the dashed lines represent an optimized gate (see text for details). (b) Variation of the barrier height $\xi(t)$ (green dashed line) and of the energies of the first six instantaneous eigenstates of the double well potential (solid lines) in the case of the optimized gate.

The gate operation is followed by solving the time-dependent Schrödinger equation along the double-well direction x' for the wave packet $\Psi(x_1', x_2', t)$ describing the motion of the two atoms

$$i\hbar \frac{\partial}{\partial t} \Psi(x_1', x_2', t) = \mathcal{H}(x_1', x_2', t)\, \Psi(x_1', x_2', t). \tag{4}$$

The two-dimensional time-dependent Hamiltonian is written as

$$\mathcal{H}(x_1', x_2', t) = \hat{T}_{x_1'} + \hat{T}_{x_2'} + V_{2\mathrm{D}}(x_1', x_2', t)\,, \tag{5}$$

where $\hat{T}_q$ denotes the kinetic energy operator along the q-coordinate. The two-dimensional potential is given by the following sum

$$V_{2\mathrm{D}}(x_1', x_2', t) = V(x_1', t) + V(x_2', t) + V_{\mathrm{int}}(|x_2' - x_1'|, t)\,, \tag{6}$$

where $V(x', t)$ is the trapping potential (3) created by the atom chip and $V_{\mathrm{int}}(|x_2' - x_1'|, t)$ represents the averaged interaction potential between the two atoms at time t

$$V_{\mathrm{int}}(|x_2' - x_1'|, t) = 2\hbar a_0\, (\omega_{y'}\, \omega_z)^{1/2}\, \delta\,(|x_2' - x_1'|). \tag{7}$$

This last expression is obtained by averaging the three-dimensional delta function interaction potential over the lowest trap states along the y' and z directions[15]. One can note that the atom-atom interaction strength is proportional to the s-wave scattering length a_0. Since the orthogonal trapping frequencies $\omega_{y'}$ and ω_z vary slightly during the gate operation[8], the averaged interaction strength is also slightly time-dependent.

We solve the time-dependent Schrödinger equation (4) in a basis set approach by propagating the initial state of the two-atom system in time. We start with the atoms initially in one of the first four eigenstates $|gg\rangle$, $|ge\rangle$, $|eg\rangle$ or $|ee\rangle$ of the double-well potential (6) shown in Figure 2 (a). The wavefunction $\Psi(x'_1, x'_2, t)$ is then expanded as

$$\Psi(x'_1, x'_2, t) = \sum_i c_i(t) \times \varphi_i(x'_1, x'_2, t) \,, \tag{8}$$

where $\varphi_i(x'_1, x'_2, t)$ represents the wavefunctions associated to the instantaneous eigenstates of the two-dimensional potential $V_{2\text{D}}(x'_1, x'_2, t)$ of Eq. (6). Inserting this expansion into the time-dependent Schrödinger equation (4) yields the following set of first order coupled ordinary differential equations for the complex coefficients $c_i(t)$

$$i\hbar \frac{d}{dt} c_i(t) = \varepsilon_i \, c_i(t) - i\hbar \sum_j c_j(t) \, V_{ij}(t) \,, \tag{9}$$

where ε_i denotes the energy of the eigenstate φ_i and $V_{ij}(t)$ is a time-dependent non-adiabatic coupling arising from the time variation of the barrier height $\xi(t)$

$$V_{ij}(t) = \langle \varphi_i | \frac{\partial}{\partial t} | \varphi_j \rangle = \frac{d\xi}{dt} \langle \varphi_i | \frac{\partial}{\partial \xi} | \varphi_j \rangle \,. \tag{10}$$

This set of equations is solved using an accurate Shampine-Gordon algorithm[16]. In order to analyze the dynamics taking place during the gate operation, it is useful to examine how the energies ε_i of the two-atom translational eigenstates $\varphi_i(x'_1, x'_2, t)$ vary with the barrier height. These quantities are responsible for the build-up of the dynamical phases φ_{00}, φ_{01}, φ_{10} and φ_{11}, and therefore of the global conditional phase $\varphi = \varphi_{00} + \varphi_{11} - \varphi_{10} - \varphi_{01}$. This variation is shown in Figure 4 (a), as a function of the barrier height expressed in terms of magnetic field $\Delta B = \xi \,/\, g_F \, \mu_B \, m_F$. During the course of the gate, the energies of the four two-qubit eigenstates vary by about 30%, while their difference $\Delta\varepsilon = \varepsilon_{00} + \varepsilon_{11} - \varepsilon_{01} - \varepsilon_{10}$, shown in Figure 4 (b), changes by four orders of magnitude. This exponential scaling is a signature of tunneling. For each barrier height, a gate duration $\tau_{\text{gate}} = \pi / \Delta\varepsilon$ can be extracted. This quantity is shown in Figure 4 (c), and it corresponds to the gate duration that one could achieve in a static gate implementation[7]. One can see from this graph that the minimum gate duration achievable with the present parameters is about 5 ms.

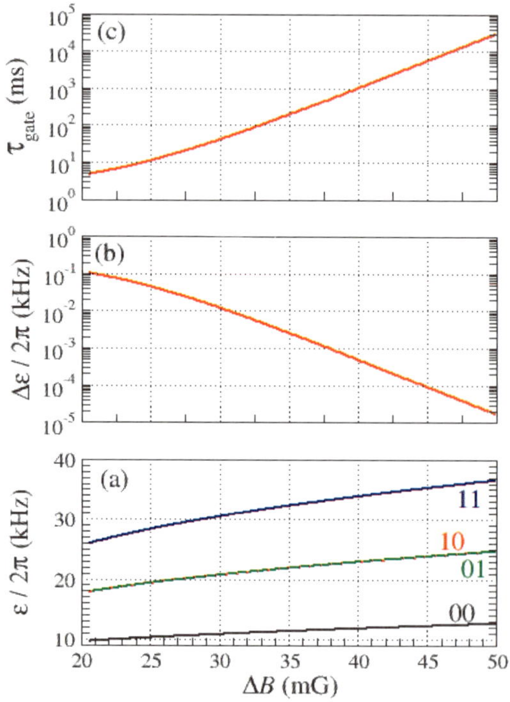

Figure 4. (a) Variation of the energies ε_{00}, ε_{01}, ε_{10} and ε_{11} of the four qubit states $|gg\rangle$ (full black line), $|ge\rangle$ (dashed green line), $|eg\rangle$ (full red line) and $|ee\rangle$ (full blue line) as a function of the barrier height expressed in terms of magnetic field ΔB. (b) Variation of $\Delta\varepsilon = \varepsilon_{00} + \varepsilon_{11} - \varepsilon_{01} - \varepsilon_{10}$ with the barrier height. (c) Variation of $\tau_{\text{gate}} = \pi/\Delta\varepsilon$ with the barrier height.

At the end of the propagation ($t = t_f$) the coefficients $c_i(t_f)$ are analyzed to calculate the infidelity of the gate

$$I = \sum_{i=|gg\rangle\cdots|ee\rangle} \left(1 - |c_i(t_f)|^2\right), \tag{11}$$

where the sum runs over all possible initial qubit states. The infidelity is therefore a measure of the deviation from adiabaticity which arises from the non-adiabatic couplings $V_{ij}(t)$. This quantity is plotted in Figure 5 (red line for the linear gate) as a function of the gate duration. It shows an oscillatory behavior partially similar to the one observed with atoms trapped in an optical lattice[17]. The succession of maxima and minima is a signature of constructive and destructive interferences between two distinct pathways of excitation of the initial qubit state. Indeed the initial state may be excited in the time intervals

$0 \leqslant t \leqslant T_0$ and $T_0 + T_1 \leqslant t \leqslant 2T_0 + T_1$, when the barrier is lowered and raised. The nature of this interference depends on the phases which develop during the gate operation[17]. The periodicity of the oscillation is simply related to the Bohr frequencies associated with the energy splitting of the two-atom eigenstates. The linear gate configuration proposed here can achieve a relatively high fidelity of about 99.6% in just 7.6 ms.

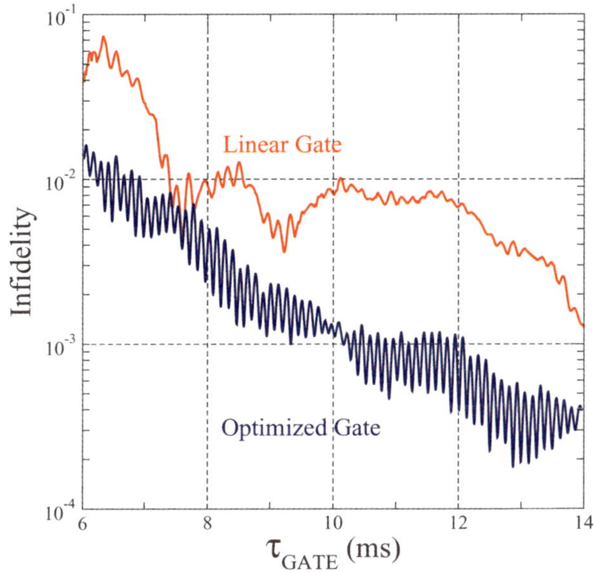

Figure 5. Infidelity of the conditional phase gate as a function of the gate duration. The red and blue lines correspond to the infidelities calculated for the linear and optimized gates respectively.

One should also realize that in the general case the couplings between the initial qubit states and the other accessible two-atom eigenstates vary with time. These couplings effectively increase when the inter-well barrier approaches the energy of the initial state. The linear gate proposed until now is therefore far from being optimal for the maximization of the gate fidelity. We have thus implemented an optimized gate which tends to minimize these couplings during the whole gate duration. For this purpose, we impose a fast variation of the barrier height $\xi(t)$ at early times $t \ll T_0$, while this variation is much slower when $t \simeq T_0$. This is done by choosing

$$\hbar \left(\frac{d\xi}{dt} \right) = \gamma \times \mathrm{Min}_i \left| \frac{\varepsilon_i - \varepsilon_{|ee\rangle}}{\langle \varphi_i | \dfrac{\partial}{\partial \xi} | \varphi_{|ee\rangle} \rangle} \right| . \tag{12}$$

In this expression, γ is a dimensionless proportionality factor, which can be decreased to achieve larger gate durations. The first derivative with respect to time of the barrier height $\xi(t)$ is therefore chosen such that the maximum effective first-order coupling between the highest energy qubit state $|ee\rangle$ and all other states remains constant during the whole gate duration. With this approach, higher fidelities are expected when compared to a linear gate of the same duration.

The variations of $I_0(t)$ and $\alpha(t)$ for this optimized gate are shown as dashed lines in Figure 3 (a). Figure 5 shows that the optimized gate infidelity (blue line) is, on average, improved by a factor of about 6 when compared to the linear gate. As a consequence, this optimized gate can achieve fidelities of 99% in only 6.3 ms and of 99.9% in just 10.3 ms.

3. Conclusions

When neutral atoms are used, it is highly desirable to employ both the vibrational and the internal states as qubit states. Vibrational states are very promising in terms of gate performance, while internal states, in a carefully chosen magnetic field environment, are highly protected from decoherence[12]. In addition, the readout process can be achieved efficiently with internal states using fluorescence measurement techniques. Two-photon Raman processes can be used to transfer the qubit states from one representation to the other[7,14].

We have presented in this proceeding a detailed analysis of the implementation of a quantum phase gate with neutral rubidium atoms on atom chips. Our analysis is quite close to the experimental conditions and is within the reach of current technology. We have shown how to create a double well potential near the surface and studied the performance of the phase gate. We have found that a fidelity of 99.9% can be achieved in just 10.3 ms. The results presented here are a significant improvement when compared to an implementation using a static trap[7].

Finally, an important additional mechanism one has to consider in a realistic evaluation of the performance of quantum gates on atom chips, is the possibility of loss and decoherence of the qubits during the operation caused by thermal electromagnetic fields generated by the nearby, "hot" solid substrate[18,19]. Following the treatment of Henkel and Wilkens[18], we estimate the lifetimes for our example setup to 0.8 s, limiting the fidelity to 98.7% at a gate operation time of 9 ms. Reducing the thickness of the wires down to 50 nm and increasing the width of the central wire to 3 μm will increase the lifetime to over 3 s and increase the fidelity to 99.7% at a gate operation time of 11 ms. With an optimized wire geometry, fidelities of better than 99.9% should therefore be possible in realistic settings with present day atom chip technology.

Acknowledgments

M. A. Cirone, A. Negretti, T. Calarco and J. Schmiedmayer acknowledge financial support from the European Union, contract number IST-2001-38863 (ACQP). T. Calarco also acknowledges financial support from the European Union through the FP6-FET Integrated Project CT-015714 (SCALA) and a EU Marie Curie Outgoing International Fellowship, and from the National Science Foundation through a grant for the Institute for Theoretical Atomic, Molecular and Optical Physics at Harvard University and Smithsonian Astrophysical Observatory. E. Charron acknowledges the IDRIS-CNRS supercomputer center for the contract number 08/051848 and the financial support of the LRC of the CEA, under contract number DSM05-33. Laboratoire de Photophysique Moléculaire is associated to Université Paris-Sud 11. We wish to thank P. Krüger, J. Reichel and P. Treutlein for useful discussions about experimental details.

References

1. M. A. Nielsen and I. L. Chuang, *Quantum Computation and Quantum Information*, Cambridge University Press, Cambridge (2000).
2. P. W. Shor, *Proc. 35th Annual Symposium on Foundations of Computer Science* (Shafi Goldwasser, ed.), IEEE Computer Society Press (1994).
3. L. K. Grover, *Phys. Rev. Lett.* **79**, 325 (1997).
4. A. Barenco, C. H. Bennett, R. Cleve, D. P. DiVincenzo, N. Margolus, P. Shor, T. Sleator, J. A. Smolin, and H. Weinfurter *Phys. Rev. A* **52**, 3457 (1995).
5. R. Folman, P. Krüger, J. Schmiedmayer, J. Denschlag, and C. Henkel, *Adv. At. Mol. Opt. Phys.* **48**, 263 (2002).
6. E. A. Hinds, C. J. Vale and M. G. Boshier, *Phys. Rev. Lett.* **86**, 1462 (2001) ; W. Hänsel, J. Reichel, P. Hommelhoff and T. W. Hänsch, *Phys. Rev. A* **64**, 063607 (2001) ; T. Schumm, S. Hofferberth, L. M. Andersson, S. Wildermuth, S. Groth, I. Bar-Joseph, J. Schmiedmayer and P. Krüger, *Nature Physics* **1**, 57 (2005).
7. M. A. Cirone, A. Negretti, T. Calarco, P. Krüger, and J. Schmiedmayer, *Eur. J. Phys. D* **35**, 165 (2005).
8. During the gate operation, the trapping frequencies vary in the ranges 13.0-10.9 kHz, 157.5-154.3 kHz and 153.8-150.9 kHz for $\nu_{x'}$, $\nu_{y'}$ and ν_z respectivelly.

9. R. Folman, P. Krüger, D. Cassettari, B. Hessmo, T. Maier, and J. Schmiedmayer, *Phys. Rev. Lett.* **84**, 4749 (2000).

10. P. Krüger, L.M. Andersson, S. Wildermuth, S. Hofferberth, E.Haller, S. Aigner, S. Groth, I. Bar-Joseph and J. Schmiedmayer, e-print arXiv:cond-mat/0504686.

11. S. Groth, P. Krüger, S. Wildermuth, R. Folman, T. Fernholz, D. Mahalu, I. Bar-Joseph, and J. Schmiedmayer, *Appl. Phys. Lett.* **85**, 2980 (2004).

12. P. Treutlein, P. Hommelhoff, T. Steinmetz, T. W. H\"ansch, and J. Reichel, *Phys. Rev. Lett.* **92**, 203005 (2004).

13. D. M. Harber, H. J. Lewandowski, J. M. McGuirk, and E. A. Cornell, *Phys. Rev. A* **66**, 053616 (2002).

14. E. Charron, M. A. Cirone, A. Negretti, J. Schmiedmayer and T. Calarco, *Phys. Rev. A* **73**, *in press* (2006) and e-print arXiv:quant-ph/0603138

15. T. Calarco, E. A. Hinds, D. Jaksch, J. Schmiedmayer, J. I. Cirac, and P. Zoller *Phys. Rev. A* **61**, 022304 (2000).

16. L. F. Shampine and M. Gordon, *Computer solution of ordinary differential equations: the initial value problem*, W. H. Freemann and Company, San Francisco, (1975). http://www.csit.fsu.edu/~burkardt/f_src/ode/ode.f90

17. E. Charron, E. Tiesinga, F. Mies, and C. Williams, *Phys. Rev. Lett.* **88**, 077901 (2002).

18. C. Henkel and M. Wilkens, *Europhys. Lett.* **47**, 414 (1999) ; C. Henkel, S. Pötting and M. Wilkens, *Appl. Phys. B* **69**, 379 (1999) ; C. Henkel and S. Pötting, *Appl. Phys. B* **72**, 73 (2001) ; C. Henkel, *Eur. Phys. J. D* **35**, 59 (2005).

19. P. K. Rekdal, S. Scheel, P. L. Knight and E. A. Hinds, *Phys. Rev. A* **70**, 013811 (2004) ; S. Scheel, P. K. Rekdal, P. L. Knight and E. A. Hinds, e-print arXiv:quant-ph/0501149.

ULTRAFAST PULSE SHAPING DEVELOPMENTS FOR QUANTUM COMPUTATION*

S.K. KARTHICK KUMAR, DEBABRATA GOSWAMI†

Department of Chemistry and the Center for Laser Technology, Indian Institute of Technology, Kanpur—208016, India

This paper is geared to provide some logical essence of our work on various aspects quantum computing through optical approach involving the engineering aspects of ultrafast laser pulse modulation and programmability. Such an effort also includes the hunt for an appropriate physical system for quantum information as we present here.

1. Introduction

Ultrafast pulse shaping is a useful tool in coherent control of molecular dynamics and in high-speed communication technologies. Typical ultrafast pulse shaping technology uses indirect Fourier Transform techniques through the use of a spatial grating based pulse shaper where the pulse can be shaped by amplitude or phase masking in the Fourier domain [1] or by time domain schemes through the use of Mach Zehnder or a Michelson interferometer [2]. Here, we discuss a novel combination of these two typical schemes wherein a weaker femtosecond pulse in one arm of the interferometer is intensity modulated by its replica that undergoes spatial pulse shaping in the other arm of the interferometer. The intensity modulations of the output pulse was found to occur at regular intervals through the zero delay position that is independently established through sum-frequency generation between the two-arms. Such techniques at 1560nm would be of immense use for simultaneous time-domain multiplexing (TDM) and wavelength-domain multiplexing (WDM).

2. Experimental Setup

An Erbium doped femtosecond fiber laser emitting synchronized femtosecond pulses centered at two different wavelengths 780nm and 1560nm collinearly was used. Independent autocorrelation measurements establish that the laser

* This work is supported by DST, MCIT, New Delhi and International SRF program of Wellcome Trust, UK.

† Corresponding author : dgoswami@iitk.ac.ic

operates at the manufacture's specifications of 100fs pulsewidth for the 780nm and 300fs for the 1560nm wavelengths. The synchronized nature of the two wavelengths is confirmed through sum-frequency generation between the two wavelengths right at the output of the laser by focusing it into a BBO crystal which results in 520nm light. We built a Mach-Zehnder type interferometer between the two wavelengths. In one of the arms of the interferometer a spatial modulator pulse shaper was constructed while the other arm has a motorized delay line setup with the help of a retroreflector (R) mounted on a motorized translation stage (Newport MFN 20CC) (Fig. 1). Such a setup allows us to measure the cross-correlation between the two wavelengths scanning through the equal-arm position of the interferometer. The cross-correlation forms a routine diagnostic scheme in the characterization of the pulse shaper.

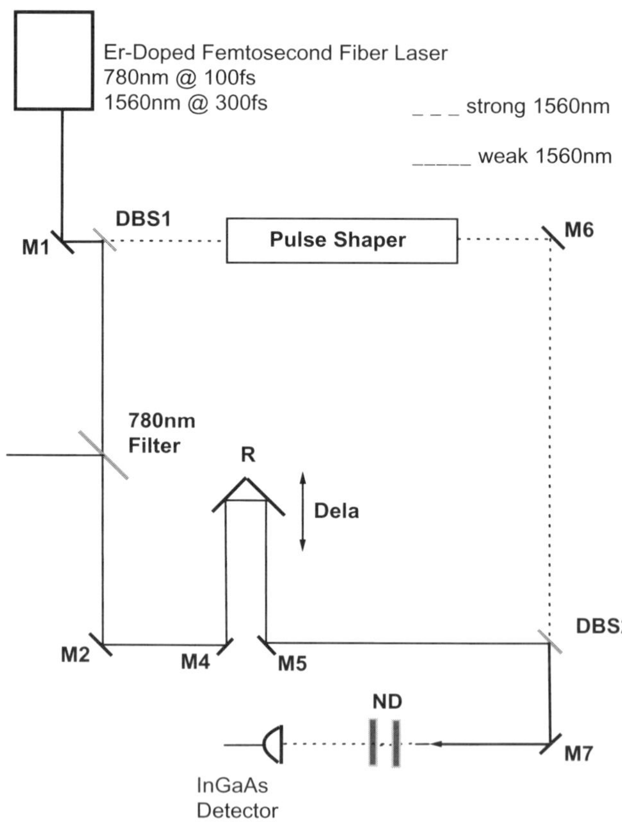

Figure 1. Experimental setup for modulating the intensity of weak pulse with a pulse shaper. DBS: Dichroic beam splitter, R: retroreflector, M: Mirror

We found that the dichroic beam splitter (DBS1) that transmits the 1560nm pulses while reflecting the 780nm pulses, also partially acts as a 90:10 beamsplitter for the 1560nm pulse. During sum-frequency generation process, this leakage is inconsequential. However, this leakage becomes critical in our design of the coupled interferometer to spatial pulse shaper design. The shaped 1560nm pulses from the pulse shaper arm is recombined with the weak 1560nm pulse from the motorized delay arm in a second dichroic beam splitter (DBS2) such that the output pulses are collinear.

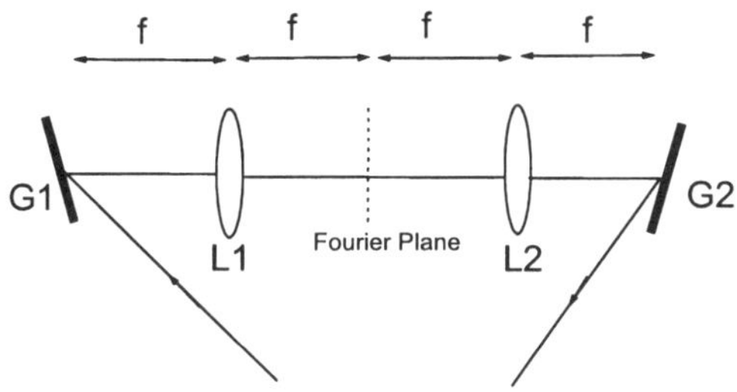

Figure 2. Pulse shaper setup in a 4f geometry. G1 and G2 are the gratings; L1 and L2 are the lenses

The pulse shaper (Fig. 2) in one arm of the interferometer has a 1200 lines/mm reflective grating G1 in dispersive mode. The dispersed incoming pulse is then focused with one of the achromat pair plano-convex lens L1 of focal length f = 30cm, which is at a distance f from the grating to complete the Forward Fourier Transform (FFT). The second of the achromat pair lens L2 thereafter starts the process of Inverse Fourier Transform (IFT) by collimating the output FFT beam and it is placed at a distance of 2f from the first lens. Thus collimated, the beam then hits a reflective grating G2 in the dispersive mode to get back the original 1560nm pulse in case no modulation is performed in the Fourier plane. The pulse is then sent to DBS2 to combine the two pulses collinearly with some loss through a second dichroic beamsplitter and monitored by an InGaAs detector (Acton Research Corp. ID441) for variable delays of the weak 1560nm pulse.

3. Results and Discussion

Fluctuations were seen in the intensity of the spatially overlapped 1560nm pulses and were observed to be changing with their temporal overlap. To establish a zero time delay for the interferometer, we generated a sum-frequency signal with 780nm pulse, by focusing the collinear the 780nm and 1560nm pulses with a fused silica lens of focal length 75mm on to the BBO for sum frequency generation at 520nm. The BBO and the lens were then removed and the change in the intensity was monitored for time delays of the weak pulse at 1560nm by using an InGaAs detector.

Fluctuations observed (Fig. 3) were of few hundred milliVolts and to record the changes in the fluctuations, the readings were taken in peak to peak mode in the oscilloscope and the step size of the time delay was 5.54μm. Our technique of quantifying amplitude of the shaped pulse signal allows us to measure all the five correlation signals, one of which is at the same position as the regular cross-correlation signal.

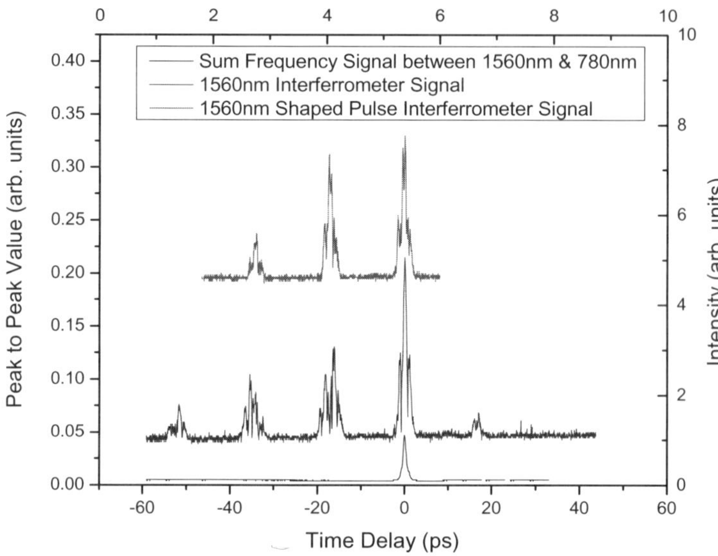

Figure 3. Plot of changes in the peak to peak value of the intensity as a function of the delay of weak pulse. The time-zero is established with the help of the sum frequency generating signal which is bottommost graph. Effect of shaping the pulse in the spatial domain is shown in the topmost pulse comb. We also find that pure phase modulation adds additional delay to the pulse train; however, for clarity in the figure we only show representative three data-sets.

Each of these time copies carry the expected effects of the shaped pulse in terms of its relative amplitude and phase content with respect to the unshaped pulse. Such interferometric results can be correlated by use of the interferrometric correlation signal:

$$|E(t)|^2 \propto \int E_1(t)E_2(\tau - t)d\tau \tag{1}$$

The phase of the electric field can be expanded as the Taylor expansion [3]:

$$\begin{aligned} E_1(t) &= FT\{\varepsilon_1(\omega)exp[-i(\omega t + \phi(\omega))]\} \\ \phi(\omega) &= b_0 + b_1\omega + b_2\omega^2 + b_3\omega^3 + \ldots \end{aligned} \tag{2}$$

It is possible to fit the experimental data quite accurately with such a simple model wherein the sign of b_1 needs to be negative. However, it is important to realize that the time axes of these data-sets are measured in relation to the spatial shaping arm containing the 1560nm as shown in Fig.1. So, if the interference is happening due to a change in arm length of the reference 780nm arm, the time axis is read in the negative sense. The overall result as we show in Fig.3 is therefore, in fact, a combination of both the issues. Such finely time-delayed shaped pulses would be of use in interesting control and molecular switching experiments.

4. Qubit Realization and Optical Network

A typical visualization of a quantum computer network would have nodes consisting of quantum storage devices, where information can be stored for very long times either in ground in or some metastable excited states of atoms, molecules or ions. The quantum information can be transferred from one node of the network to the other using photons. The nodes would carry out the required computations and also serve as a storage or memory unit. The storage time is limited by decoherence time. Transferring quantum information between the two nodes without allowing for decoherence is very difficult. There are already some proposals in quantum communication to transmit and exchange quantum information between distant users, which includes distribution of quantum secure key information for secure communication. Teleportation allows an arbitrary unknown quantum state to be conveyed from one distant part to another with perfect fidelity by the establishment of a maximal entangled state of two distant quantum bits. However, the bottleneck for communication between distant users is the scaling of the error probability with the length of the channel connecting the users. The error results from amplitude and phase damping.

Our proposal in realizing such schemes would achieve quantum computing at nodes of the quantum networks and send photons through free space or through standard optical fibers for networking (Fig.4). In such a scheme we would exploit the low cross-talk between two light signals, capacity to carry enormous amount of information encoded on the optical pulse, combined with its ability to induce the desired transitions in the target systems. Ultrafast optically shaped pulses enable us to get to our goals. Appropriately shaped optical pulse has been successfully used to induce selective excitation to control the yield of the desired product at the end of the chemical reaction either under single or multi-photon condition [1]. Such designed pulses also have potential to control decoherence in large molecules by controlling intra-molecular vibrational relaxation. The quantum interaction with the incident shaped pulse is in effect as long as the pulse is present irrespective of the dephasing mechanism. This would add to the high fidelity of the systems, since the presence of the environment, which would otherwise destroy the coherence, would not affect the operation. Shaped optical pulses have furthermore been used to demonstrate the quantum logic gate (CNOT) in such systems [3].

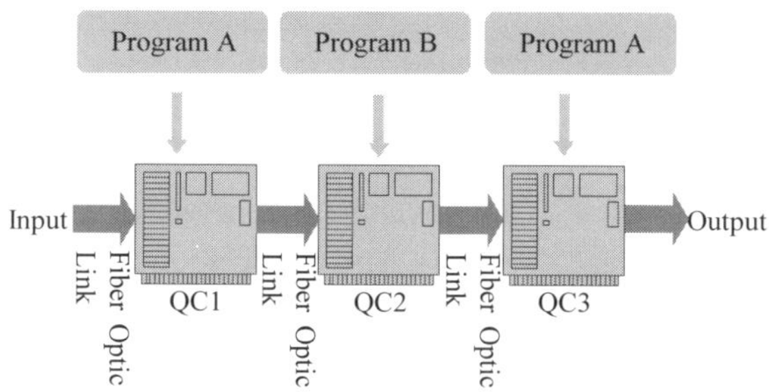

Figure 4. Schematic of the AOM pulse shaper based fiber- optic networked distributed quantum-computing framework as discussed in the text. For clarity purposes we show the schematic with only three quantum computers (QC).

The shaped pulses can be split into a number of different paths which can carry different train of pulses at different timescales. This would provide leverage to control the various nodes where molecular systems interact with shaped pulses to carry out various instructions and perform quantum computing activity at each node during the pulse duration. This could enable the processing of different quantum computational steps at various nodes simultaneously, which implies that we can parallelize our code and distribute the task over the network. At the end of the computation we can read out the results by sending in a "read pulse" and recombining the results. Essentially, this is distributed quantum computing over the network using shaped pulses. Currently, 106 bits can be transmitted/encoded in a single burst of light [4] with the present day optical pulse shaping technology as shown in Fig. 3. The repetition rate from the laser source is about 50 to 100 MHz. Thus, one would be able to use terabit/sec bit of communication channel through the existing infrastructure available with the optical community. Once such a quantum computer is available at remote site, these packets acting as "quantum software" can be transferred through high-speed communication channels. Thus, it is possible to carry out quantum computation at a remote distance with the proposed scheme of shaped pulses for terabit/sec communication and molecular control (Fig.4).

5. Experimental Developments

One of our goals towards Quantum Computing also involves the hunt for molecular systems that could effectively be used through optical interactions for optical logic. Towards this goal, several developments in hunting molecular systems have been quite successful and several literature reports have been made. We have, for example, developed a very sensitive femtosecond z-scan apparatus that has the capability to measure extremely low as well as very high cross-section (σ_2) values and nonlinear index of refraction (n_2). A general trend is evolving in the molecular structure to nonlinearity measurements. In these results, we have taken the advantage of using both closed aperture z-scan as well as open-aperture z-scan data. A typical z-scan apparatus and an open-aperture z-scan data is depicted below in Fig. 5. Since the incident laser used is 100fs wavelength tunable MIRA Ti:Sapphire (Coherent Inc., USA), this experimental arrangement also enables wavelength tunable scans.

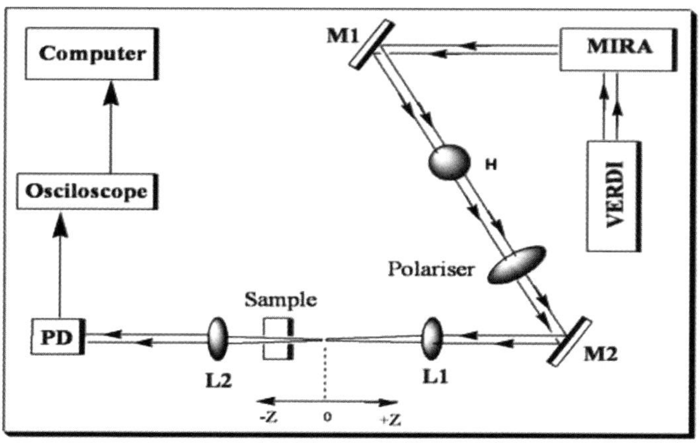

Figure 5. Experimental setup for the z-scan experiments. An iris in between lens L2 & photodiode (PD) with ~50% transmission aperturing would transform an open z-scan setup into a closed z-scan.

In this same setup as shown in Fig.5, if we introduce another laser pulse exactly overlapped in time and space, we can then measure the effect of nonlinearity at one wavelength by the other. We have demonstrated this effect [5] through the combination of 1560nm with 780nm output from the Erbium-doped fiber laser (IMRA Inc. USA). Interestingly, such sensitive results could also indicate the transient heating and thermal lensing effects that are generated as a result of such "pump-probe" z-scan measurements as shown in Fig.6 below.

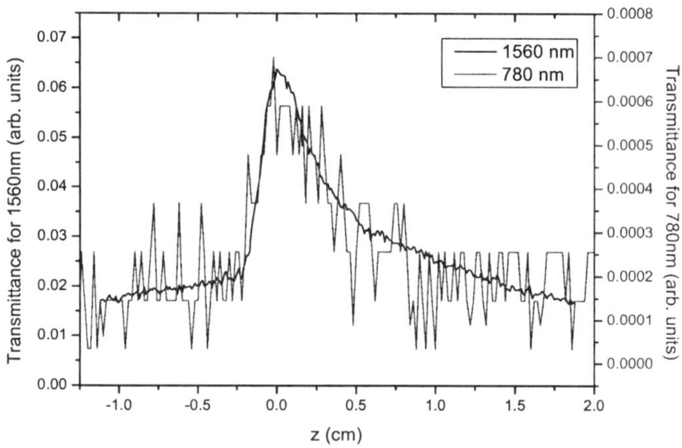

Figure 6. Measured *z*-scan transmittance of 80 fs pulses of $\lambda = 780$ nm as a probe through a 1.6 mm thick double distilled water being irradiated with 95 fs of $\lambda = 1560$ nm as pump (raw transmittance data is presented to illustrate sensitivity of our measurements). We plot the open aperture data for 1560 nm alone on this same plot to show that both essentially have same features except their two orders of magnitude difference in their signal levels and hence the same fit would suffice.

6. Conclusion

The experimental results presented here are fueled by either the desire to enhance the capabilities of femtosecond pulse shaping and in the hunt for good physical systems that are potentially good candidates for handling quantum information. A key concept in understanding this is the notion of quantum noise or decoherence that corrupts desired evolution of the system. The length of the longest possible quantum computation is roughly given by the ratio of τ_q (the time for which a system remains quantum-mechanically coherent) to τ_{op} (the time it takes to perform elementary unitary transformations, which involve at least two qubits). These two times are actually related to each other in many systems, since they are both determined by the strength of coupling of the system to the external world. Nevertheless, $\lambda = \tau_{op}/\tau_q$ can vary over surprisingly wide range in the kind of systems where quantum computation has been achieved [6].

We demonstrated in the first section that the enhanced capabilities of femtosecond pulse shaping can in fact lead to time-comb pulses that would have various applications in information technology and coherent control issues. As part of sensitivity enhancement and high finesse experiments, we also demonstrate the capability to measure minute localized temperature jump through pump-probe z-scan experiments. Further studies on such developments for high finesse experiments are currently in progress in the authors' laboratory.

References

1. D. Goswami, *Physics Reports* **374**, 385-483 (2003).
2. D. Goswami, C. W. Hillegas, J. X. Tull, and W. S. Warren, *Femtosecond Reaction Dynamics*, **42**, 291-298 (1994).
3. D. Goswami, *Phys. Rev. Lett.* **88**, 177901-1—177901-4 (2002).
4. W. Yang, M.R. Fetterman, D. Goswami, W.S. Warren, *J. Opt. Comm.* **22**(1), 2001, 694-697 (2001).
5. D. Goswami, *Optics Comm.* **261**, 158-162 (2006).
6. M.A. Nielsen, I.L. Chuang, *Quantum Computing and Quantum Information*. Cambridge University Press: Cambridge, UK, 2000.

Quantum information transfer in atom-photon interactions in a cavity

A. S. Majumdar*, N. Nayak and B. Ghosh

S. N. Bose National Centre for Basic Sciences, Salt Lake, Kolkata 700 098, India
**E-mail: archan@bose.res.in*

We consider the interaction of one or more two-level atom(s) with photons inside the controlled environment of a microwave cavity. The entanglement generated between the atoms and the cavity photons has several interesting consequences. Robust atom-atom entanglement can be created through this mechanism without the direct overlap between the atomic wave functions. The monogamous nature of entanglement is clearly exhibited. Cavity dissipation is monitored in a controlled manner, and the transfer of quantum information between the atoms, cavity, and the external reservoir is studied. Practical realization of several information theoretic concepts is achieved through the micromaser.

1. Introduction

Composite quantum systems are usually endowed with quantum entanglement. A pair of particles is said to be entangled in quantum mechanics if its state cannot be expressed as a product of the states of its individual constituents. This was first noted by Einstein, Podolsky and Rosen in 1935 [1]. Entangled states have nonclassical and nonlocal properties, such as the violation of Bell's inequality [2]. The preparation and manipulation of these states leads to a better understanding of basic quantum phenomena. A striking example is that of the GHZ states [3] which have been used for tests of quantum nonlocality [4]. At the practical level entanglement has become a fundamental resource in quantum information processing [5]. There has been rapid theoretical development and experimental realization of different features of quantum entanglement in recent years [6].

Unlike classical correlations, quantum entanglement can not be freely shared among many quantum systems. It has been observed that a quantum system being entangled with another one limits its possible entanglement with a third system. This is known as the "monogamous nature of entanglement" which was first proposed by Bennett [7]. If a pair of two-level quantum systems A and B have a perfect quantum correlation, namely, if they are in a maximally entangled state $\Psi^- = (|01\rangle - |10\rangle)/\sqrt{2}$, then the system A cannot be entangled to a third system C. This indicates that there is a limitation in the distribution of entanglement, and several efforts have been devoted to capture this unique property of "monogamy of quantum entanglement" in a quantitative way for tripartite and multipartite systems [8–10]. Another distinctive property of quantum entanglement for multipartite systems is the possibility of entanglement swapping between two or more pairs of qubits. Using this property, two parties that never interacted in their history can be entangled [11]. There may indeed exist a deeper connection between the characteristics of "monogamy" and entanglement swapping since the features of the distribution and transfer of quantum information is essentially reflected in the both these properties.

Entanglement has been widely observed within the framework of atom-photon interactions such as in cavity quantum electrodynamics. Several experiments have been carried out in recent years where entangled states have been created and verified. Practical realization of various features of quantum entanglement are obtained in quantum optical devices consisting of optical and microwave cavities [12]. An example that could be highlighted is the generation of a maximally entangled state between two modes in a single cavity using a Rydberg atom coherently interacting with each mode in turn [13]. For practical implementation of quantum information protocols useful in communication and computation [6], entanglement has to be created and preserved between qubits that are well separated. A recent experimental breakthrough has been obtained by entangling two distant atomic qubits by their interaction with the same photon [14]. From the viewpoint of information processing, quantification of entanglement is an important aspect. Recently some studies have been performed to quantify the entanglement that is obtained in atom-photon interactions in cavities [15,16].

An important attribute of real devices in the ubiquitous presence of dissipative effects in them. These have to be monitored in order for the effects of quantum correlations to survive till detection. The consequences of cavity leakage on information transfer in the micromaser has been quantified recently [16]. Other characteristics of entanglement such as its "monogamous" nature, and also its exchange or swapping are affected by dissipative processes [17]. Atom-photon interactions in cavities are a sound arena for the quantitative investigations of different aspects of quantum entanglement in realistic situations. In this article we discuss the entanglement generated between atoms and photons, and also between two atoms mediated by the cavity photon field. Our system consists of single-mode cavities which are empty initially, and Rydberg atoms passing through them. We show the entanglement generated could be quantified both in the case of ideal cavities, and in the presence of dissipative effects like cavity photon leakage. We discuss two features of atom-photon interactions in cavity QED systems in detail, viz., the quantitative nature of information transfer, and the monogamous character of entanglement.

2. The generation of atomic entanglement in cavities

A composite two-component state $|\Psi\rangle_{AB}$ in quantum mechanics is said to be entangled if

$$|\Psi\rangle_{AB} \neq |\chi\rangle_A \otimes |\phi\rangle_B \tag{1}$$

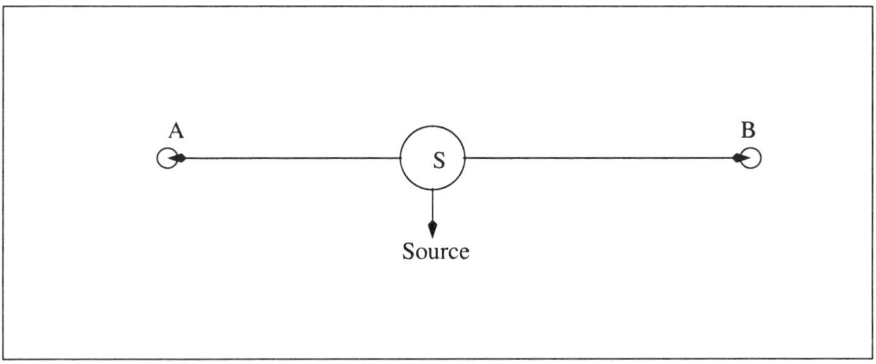

Fig. 1. Entanglement between two particles can be created by their interaction or common origin, e.g., two spin-1/2 particles are emitted from a common source.

The entangled states of two spin-1/2 particles may take forms as

$$\begin{aligned} |\phi^{\pm}\rangle_{AB} &= \frac{1}{\sqrt{2}}\left(|\uparrow_z, \uparrow_z\rangle_{AB} \pm |\downarrow_z, \downarrow_z\rangle_{AB}\right), \\ |\psi^{\pm}\rangle_{AB} &= \frac{1}{\sqrt{2}}\left(|\uparrow_z, \downarrow_z\rangle_{AB} \pm |\downarrow_z, \uparrow_z\rangle_{AB}\right) \end{aligned} \tag{2}$$

These above states are maximally entangled states and they are a fundamental resource in quantum information processing A two-level atom (which is formally analogous to a spin-1/2 system) interacting with a single resonant cavity mode results in entanglement between the atom and the cavity photon. The resonant interaction between atom and photon inside a cavity is governed by the Jaynes-Cummings model.

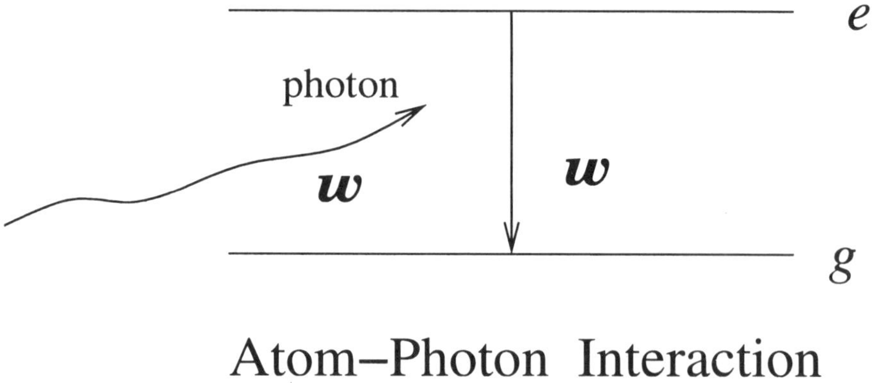

Fig. 2. A schematic diagram of a two level atom inside a single mode cavity.

The Jaynes-Cummings (JC) model [18] is one of the simplest examples of two interacting quantum systems. It is one of the most studied models in quantum optics because it is an exactly solvable model. Entanglement of the optical field with matter in the JC model has been studied earlier [19]. Here we will discuss the entanglement between atoms mediated by the optical field, where the light-atom interaction is governed by the JC model. The JC model conists of a two-level atom coupled to a single-mode radiation field inside a cavity. A two level atom is formally analogous to a spin-1/2 system. Let us denote the upper level of the atom as $|e\rangle$ and the lower level as $|g\rangle$ and the spin (atomic) raising and lowering operators can be defined as $\sigma^+ = |e\rangle\langle g|$ and $\sigma^- = |g\rangle\langle e|$, respectively, with the commutation relation

$$\begin{aligned}[\sigma^+, \sigma^-] &= |e\rangle\langle e| - |g\rangle\langle g| \\ &= \sigma_z . \end{aligned} \tag{3}$$

A quantum mechanical field can be represented as (for the present purpose, we consider a single mode field)

$$E(t) = \frac{1}{2}[ae^{-i\omega t} + a^\dagger e^{i\omega t}] \tag{4}$$

apart from a mode function which we omit here since it is not required for the present discussion. Here a and $a^\dagger$ are annihilation and creation operators, respectively. The graininess of the radiation field is represented by the photon number state $|n\rangle$, $n = 0, 1, 2, \ldots\ldots$, such that $a|n\rangle = \sqrt{n}|n-1\rangle$

and $a^\dagger|n\rangle = \sqrt{n+1}|n+1\rangle$. It is an eigenstate of the number operator $\hat{n} = a^\dagger a$ given by

$$\hat{n}|n\rangle = n|n\rangle \tag{5}$$

The field in Eq.(4) can be represented by a quantum mechanical state vector $|\psi\rangle$ which is a linear superposition of of the number states $|n\rangle$, i.e.,

$$|\psi\rangle = \sum_{n=0}^{\infty} c_n|n\rangle \tag{6}$$

where c_n is, in general, complex and gives the probabilty of the field having n photons by the relation

$$P_n = \langle n|\psi\rangle\langle\psi|n\rangle = |c_n|^2 \tag{7}$$

The field obeys quantum statistics and its average photon number is given by

$$< n >= \sum_{n=0}^{\infty} nP_n \tag{8}$$

with the intensity of the field $I \propto < n >$. The statistics brings in a quantum mechanical noise which is represented by the variance

$$V = \frac{< n^2 > - < n >}{< n^2 >}. \tag{9}$$

$V = 1$ is for coherent state field and $V < 1$ signifies a non-classical state. The parameters $< n >$ and V give a fair description of the quantum mechanical nature of the radiation field.

The Hamiltonian of the joint atom-field system can be written in the rotating wave approximation [20] as,

$$H = \frac{\omega\sigma_z}{2} + a^\dagger a\omega + g(\sigma^+ a + \sigma^- a^\dagger), \tag{10}$$

where $a^\dagger$ and a are usual creation and destruction operators of the radiation field. (Here we have set $\hbar = 1$). In the interaction picture, the equation of motion defining the system is

$$i\frac{\partial}{\partial t}|\psi_{a-f}(t)\rangle_I = H^I_{int.}|\psi_{a-f}(t)\rangle_I, \tag{11}$$

where the Hamiltonian reduces to

$$H_I = g(\sigma^+ a + \sigma^- a^\dagger). \tag{12}$$

We consider a micromaser system in which two atoms prepared in excited state are passing through a single mode cavity one after the other

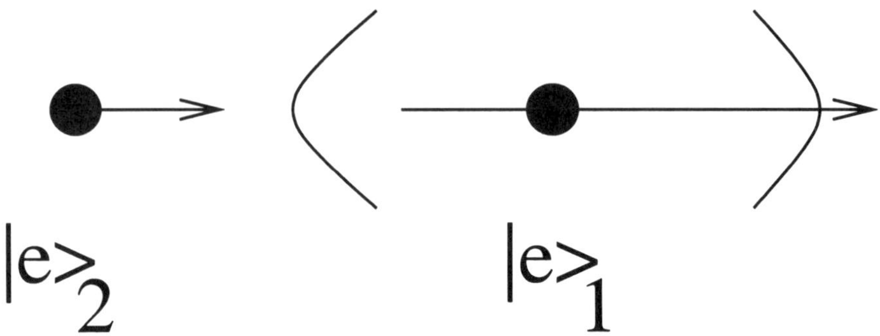

Fig. 3. Two two-level atoms prepared initially in their upper (excited) states pass through the single mode cavity one after the other. There is no spatial overlap between the two atoms.

With the passage of the two atoms, one after the other, the joint state of both the atoms and the field at some instance t may be denoted by $|\Psi(t) >_{a-a-f}$. The corresponding atom-atom-field pure density state is

$$\rho(t) = |\Psi(t)\rangle\langle\Psi(t)| \tag{13}$$

The state of the two atoms emerging from the cavity is not separable, as has been observed through the violation of Bell's inequality [21]. In order to quantify the entanglement between the two atoms, the field variables have to be traced out. The reduced mixed density state of two atoms after taking trace over the field is

$$\rho(t) = \mathrm{T}r_{\text{field}}|\Psi(t) >_{a-a-f.a-a-f}< \Psi(t)| \tag{14}$$

Entanglement within pure states of bipartite system can be measured by the Von Neumann entropy of the reduced density matrices. Various measures have been introduced to quantify bipartite entanglement of mixed states but none of them satisfies all the requisite criteria for an axiomatically sound measure [22]. While the entanglement for a mixed state can be measured as the average entanglement of its pure state decompositions, the existence of an infinite number of such decompositions makes their minimization over this set a nontrivial task. Hill and Wooters [23] carried out such a procedure for bipartite, $2\otimes 2$ systems and showed that a quantity 'entanglement of formation' is a measure of entanglement. The entanglement of formation has since turned out to be a popular measure for computing atomic entagelement in quantum optical systems [15–17]. The entanglement of formation for a bipartite density operator ρ is given by

$$E_F(\rho) = h\left(\frac{1+\sqrt{1-C^2(\rho)}}{2}\right), \tag{15}$$

where C is called the concurrence defined as

$$C(\rho) = \max(0, \sqrt{\lambda_1} - \sqrt{\lambda_2} - \sqrt{\lambda_3} - \sqrt{\lambda_4}), \tag{16}$$

where the λ_i are the eigenvalues of $\rho_{12}(\sigma_y \otimes \sigma_y)\rho_{12}^*(\sigma_y \otimes \sigma_y)$ in descending order, and

$$h(x) = -x\log_2 x - (1-x)\log_2(1-x) \tag{17}$$

is the binary entropy function. The entanglement of formation is a monotone of the concurrence.

In order to demonstrate the quantitative nature of information transfer in cavity-QED systems, it is useful to compute the Shannon entropy of the steady-state cavity defined as

$$S(\rho_f^{(ss)}) = -\sum_{n=1}^{\infty} P_n^{(ss)} \log P_n^{(ss)} \tag{18}$$

using the expression for the steady-state photon number distribution $P_n^{(ss)}$ given in Ref. [24], with the normalysation $\sum_n P_n = 1$. The expression for the density operator of the cavity field after passage of the two experimental atoms has been previously derived [21,24]. We denote this cavity state as $\rho_f^{(2)}$, given by

$$\rho_f^{(2)} = \mathcal{A}\mathcal{A}\rho_f^{(ss)}\mathcal{A}^\dagger\mathcal{A}^\dagger + \mathcal{D}\mathcal{D}\rho_f^{(ss)}\mathcal{D}^\dagger\mathcal{D}^\dagger + 2\mathcal{A}\mathcal{D}\rho_f^{(ss)}\mathcal{D}^\dagger\mathcal{A}^\dagger \tag{19}$$

where the field operators $\mathcal{A}$ and $\mathcal{D}$ are defined as

$$\begin{aligned} \mathcal{A} &= \cos(gt\sqrt{a^\dagger a+1}) \\ \mathcal{D} &= -ia^\dagger\frac{\sin(gt\sqrt{a^\dagger a+1})}{\sqrt{a^\dagger a+1}} \end{aligned} \tag{20}$$

The expression fore the Shannon entropy is given by

$$S(\rho_f^{(2)}) = -\sum_{n=1}^{\infty} P_n^{(2)} \log P_n^{(2)} \tag{21}$$

and it can be computed through the corresponding photon number distribution $P_n^{(2)} = \langle n|\rho_f^{(2)}|n\rangle$. It is observed that the streaming atoms leave an imprint on the the photon distribution function, as reflected by its reduced Shannon entropy [16].

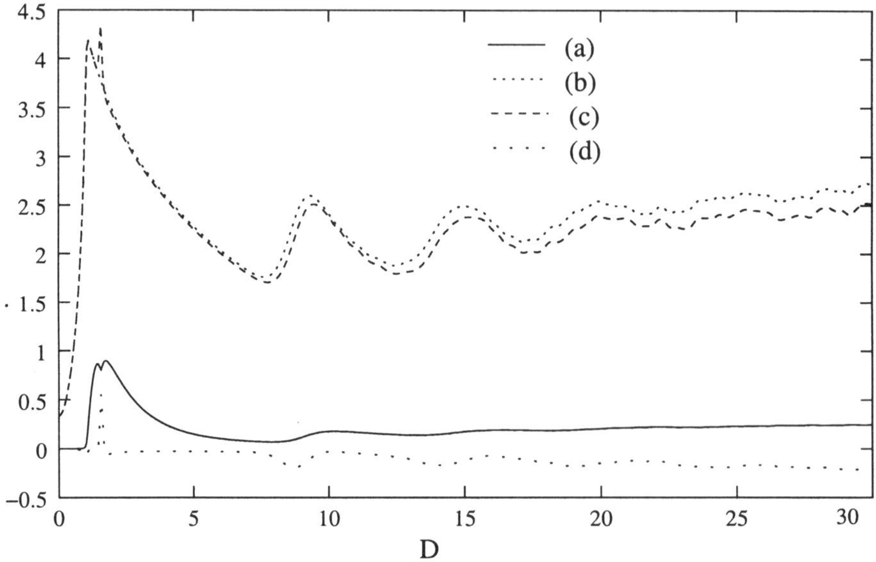

Fig. 4. The entanglement of formation E_F (a), the Shannon entropies of the cavity, $S(\rho_f^{(ss)})$ (b), $S(\rho_f^{(2)})$ (c), and their difference $S(\rho_f^{(2)}) - S(\rho_f^{(ss)})$ (d) are plotted with respect to the cavity pump parameter D.

A striking feature of our results [16] is the remarkable correspondence between the difference $S(\rho_f^{(2)}) - S(\rho_f^{(ss)})$ and the entanglement of formation $E_F(\rho_a)$ for the atoms. In Fig4 the four functions, $S(\rho_f^{(ss)})$, $S(\rho_f^{(2)})$, $S(\rho_f^{(2)}) - S(\rho_f^{(ss)})$, and $E_F(\rho_a)$, are plotted respectively, versus the cavity pump parameter D. The difference of the cavity entropy before and after the passage of the atoms is transported towards constructing the atomic entanglement. The quantitative nature of information transfer is clearly revealed even in the presence of dissipation.

3. The monogamous nature of entanglement

Here we consider two cavities which can be maximally entangled by sending a single circular Rydberg atom prepared in the excited state through two identical and initially empty high-Q cavities (C_1 and C_2) [25]. The initial state of the two-cavity entangled system can be written as

$$|\Psi\rangle_{C_1C_2} = \frac{1}{\sqrt{2}}(|0_1 1_2\rangle + |1_1 0_2\rangle), \tag{22}$$

where the index 1 and 2 refer to the first and second cavity, respectively. In this set-up we consider the passage of a two-level Rydberg atom A_1 prepared in the ground state $|g\rangle$ through the cavity C_1. We are considering the resonant interaction between the two-level atom and cavity mode frequency. The dynamics of the atom-photon interaction is governed in the ideal cavity (no dissipation) case by the equation

$$\dot{\rho} = -i[H_I, \rho] \tag{23}$$

with joint three-party initial ($t = 0$) state corresponding to

$$|\Psi(t=0)\rangle_{C_1C_2A_1} = \frac{1}{\sqrt{2}}(|0_1 1_2\rangle + |1_1 0_2\rangle) \otimes |g_1\rangle \tag{24}$$

Hence, a two-level atom entering the empty cavity in the upper state ($|e\rangle$) evolves to

$$\begin{aligned} |\Psi_e(t)\rangle &= e^{-iH_I t}|e, 0\rangle \\ &= \cos(gt)|e, 0\rangle + \sin(gt)|g, 1\rangle \end{aligned} \tag{25}$$

at some time t, and similarly, a two-level Rydberg atom entering the one photon cavity in the ground state evolves to

$$\begin{aligned} |\Psi_g(t)\rangle &= e^{-iH_I t}|g, 1\rangle \\ &= \cos(gt)|g, 1\rangle - \sin(gt)|e, 0\rangle \end{aligned} \tag{26}$$

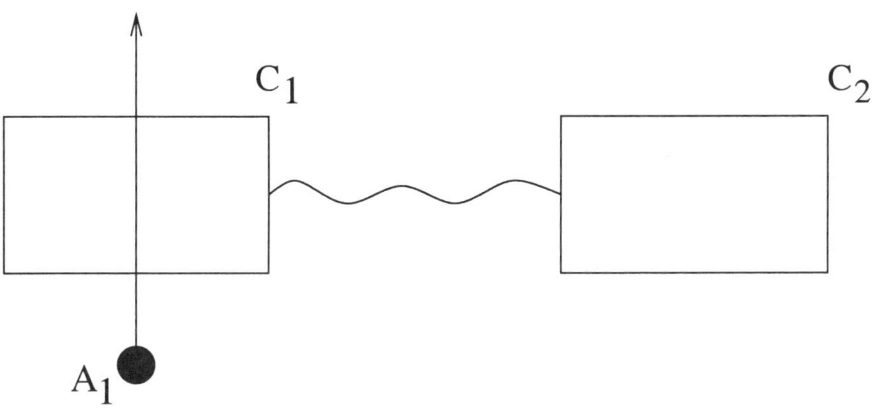

Fig. 5. A two-level Rydberg atom prepared in the ground state is passing throuh the one of the maximally entangled cavities C_1

Let us now discuss the above scenario in the presence of cavity dissipation. Since the lifetime of a two-level Rydberg atom is usually much longer compared to the atom-cavity interaction time, we can safely neglect the atomic dissipation. The dynamics of the flight of the atom is governed by the evolution equation

$$\dot{\rho} = \dot{\rho}|_{\text{atom-field}} + \dot{\rho}|_{\text{field-reservoir}} \tag{27}$$

where the strength of the couplings are given by the parameters κ (the cavity leakage constant) and g (the atom-field interaction constant). At temperature $T = 0K$ the average thermal photon number is zero, and hence one has [26]

$$\dot{\rho}|_{\text{field-reservoir}} = -\kappa(a^\dagger a\rho - 2a\rho a^\dagger + \rho a^\dagger a). \tag{28}$$

When $g \gg \kappa$, it is possible to make a secular approximation [27] while solving the complete evolution equation in order to get the density elements of $\rho(t)_{C_1C_2A_1}$. We also work under a further approximation (that is justified when the cavity is close to $0K$) that the probability of getting two or more photons inside the cavities is zero, or in other words, a cavity always remains in the two level state comprising of $|0>$ and $|1>$. The tripartite (mixed) state is then obtained to be

$$\begin{aligned} \rho(t)_{C_1C_2A_1} &= \alpha_1|0_1 1_2 g_1\rangle\langle 0_1 1_2 g_1| \\ &+\alpha_2|1_1 0_2 g_1\rangle\langle 1_1 0_2 g_1| \\ &+\alpha_3|0_1 0_2 e_1\rangle\langle 0_1 0_2 e_1| \\ &+\alpha_4|0_1 1_2 g_1\rangle\langle 1_1 0_2 g_1| \\ &+\alpha_4|1_1 0_2 g_1\rangle\langle 0_1 1_2 g_1| \\ &+\alpha_5|1_1 0_2 g_1\rangle\langle 0_1 0_2 e_1| \\ &-\alpha_5|0_1 0_2 e_1\rangle\langle 1_1 0_2 g_1| \\ &+\alpha_6|0_1 0_2 e_1\rangle\langle 0_1 1_2 g_1| \\ &-\alpha_6|0_1 1_2 g_1\rangle\langle 0_1 0_2 e_1|, \end{aligned} \tag{29}$$

where the α_i are given by

$$\alpha_1 = (1 - \frac{e^{-\kappa_1 t}}{2})e^{-2\kappa_2 t},$$
$$\alpha_2 = (\cos^2 gt)e^{-\kappa_1 t}(1 - \frac{e^{-2\kappa_2 t}}{2}),$$
$$\alpha_3 = (\sin^2 gt)e^{-\kappa_1 t}(1 - \frac{e^{-2\kappa_2 t}}{2}),$$
$$\alpha_4 = \frac{(\cos gt)e^{-\frac{\kappa_1 t}{2}}e^{-\kappa_2 t}}{2},$$
$$\alpha_5 = i(\sin 2gt)e^{-\kappa_1 t}(1 - \frac{e^{-2\kappa_2 t}}{2}),$$
$$\alpha_6 = i(\frac{e^{-\frac{\kappa_1 t}{2}}\sin gt}{2} - \frac{\kappa_1 e^{-\frac{\kappa_1 t}{2}}\cos gt}{4g} + \frac{\kappa_1}{4g})e^{-\kappa_2 t},$$

κ_1 and κ_2 are the leakage constants for cavity C_1 and C_2 respectively. The reduced density states of the pairs C_1C_2, C_2A_1, C_1A_1 are thus given by

$$\begin{aligned}\rho(t)_{C_1C_2} &= \mathrm{Tr}_{A_1}(\rho_{C_1C_2A_1}),\\ &= \alpha_1|0_1 1_2\rangle\langle 0_1 1_2| + \alpha_2|1_1 0_2\rangle\langle 1_1 0_2|\\ &+ \alpha_3|0_1 0_2\rangle\langle 0_1 0_2| + \alpha_4|0_1 1_2\rangle\langle 1_1 0_2|\\ &+ \alpha_4|1_1 0_2\rangle\langle 0_1 1_2|.\end{aligned} \tag{30}$$

$$\begin{aligned}\rho(t)_{C_2A_1} &= \mathrm{Tr}_{C_1}(\rho_{C_1C_2A_1}),\\ &= \alpha_1|1_2 g_1\rangle\langle 1_2 g_1| + \alpha_2|0_2 g_1\rangle\langle 0_2 g_1|\\ &+ \alpha_3|0_2 e_1\rangle\langle 0_2 e_1| - \alpha_6|1_2 g_1\rangle\langle 0_2 e_1|\\ &+ \alpha_6|0_2 e_1\rangle\langle 1_2 g_1|.\end{aligned} \tag{31}$$

$$\begin{aligned}\rho(t)_{C_1A_1} &= \mathrm{Tr}_{C_2}(\rho_{C_1C_2A_1}),\\ &= \alpha_1|0_1 g_1\rangle\langle 0_1 g_1| + \alpha_2|1_1 g_1\rangle\langle 1_1 g_1|\\ &+ \alpha_3|0_1 e_1\rangle\langle 0_1 e_1| + \alpha_5|1_1 g_1\rangle\langle 0_1 e_1|\\ &- \alpha_5|0_1 e_1\rangle\langle 1_1 g_1|.\end{aligned} \tag{32}$$

and one can obtain the respective concurrences. These, namely, $C(\rho(t)_{C_1C_2})$, $C(\rho(t)_{C_1A_1})$, and $C(\rho(t)_{C_2A_1})$ are plotted with respect to the Rabi angle gt in Figure6. The monogamous nature of entanglement is clearly exhibited by the respective concurrences. Though dissipation reduces the magnitude of entanglement, the 'complementarity' between $C(\rho(t)_{C_1C_2})$ and $C(\rho(t)_{C_2A_1})$ is maintained even with cavity leakage.

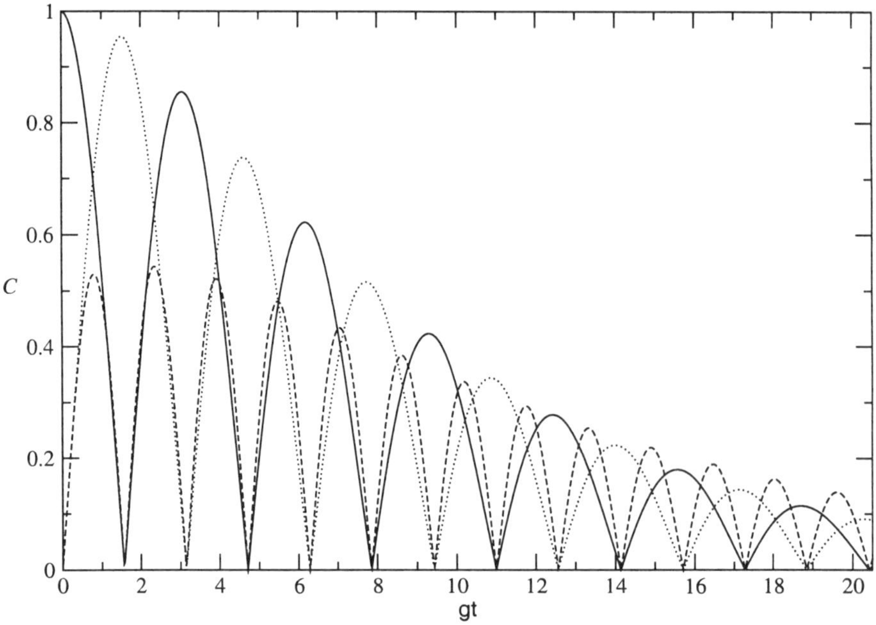

Fig. 6. $C(\rho(t)_{C_1C_2})$ (solid line), $C(\rho(t)_{C_2A_1})$, (dotted line), $C(\rho(t)_{C_1A_1})$ (broken line) plotted with respect to the Rabi angle gt. $\frac{\kappa_1}{g} = \frac{\kappa_2}{g} = 0.1$.

The CKW inequality [9] for the tripartite pure state $\rho(t)_{C_2C_1A_1}$ is given by $C^2_{C_2C_1} + C^2_{C_2A_1} \leq C^2_{C_2(C_1A_1)}$ To verify the CKW inequality for the mixed state $\rho(t)_{C_1C_2A_1}$, one has to average $C(\rho(t)_{C_2(C_1A_1)})$ over all pure state decompositions [10]. One may however, adopt a point of view that for small κ $C(\rho(t)_{C_2(C_1A_1)}) \approx 2\sqrt{\mathrm{det}\rho_{C_2}}$. Note that this result holds exactly for a pure state [9]. Nevertheless, for a small value of κ and for a bipartite photon field, one stays very close to a pure state. It can be verrified that the corresponding CKW inequality always holds under the above approximation [17]. An interesting feature of the entanglement obtained between the atom $A1$ and the cavity $C1$ through which it interacts directly is that the concurrence $C(\rho(t)_{A_1C_1})$ increases for increasing cavity loss [17]. This happens because the cavity leakage reduces the intial entanglement between C_1 and C_2, and hence makes room for the subsequent entanglement between C_1 and A_1 to form. The dissipative mechanism is thus a striking confirmation of the monogamous character of entanglement.

4. Conclusions

In this article we have discussed two important and interesting features of quantum entanglement in atom-photon interactions inside cavities, viz., the monogamy [7] of entanglement, and the quantitative nature of information transfer. We have seen that it is possible to demonstrate the transfer of information through the micromaser set-up [16]. The initial joint state of two successive atoms that enter the cavity is unentangled. Interactions mediated by the cavity photon field results in the final two-atom state being of a mixed entangled type which violates Bell-type inequalities [21]. The information content of the final two-atom state characterised by its entanglement of formation emanates from the loss of information of the cavity field quantified by the reduction of Shannon entropy. Dissipation results in a part of the Shannon entropy, or information content of the cavity to be lost to the environment. Alternatively, the final entangled state of the two atoms could be viewed as arising through the interaction of two separate atoms with the common, but suitably tailored "environment", the role of which is played by the cavity field. In this respect, the physics of entanglement generation in the micromaser is a special case of 'environment induced entanglement' [28].

We have also used the set-up of two initialy entangled cavites [25] and a single Rydberg atom passing through one of them to discuss the quantitative manifestation of a monogamy inequality [9] in atom-photon interactions. The unavoidable photon leakage exists in all real cavities used for the practical realization of quantum information transfer. The effects of such dissipation have been investigated on the monogamous nature of the entanglement between the two cavities, on one hand, and the atom and the second cavity on the other. It has been observed have that the essential monogamous character is preserved even with cavity dissipation [17]. It is further possible to see that the entanglement between the atom and the cavity through which it passes increases with larger dissipation, a feature that could be understood by invoking the monogamous character of entanglement. Further studies on different quantitative manifestations of information transfer in atom-photon interactions in the presence of dissipative effects inside cavities might be useful for the construction of realistic devices for implementing various information theoretic protocols [6].

References

1. A. Einstein, B. Podolsky and N. Rosen, Phys. Rev. **47**, 777 (1935).
2. J. S. Bell, Physics **1**, 195 (1964).

3. D. M. Greenberger, M. A. Horne, and A. Zeilinger, Am. J. Phys. **58**, 1131 (1990).
4. J. W. Pan, D. Bouwmeester, M. Daniell, H. Weinfurter and A. Zeilinger, Nature (London) **403**, 515 (2000).
5. C. H. Bennett, G. Brassard, C. Crpeau, R. Jozsa, A. Peres, and W. K. Wootters Phys. Rev. Lett. **70**, 1895 (1993); M. Zukowski, A. Zeilinger, M. A. Horne and A. K. Ekert, Phys, Rev. Lett. **71**, 4287 (1993).
6. See, for example, M. A. Nielsen and I. L. Chuang, *Quantum Computation and Information* (Cambridge University Press, Cambridge, England, 2000).
7. Bennett C. H., Lecture course in the School on Quantum Physics and Information Processing, TIFR, Mumbai, 2002 (http://qpip-server.tcs.tifr. res.in/qpip/HTML/Courses/Bennett/TIFR2.pdf).
8. V. Buzek, V. Vedral, M. B. Plenio, P. L. Knight and M. Hillery, Phys. Rev. A**55**, 3327 (1997); D. Bruβ, Phys. Rev. A **60**, 4344 (1999); W. Dür, G. Vidal and J. I. Cirac, Phys. Rev. A 62, 062314 (2000); M. Koashi, V. Buzk and N. Imoto, Phys. Rev. A **62**, 050302 (2000); K. A. Dennison and W. K. Wootters, Phys. Rev. A **65**, 010301R (2001); B. M. Terhal, quant-ph/0307120; M. Koashi and A. Winter, Phys. Rev. A**69**, 022309 (2004).
9. V. Coffman, J. Kundu and W. K. Wootters, Phys. Rev. A **61**, 052306 (2000).
10. T. J. Osborne and F. Verstraete, e-print quant-ph/0502176 (2005).
11. J. W. Pan, D. Bouwmeester, H. Weinfurter, and A. Zeilinger, Phys. Rev. Lett. **80**, 3891 (1998); E. S. Guerra and C. R. Carvalho, quant-ph/0501078.
12. J. M. Raimond, M. Brune and S. Haroche, Rev. Mod. Phys. **73**, 565 (2001).
13. A. Rauschenbeutel, P. Bertet, S. Osnaghi, G. Nogues, M. Brune, J. M. Raimond and S. Haroche, Phys. Rev. A **64**, 050301 (2001).
14. D. N. Matsukevich, T. Chanelire, S. D. Jenkins, S.-Y. Lan, T. A. B. Kennedy, and A. Kuzmich, Phys. Rev. Lett. **96**, 030405 (2006).
15. P. Masiak, Phys. Rev. A **66**, 023804 (2002); M. S. Kim, Jinhyoung Lee, D. Ahn and P. L. Knight, Phys. Rev. A **65**, 040101(R) (2002); L. Zhou, H. S. Song and C. Li, J. Opt. B: Quantum Semiclass. Opt. **4**, 425 (2002); T. Tessier, A. Delgado, I. Fuentes-Guridi, and I. H. Deutsch, Phys. Rev.A **68**, 062316 (2003).
16. A. Datta, B. Ghosh, A. S. Majumdar and N. Nayak, Europhys. Lett. **67**, 934 (2004).
17. B. Ghosh, A. S. Majumdar and N. Nayak, Int. J. Quant. Inf. (2006).
18. E. T. Jaynes, F. W. Cummings, Proc. IEEE **51**, 89 (1963).
19. S. J. D. Phoenix and P. L. Knight, Annals of Physics **186**, 381 (1988); E. Boukobza and D. J. Tannor, quant-ph/0505119.
20. See, for instance, W. H. Louisell, *Quantum Statistical Properties of Radiation* (Wiley, New York, 1973).
21. A. S. Majumdar and N. Nayak, Phys. Rev A **64**, 013821 (2001).
22. M. Keyl, Phys. Rep. **369**, 431 (2002).
23. S. Hill and W. K. Wootters, Phys. Rev. Lett. **78**, 5022 (1997); W. K. Wootters, Phys. Rev. Lett. **80** 2245 (1998).
24. N. Nayak, Opt. Commun. **118**, 114 (1995).
25. L. Davidovich, N. Zagury, M. Brune, J.M. Raimond, and S. Haroche, Phys.

Rev. A **50**, R895 (1994); V. Giovannetti, D. Vitali, P. Tombesi, and A. Ekert, Phys. Rev. A**62**, 032306 (2000); A. Rauschenbeutel, P. Bertet, S. Osnaghi, G. Nogues, M. Brune, J.M. Raimond, and S. Haroche, Phys. Rev. A**64**, 050301 (2001).

26. See, for instance, G. S. Agrawal, in *Springer Tracts in Modern Physics*, Vol.70, (Springer-Verlag, Berlin & New York, 1974).
27. S. Haroche and J. M. Raimond, in "Advances in atomic and molecular physics", Vol. 20, eds. D. R. Bates and B. Bederson (Academic, New York, 1985).
28. A. Beige, D. Braun, B. Tregenna and P. L. Knight, Phys. Rev. Lett. **85**, 1762 (2000); D. Braun, Phys. Rev. Lett. **89**, 277901 (2002).

Liouville density evolution in billiards and the quantum connection

Debabrata Biswas

Theoretical Physics Division, Bhabha Atomic Research Centre, Mumbai 400 085, INDIA

The study of classical Liouville density arises naturally in chaotic systems where a probabilistic treatment is more appropriate. The evolution of a density projected on to configuration space has a quantum connection in billiard systems when the initial density is isotropic in momentum. We outline here a sketch of this connection and demonstrate that quantum states can be used to evolve classical projected densities.

Keywords: projected-density, billiards, classical quantum correspondence

1. Introduction

The study of classical Liouville density arises naturally in chaotic systems which generically possess an infinite hierarchy of phase space structures and display sensitivity to small changes in initial conditions. The finite precision inherent in measurements or computations puts a limit on phase space resolution and makes trajectory calculations approximate at shorter times. Every "finite resolution" initial condition actually stands for an entire ensemble of trajectories with possibly very different long-time behaviours. Thus, at longer times, trajectory calculations can even be meaningless. The phase space density provides an alternate route for the study of classical dynamics that is both interesting and practical. Its evolution is governed by the Perron-Frobenius (PF) operator, $\mathcal{L}^t$,

$$\rho(\boldsymbol{x},t) = \mathcal{L}^t \circ \rho(\boldsymbol{x}) = \int \delta(\boldsymbol{x} - \boldsymbol{f}^t(\boldsymbol{x}'))\rho(\boldsymbol{x}')d\boldsymbol{x}' \tag{1}$$

where $\rho(\boldsymbol{x})$ refers to the phase space density at time $t = 0$, $\boldsymbol{x} = (\boldsymbol{q}, \boldsymbol{p})$ is a point in phase space and $\boldsymbol{f}^t(\boldsymbol{x})$ is its position at time t. Equivalently, one may solve the Liouville equation

$$\frac{\partial \rho}{\partial t} = \{H, \rho\} \tag{2}$$

to determine the evolution of the density. Here $\{\ ,\ \}$ refers to the Poisson bracket. A knowledge of the spectral decomposition of $\mathcal{L}^t$ allows one to evaluate correlations, averages and other quantities of interest [1–3].

The problem we shall address here deals with the evolution of an initial phase space density projected on to the configuration space. Imagine therefore a blob of initial conditions in configuration space (as shown in Figs. (2a) and (4a)) moving outwards isotropically. We shall be interested here in the evolution of the blob using the eigenvalues and eigenfunctions of the appropriate evolution operator for the projected-density.

We shall restrict ourselves to billiard systems due to the interesting properties of the eigenvalues and eigenfunctions. In a billiard, a point particle moves freely on a plane inside an enclosure and reflects specularly from the boundary. Depending on the shape of the boundary, billiards exhibit the entire range of behaviour observed in other dynamical systems. The quantum billiard problem consists of determining the eigenvalues and eigenfunctions of the Helmholtz equation

$$\nabla^2 \psi(\boldsymbol{q}) + k^2 \psi(\boldsymbol{q}) = 0 \tag{3}$$

with $\psi(\boldsymbol{q}) = 0$ (Dirichlet) or $\hat{\boldsymbol{n}}.\nabla\psi = 0$ (Neumann boundary condition; $\hat{\boldsymbol{n}}$ is the unit normal) on the boundary. Apart from being a paradigm in the field of classical and quantum chaos, billiards have relevance in a variety of contexts including the motion of ultra-cold atoms confined by laser beams (the so called "optical billiards") [4,5]. The Helmholtz equation describing the quantum billiard problem also describes acoustic waves, modes in microwave cavities and has relevance in studies on "quantum wells", "quantum corrals", mesoscopic systems and nanostructured materials. The Neumann boundary condition has important manifestation in acoustic waves, surface water waves, TE modes in cavities and modes of a drum with stress-free boundaries [6] as well as excitations of Bose-Einstein condensates in billiards [7].

In the following, we shall first arrive at the projected evolution operator $\mathcal{L}_P^t$, and then sketch out the connection between the eigenvalues and eigenfunctions of $\mathcal{L}_P^t$ and the quantum Neumann energy eigenstates. Finally, we shall provide numerical evidence of this connection by demonstrating that an arbitrary classical projected density can be evolved using information about the quantum eigenstates.

2. The projected Perron-Frobenius Operator for Billiards: Eigenvalues and Eigenfunctions

As stated above, the density, projected on to the configuration space, is the object of study in this paper. This can be obtained by integrating out the momentum $\boldsymbol{p}$. The time evolution of the projected-density, $\rho(\boldsymbol{q})$, can be expressed as

$$\mathcal{L}_P^t \circ \rho(\boldsymbol{q}) = \int d\boldsymbol{p}\ \mathcal{L}^t \circ \rho(\boldsymbol{q}, \boldsymbol{p}) \tag{4}$$

where the subscript P refers to the projected operator. When the initial density is uniformly distributed in momentum,

$$\rho(\boldsymbol{q}, \boldsymbol{p}) = \rho(\boldsymbol{q})/\mu_p \tag{5}$$

where $\mu_p = \int_{\Gamma_p} d\boldsymbol{p}$ is a measure of momentum space volume subject to the energy conservation constraint (denoted by Γ_p). The evolution operator for the projected density then takes a particularly simple form [8]:

$$\mathcal{L}_P^t \circ \rho(\boldsymbol{q}) = \int d\boldsymbol{q}'\ \rho(\boldsymbol{q}')\ K_P(\boldsymbol{q}, \boldsymbol{q}', t) \tag{6}$$

where the kernel K_P is [9]

$$K_P(\boldsymbol{q}, \boldsymbol{q}', t) = \frac{1}{\mu_p} \int d\boldsymbol{p}' d\boldsymbol{p}\ \delta(\boldsymbol{x} - \boldsymbol{f}^t(\boldsymbol{x}')). \tag{7}$$

Specializing now to the case of billiards, note that the motion inside a billiard enclosure is free, so that, the magnitude of the momentum is conserved. Restriction to the constant energy surface thus implies that the projection can be achieved by integrating over the directions, φ that $\boldsymbol{p}$ makes with a given axis. The kernel can thus be expressed as

$$K_P(\boldsymbol{q}, \boldsymbol{q}', t) = \frac{1}{2\pi} \int d\varphi\ \delta(\boldsymbol{q} - \boldsymbol{q}'^t(\boldsymbol{q}'; p, \varphi)) \tag{8}$$

so that the time evolution of the projected-density can be determined using Eq. (6) and Eq. (8). Note that Eq. (6) preserves positivity provided the initial density, $\rho(q, 0)$ is positive.

For general 2-dimensional billiards, there are two approaches for determining the eigenmodes of $\mathcal{L}_P^t$ [10,11]. Both rely on polygonalization of the billiard boundary [12,13] and the argument that as the number of sides of the polygon is increased, the modes of the smooth billiard are approximated better. In the first approach [14], the trace of the $\mathcal{L}_P^t$ is related to the trace of the energy dependent quantum Neumann propagator while in the second, a plane wave expansion enables one to conclude that there exists a correspondence between the quantum Neumann eigenmodes and the modes of $\mathcal{L}_P^t$ [11,15]. We merely state the result here and refer the interested reader to [10,11]. For $t > 0$,

$$\mathcal{L}_P^t \circ \psi_n(\boldsymbol{q}) = f(k_n vt)\ \psi_n(\boldsymbol{q}) \tag{9}$$

where $E_n = k_n^2$ are well approximated by the quantum Neumann energy eigenvalues, $f(x) = \sqrt{2/\pi x}\cos(x - \pi/4)$ is the asymptotic form of $J_0(x)$ and $\psi_n(q)$ are well approximated by the quantum Neumann eigenfunctions. For some integrable systems, it is possible to show that $\{E_n\}$ and $\{\psi_n\}$ are the exact quantum Neumann eigenvalues and eigenfunctions.

Note that the correspondence is not necessarily exact except for some integrable geometries where this can be established rigorously. Nevertheless, we aim to use the exact quantum Neumann eigenfunctions to expand an arbitrary initial projected classical density, $\rho(\boldsymbol{q})$, and evolve it using the eigenvalues $f(\sqrt{E_n}vt)$ with $\{E_n\}$ as the exact quantum Neumann eigenvalues. Due to the assumption of uniformity in momentum in arriving at the kernel, the evolution under $\mathcal{L}_P^t$ should be such that from each point, the particles move out uniformly in all directions.

3. The quantum connection: numerical evidence

There are two levels at which numerical verification of the quantum connection can be provided. In the first instance, a few individual eigenfunctions of $\mathcal{L}_P^t$ and quantum Neumann eigenfunctions can be compared. For details of how the classical Neumann eigenfunctions can be extracted, we refer the interested reader to [10]. Here, we shall use the quantum Neumann energy eigenfunctions to evolve an arbitrary classical density $\rho(\boldsymbol{q})$ and compare this using the actual evolution of trajectories [8].

The completeness of the quantum Neumann eigenfunctions allows us to expand $\rho(\boldsymbol{q})$ as:

$$\rho(\boldsymbol{q}) = \sum_{\boldsymbol{n}} C_{\boldsymbol{n}} \psi_{\boldsymbol{n}}(\boldsymbol{q}) \tag{10}$$

where

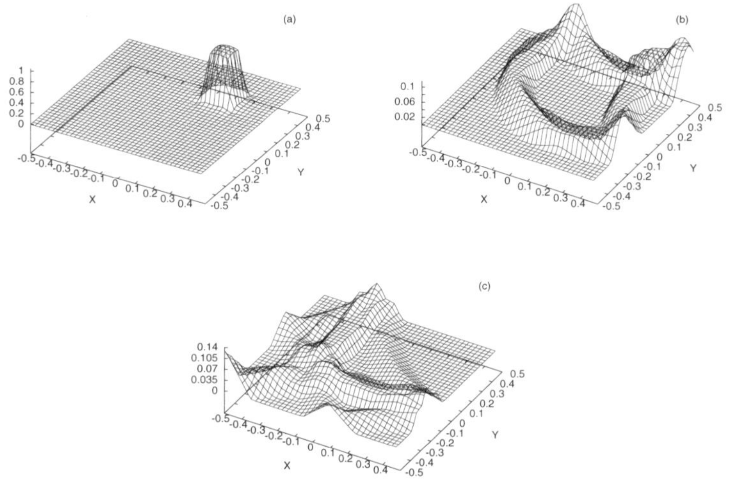

Fig. 1. The evolution of a projected-density, $\rho(\boldsymbol{q})$, in the rectangular billiard is computed using the exact quantum Neumann eigenfunctions as the eigenfunctions of the projected Perron-Frobenius operator, $\mathcal{L}_P^t$. The initial density is shown in (a) while (b) and (c) show the density at t=1.0 and 2.0 respectively. The evolution reflects isotropy in momentum.

$$C_{\boldsymbol{n}} = \int d\boldsymbol{q}\ \rho(\boldsymbol{q})\psi_{\boldsymbol{n}}^*(\boldsymbol{q}). \tag{11}$$

In the above, it is assumed that $\int\ d\boldsymbol{q}\ \ \psi_{\boldsymbol{n}}^*(\boldsymbol{q})\psi_{\boldsymbol{m}}(\boldsymbol{q}) = \delta_{\boldsymbol{m},\boldsymbol{n}}$. On using Eq. (9), we have

$$\mathcal{L}_P^t \circ \rho(\boldsymbol{q}) = \sum_{\boldsymbol{n}} C_{\boldsymbol{n}} J_0(\sqrt{E_{\boldsymbol{n}}}vt)\psi_{\boldsymbol{n}}(\boldsymbol{q}). \tag{12}$$

In order to account for the behaviour at small t, we use the Bessel function J_0 as $f(x)$ is an asymptotic form of $J_0(x)$. Note that the evolution of the projected-density, $\rho(\boldsymbol{q})$, under Eq. (12) should reflect isotropy in momentum.

To demonstrate that Eq. (12) indeed gives us the correct classical evolution, we choose points (initial conditions) in $\boldsymbol{q}$-space distributed according to the initial density $\rho(\boldsymbol{q})$ and evolve them in time using classical equations of motion. The evolution of these points give us the density $\rho(\boldsymbol{q}, t)$ at any time t. This is compared with the density obtained from Eq. (12).

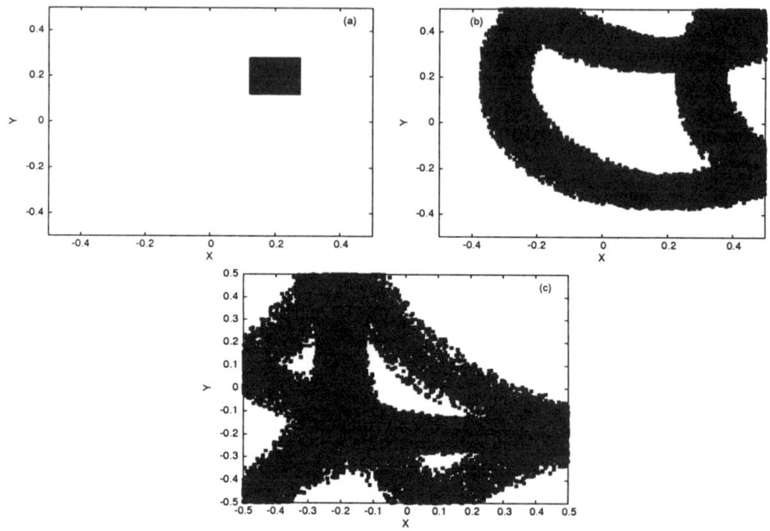

Fig. 2. The initial points chosen according to the density in Fig. 1a is shown in Fig. 2a. These are evolved classically using momentum direction (φ) distributed uniformly in $(0, 2\pi)$. The distribution of points at time 1.0 and 2.0 can be compared with the densities in Fig. 1b and 1c respectively.

In both the examples considered below, we shall consider a hat function as the initial density for convenience. The function takes the value unity within a square strip of side ΔL centred at a point $\boldsymbol{q}$ and is zero outside. Thus, all initial conditions lie uniformly within the square strip. For convenience, we choose the velocity $v = 1$.

As a first example, consider a particle in a rectangular box of side $L_1 = L_2 = 1.0$. Fig. 1a shows the initial density centred at $\boldsymbol{q} = (0.2, 0.2)$ with $\Delta L = 0.15$. The first five thousand quantum Neumann states have been used to expand the initial density. Figs. (1b) and (1c) are the densities at time $t = 1$ and 2 respectively as computed using Eq. (12). For comparison, Fig. (2) is a 2-dimensional view of the density evolved using classical trajectories with their momentum distributed isotropically. At time $t = 0$, the points lie in a square patch as shown in Fig. (2a) while Figs. (2b) and (2c) show the points as they evolve from Fig. (2a) at times $t = 1$ and 2 respectively. Note that the evolution in Fig. 1 reflects isotropy in momentum as expected.

We next consider the chaotic Stadium billiard. In this case, the quantum Neumann eigenfunctions have been computed numerically using the

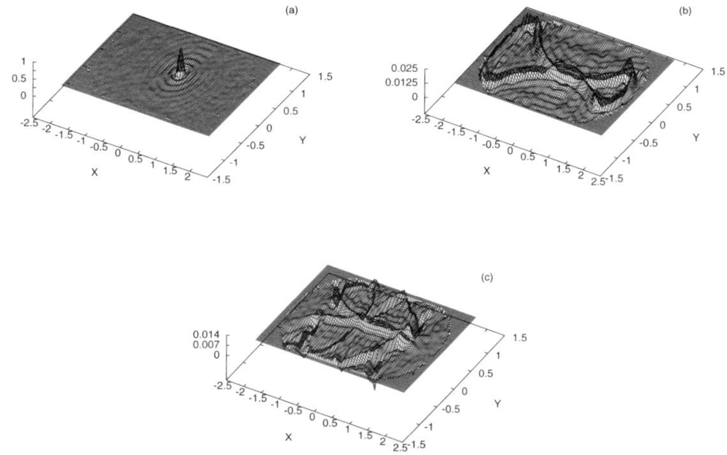

Fig. 3. As in Fig. 1 for the stadium billiard at times t=0, t=2 and t=4.

boundary integral method. We use the first thousand eigenfunctions to expand the initial density and evolve it using Eq. (12). Fig. 3 is similar to Fig. 1 but for times t=0,2 and 4 while Fig. 4 (similar to Fig. 2) shows the classical (trajectory) evolution of the initial points at these times. At longer times, both evolutions lead to a uniform distribution in $\boldsymbol{q}$ space (see [10] for a figure).

There are two possible sources of errors in this prescription of evolving the classical density numerically. The first of these is the truncation of the basis. The second stems from the fact that the quantum Neumann eigenfunctions and the eigenvalues $f(\sqrt{E_n}tv)$ are generally approximate eigenfunctions and eigenvalues of $\mathcal{L}_P^t$. Of the above examples, the rectangular billiard has only the truncation problem to contend with. In case of the stadium, apart from the above mentioned sources of errors, it must also be noted that the quantum eigenfunctions are not known analytically and their numerical evaluation introduces another source of error. Notwithstanding these difficulties, the two evolutions agree reasonably well.

It is worth noting that the determination of exact eigenstates of the Perron-Frobenius operator (or $\mathcal{L}_P^t$) is generally nontrivial while quantum states are easier to determine. The usefulness of the results presented here thus lies in using quantum states to evolve classical configuration space densities. In conclusion, we have clearly demonstrated that for both the

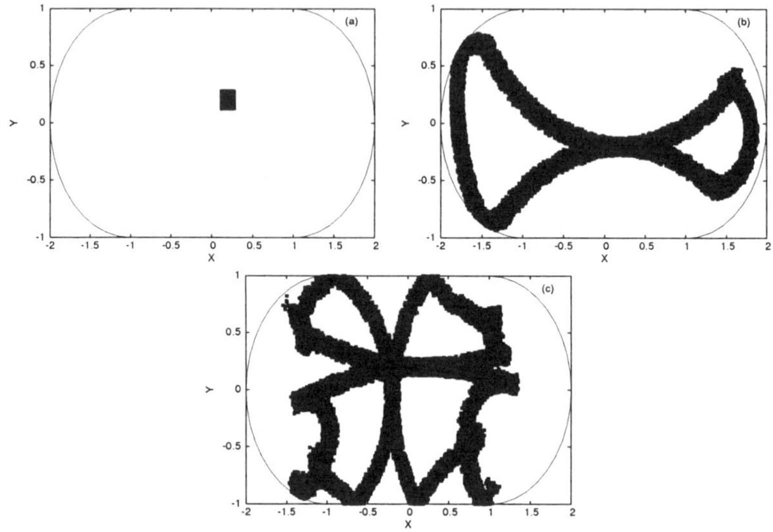

Fig. 4. As in Fig. 2 for the stadium billiard at times t=0 , t=2 and t=4.

rectangular and stadium billiards, the evolution of the initial projected-density using quantum Neumann eigenfunctions as eigenfunctions of $\mathcal{L}_P^t$, captures the classical evolution reasonably well.

References

1. P. Cvitanovic et al, *Chaos: classical and quantum*, http://chaosbook.org.
2. A. Lasota and M. MacKey, *Chaos, Fractals, and Noise; Stochastic Aspects of Dynamics*, Springer-Verlag, Berlin, 1994.
3. P. Gaspard, *Chaos, scattering and statistical Mechanics*, Cambridge University Press, 1998.
4. V. Milner, J.L. Hanssen, W. C. Campbell and M. G. Raizen,Phys. Rev. Lett. **86**, 1514 (2001).
5. N. Friedman, A. Kaplan, D. Carasso, N. Davidson, Phys. Rev. Lett. **86**, 1518 (2001).
6. T. A. Driscoll and H. P. W. Gottlieb, Phys. Rev. E **68**, 016702 (2003); H.-J. Stockmann, *Quantum Chaos - an Introduction*, Cambridge University Press, Cambridge, 1999.
7. C. Zhang, J. Liu, M. G. Raizen and Q. Niu, Phys. Rev. Lett. **93**, 074101 (2004).
8. D. Biswas, "Evolving classical projected-densities in billiards with quantum states", nlin.CD/0603061; Mod. Phys. Lett. B **20**, 795 (2006).
9. Initial densities of the form $\rho(\boldsymbol{q},\boldsymbol{p}) = f(\boldsymbol{q})g(\boldsymbol{p})$ can be treated similarly and result in a projected kernel $K_P(\boldsymbol{q},\boldsymbol{q}',t) = \int\ d\boldsymbol{p}'d\boldsymbol{p}\ \delta(\boldsymbol{x} - \boldsymbol{f}^t(\boldsymbol{x}'))g(\boldsymbol{p}')$.

10. D. Biswas, Phys. Rev. Lett. **93**, 204102 (2004).
11. D. Biswas, Pramana - J. Phys, **64**, 563 (2005).
12. J. L. Vega, T. Uzer and J. Ford, Phys. Rev. E **48**, 3414 (1993).
13. D. Biswas, Phys. Rev. E **61**, 5073 (2000).
14. D. Biswas, Phys. Rev. E **63**, 016213 (2001).
15. D. Biswas, Phys. Rev. E **67**, 026208 (2003).

MRCPA: Theory and Application to Highly Correlating System

Kiyoshi Tanaka

Advancesoft, Center for Collaborative Research, The University of Tokyo, 4-6-1, Komaba, Meguro-ku, Tokyo, 153-8904, Japan, and
Institute of Industrial Science, The University of Tokyo, 4-6-1, Komaba, Meguro-ku, Tokyo, 153-8904, Japan, and
[1]Division of Chemistry, Graduate School of Science, Hokkaido University, Kita 10, Nishi 8, Kita-ku, Sapporo, 060-0810, Japan

Abstract

This contribution primarily discusses multi-reference coupled pair approximation which is a non-variational variants of the multi-reference single and double excitation configuration interaction (MRSDCI) method. This type of theory takes advantage of simplicity and flexibility of configuration interaction method. However they incorporate more or less the knowledge from cluster expansion method. In the early stage of this note, the strategy why we selected and developed non-variational variant of MRSDCI, and further why we select the state-universal approach rather than the state-selective (state-specific) approach. A brief sketch of MRCPA approach will be shown and discussed. Some important recent results will be presented to demonstrate the characteristic of the method.

---

[1]This is the author's previous address. The works presented in this article was carried out when the author worked in Hokkaido University.
$E - mailaddress$: k-tanaka@fsis.iis.u-tokyo.ac.jp

1.1 Introduction

During the last decade, remarkable progress of many electron theory has been made in the field of quantum molecular electronic structure theory. The size of the systems and complexity of the problems have been increasing as well as the improvement in the reliability of the results from theoretical studies. Such progress owes a lot not only to the development of new extensive computational methods but also to the rapid development of computational power. The many electronic structure theories pursue accuracy of the total energies and wavefunctions of small molecules and applicablity to large scale molecules, or nano scale systems.

It is commonly well recognized that inclusion of electron correlation is significant in describing the electronic structure of molecules. As a standard computational procedure of post Hartree-Fock calculations, the configuration interaction (CI) method is well-established.

Cluster expansion method is expected to provide more accurate electronic energies than CI and to be significant in studying systems of a larger number of correlating electrons and highly correlating systems. Many groups, nowadays, devote themselves to study cluster expansion methods and to device various approximations [1, 2]. Unlike truncated CI, the cluster expansion method and the closely related many body perturbation theory (MBPT) without exception have the advantage of size-consistency, or size-extensivity at each level of approximation [3–7], whereas these methods are non-variational. Since applicability of these methods based on a single reference function to real molecular systems is relatively limited, efforts to develop the multi-reference based methods have been made [8–22] by several groups. The framework of the cluster expansion theory or MBPT is much more complicated than CI, and its computational procedure includes much more logics than CI.

The CI method is variational and the calculated energies are upper bound to the true energies of the respective states. The expansion of the wavefunction by the CI method converges slowly to the convergence limit of the full CI wavefunction against increasing the length of the expansion. Since the number of configuration state functions (CSF's) which expand a CI wavefunction must be limited, careful preparation of a set of CSF's is significant. Almost all approaches generate a set of CSF's which are composed of singly and doubly excited CSF's from several leading CSF's, i.e. multi-reference single and double excitation CI (MRSDCI). The scheme works well and has contributed enormously to predicting various properties of atoms and molecules [23]. The advantage of the scheme is as follows;

- several excited states as well as the ground state including electron correlation,
- avoided crossing among adiabatic potential surfaces,
- primarily based on simple linear algebra and,
- extensive development of efficient algorithms.

In spite of the magnificent success of the MRSDCI scheme, the well known difficulty, size-inconsistency and size-inextensivity [3], is involved in the wavefunction; the accuracy of the wavefunction deteriorates as the number of correlating electrons increases. For the last fifteen years, a few approaches which could be considered as intermediate between the CI methods and cluster expansion methods were proposed. They fall basically under the category of the cluster expansion method, but they could be considered as non-variational variants of CI from a technical point of view. It is accomplished by making use of sophisticated algorithms developed in the field of CI and incorporating significant effects found in the cluster expansion. The coupled electron pair approximation (CEPA; see review by Kutzelnigg [7] and Ahlriches and Scharf [24]) fall under this category, although it is based on a single reference function. An advantage of the scheme is that we can partly utilize the extensively developed CI computational procedure. Implementation is easy since the scheme is simple and flexible. Several CEPA type non-variational variants of multi-reference single and double excitation CI (MRSDCI) method [25–33] have been developed extensively as a substitution or approximation to the multi-reference based cluster expansion method. These methods incorporate more or less knowledge obtained by the cluster expansion theory for the sake of incorporating the higher order correlation and materializing size-consistency, or size-extensivity approximately.

These methods employ state-specific representation except for the present multireference coupled pair approximation (MRCPA) [27]. This is strongly connected with the fact that most of cluster expansion methods employ state-specific approach and the equations to be solved are comparatively stable [34, 35]. Our primary strategy of developing MRCPA was to employ state-universal representation which could be easily applicable to studying electronic excited states, though the state-universal approach is employed only a little in cluster expansion approach [2, 11, 36].

The formal theory will be shown in the next section and some results characterizing the method will be shown in the end.

1.2 Non-variational variants of MRSDCI

Multi-reference version of the second order perturbation theory is one of the simplest schemes for estimating electron correlation. This type of approach needs Hamiltonian matrix in the reference space, those elements between reference space and excited space and only zeroth-order diagonal elements of the excited space. It was shown that the Rayleigh-Schrödinger type second order perturbation method [38] is a better approach than Brillouin-Wigner type B_k one [39] from the view point of size-consistency. The MR Mϕller-Plesset type perturbation theory [40, 41] is in line with this approach.

We will start with a single closed shell reference function, Φ_0. Correlated wavefunction is expressed by $\Psi = \exp(T_2)|\Phi_0>$, for simplicity, where $T_2 = \sum_p T_p \gamma_p$. The index p specifies an electron pair excitation. A linearized version of the cluster expansion has the following equations;

$$E =< \Phi_0|(H + [H, T_2])\Phi_0 >, \tag{1.1}$$

$$< \Phi^*|(H + [H, T_2])\Phi_0 >= 0, \tag{1.2}$$

where Φ^* represents a set of excited functions. Those equations are similar to those of single reference SDCI. It was found that the method overestimated correlation energy and development to include higher order correction was tried within the framework of CI-like treatment. That is so called CEPA type methods [7, 24, 42]. The formalism is based on the cluster expansion method fundamentally as is shown below. Then the following two equations are obtained;

$$E =< \Phi_0|H\Phi_0 > + \sum_q < \Phi_0|HT_q\Phi_0 > \gamma_q, \tag{1.3}$$

$$E\gamma_p =< \Phi_0|T_p^\dagger H\Phi_0 > + \sum_q < \Phi_0|T_p^\dagger HT_q\Phi_0 > \gamma_q$$
$$+ \sum_{q,q'}^{q\neq q'} \frac{1}{2} < \Phi_0|T_p^\dagger HT_qT_{q'}\Phi_0 > \gamma_q\gamma_{q'}. \tag{1.4}$$

The combination of the two equations leads to

$$< \Phi_0|T_p^\dagger H\Phi_0 >= \sum_q \{< \Phi_0|H\Phi_0 > \delta_{pq} - < \Phi_0|T_p^\dagger HT_q\Phi_0 >\}\gamma_q$$
$$+ \sum_q < \Phi_0|HT_q\Phi_0 > \gamma_q\gamma_p - \sum_{q,q'}^{q\neq q'} \frac{1}{2} < \Phi_0|T_p^\dagger HT_qT_{q'}\Phi_0 > \gamma_q\gamma_{q'}. \tag{1.5}$$

The last two terms in the right hand side of the last equation almost cancel and an important part of the rest is composed of so called exclusion principle violating (EPV) terms. Several approximations take a part of the terms and reformulate them so as to fit in the framework of computational procedure close to SDCI as follows;

$$< \Phi_0|T_p^\dagger H\Phi_0 >= \sum_q \{(< \Phi_0|H\Phi_0 > +\Delta_q)\delta_{pq} - < \Phi_0|T_p^\dagger HT_q\Phi_0 >\}\gamma_q, \tag{1.6}$$

where Δ_q depends on γ and takes different form depending on the approximations and it is discussed extensively by Ahlrichs and Scharf [24]. The CEPA type methods were successfully used in a number of research works.

As for the MR theory, the state-selective linear cluster expansion was implemented and applied to some molecules [10, 18–20]. The multi-reference based CEPA type approaches were also developed extensively by various groups [25–31] within the last 15 years. Most of them are state-selective approach except for the present approach [27] which belongs to a category of the state-universal approach. We are going to present the CSF based multi-reference coupled pair approximation (MRCPA) proposed by the present authors and to compare it with the state-selective methods. The size-consistency of the present method will be discussed and computational results are compared with the experimental results and the results obtained by the other theoretical methods.

1.2.1 *MRCPA method*

Assuming we have a complete set of CSF's, we partition the space into three subspaces: P space composing of reference functions Φ_a^P, Q space including all single and double excitations Φ_i^Q from reference functions and R space including all other functions Φ_s^R. It is assumed that one-to-one mapping between the reference space and the exact manifold holds, as was discussed in the state-universal formalism of MR cluster expansion theory [11]. The exact wavefunction may be expressed as

$$\Psi_\nu = \sum_{a\in P} \Phi_a^P c_{a\nu}^P + \sum_{i\in Q} \Phi_i^Q c_{i\nu}^Q + \sum_{s\in R} \Phi_s^R c_{s\nu}^R. \tag{1.7}$$

The wavefunction is rewritten in the following form;

$$\Psi_\nu = \sum_{a\in P} (\Phi_a^P + \sum_{i\in Q} \Phi_i^Q C_{ia} + \sum_{s\in R} \Phi_s^R D_{sa}) c_{a\nu}^P. \tag{1.8}$$

The matrices **C** and **D** are uniquely defined [27](a) by the following equation, because the matrix $\mathbf{c}^P$ is regular.

$$c^Q_{i\nu} = \sum_{a\in P} C_{ia} c^P_{a\nu}, \qquad c^R_{s\nu} = \sum_{a\in P} D_{sa} c^P_{a\nu}. \tag{1.9}$$

This is expressed by the wave operator formalism [27](b)

$$|\Psi_\nu >= \sum_{a\in P} \hat{U} |\Phi^P_a > c^P_{a\nu}. \tag{1.10}$$

and $\hat{U}$ is defined as follows;

$$\hat{U} = \sum_{a\in P} (|\Phi^P_a > + \sum_{i\in Q} |\Phi^Q_i > C_{ia} + \sum_{s\in R} |\Phi^R_s > D_{sa}) < \Phi^P_a|. \tag{1.11}$$

Projection operators $\hat{P}$, $\hat{Q}$ and $\hat{R}$ are defined as follows;

$$\hat{P} = \sum_{a\in P} |\Phi^P_a >< \Phi^P_a|, \quad \hat{Q} = \sum_{i\in Q} |\Phi^Q_i >< \Phi^Q_i|, \quad \hat{R} = \sum_{s\in R} |\Phi^R_s >< \Phi^R_s|. \tag{1.12}$$

The matrices **C** and **D** are obtained by the Rayleigh-Schrödinger perturbation theory by defining unperturbed part of Hamiltonian H_0 and perturbed part H' as

$$H = H_0 + H', \tag{1.13}$$

$$H_0 = \hat{P}H\hat{P} + \hat{Q}H\hat{Q} + \hat{R}H\hat{R}, \tag{1.14}$$

$$H' = \hat{P}H\hat{Q} + \hat{Q}H\hat{P} + \hat{Q}H\hat{R} + \hat{R}H\hat{Q}. \tag{1.15}$$

The Bloch equation, $H\hat{U} = \hat{U}H\hat{U}$, which determine $\hat{U}$, (in other words matrices **C** and **D**), leads to

$$[\hat{U}, H_0] = H'\hat{U} - \hat{U}H'\hat{U}. \tag{1.16}$$

The wave operator $\hat{U}$ is expanded in order of the perturbation

$$\hat{U} = \sum_n \hat{U}^{(n)}, \tag{1.17}$$

where $\hat{U}^{(0)} = \hat{P}$, since the reference space gives the zeroth-order wavefunction. The higher order terms are given by

$$\hat{U}^{(n)} = \sum_{a\in P} (\sum_{i\in Q} |\Phi^Q_i > C^{(n)}_{ia} + \sum_{s\in R} |\Phi^R_s > D^{(n)}_{sa}) < \Phi^P_a|. \tag{1.18}$$

Then the order dependent Bloch equation is given by

$$[\hat{U}^{(n)}, H_0] = H'\hat{U}^{(n-1)} - \sum_{k=1}^{n-1} \hat{U}^{(n-k)} H' \hat{U}^{(k-1)}. \quad n \geq 1 \tag{1.19}$$

The lower order equations up to the third order are given as follows;

$$[\hat{U}^{(1)}, H_0] = \hat{Q}H\hat{P}, \tag{1.20}$$

$$[\hat{U}^{(2)}, H_0] = \hat{R}H\hat{Q}\hat{U}^{(1)}, \tag{1.21}$$

$$[\hat{U}^{(3)}, H_0] = \hat{Q}H\hat{R}\hat{U}^{(2)} - \hat{U}^{(1)}\hat{P}H\hat{Q}\hat{U}^{(1)}. \tag{1.22}$$

The wave operator $\hat{U}$ is lead to the following form

$$\hat{U} = \{1 + \sum_k (\hat{Q}\hat{U}^{(2k-1)} + \hat{R}\hat{U}^{(2k)})\}\hat{P}, \tag{1.23}$$

with the matrices $\mathbf{C}^{(2k)} = 0$ and $\mathbf{D}^{(2k-1)} = 0$ owing to the present partitioning of unperturbed and perturbing terms. The effective Hamiltonian, H_{eff}, is given by

$$H_{eff} = \hat{P}H\hat{P} + \sum_k \hat{P}H\hat{U}^{(2k-1)}. \tag{1.24}$$

Assuming that $\hat{P}H\hat{P}$ is diagonalized, Eqs. (1.20) - (1.22) lead to

$$\sum_{j\in Q}\{H_{aa}\delta_{ij} - H_{ij}\}C_{ja}^{(1)} = H_{ia}. \tag{1.25}$$

$$\sum_{t\in R}\{H_{aa}\delta_{st} - H_{st}\}D_{ta}^{(2)} = \sum_{j\in Q} H_{sj}C_{ja}, \tag{1.26}$$

and

$$\sum_{j\in Q}\{H_{aa}\delta_{ij}-H_{ij}\}C_{ja}^{(3)} = \sum_{t\in R} H_{it}D_{ta}^{(2)} - \sum_{j\in Q} H_{aj}C_{ja}^{(1)}C_{ia}^{(1)} - \sum_{b\in P}^{b\neq a}\sum_{j\in Q} H_{bj}C_{ja}^{(1)}C_{ib}^{(1)}, \tag{1.27}$$

The solutions of Eq. (1.25) is equivalent to those of state-universal linearized version of MR cluster expansion theory. The scheme is now named as MRCPA(2) because the total energy is obtained in the second order perturbation treatment. In order to take the fourth order energy, we need to solve Eqs. (1.26) and (1.27) successively. It is, however, impossible to solve Eq. (1.26). By similar consideration given in the CEPA theory, we approximate the right hand side of Eq. (1.27) as follows;

$$\sum_{j\in Q}\{H_{aa}\delta_{ij} - H_{ij}\}\bar{C}_{ja}^{(3)} = -H_{ai}C_{ia}^{(1)}C_{ia}^{(1)} - \sum_{b\in P}^{b\neq a} H_{bi}C_{ia}^{(1)}C_{ib}^{(1)}. \tag{1.28}$$

The approximate third order coefficients $\{\bar{C}^{(3)}_{ja}\}$ are determined by this equation. Combination of Eqs. (1.25) and (1.28) leads to the following equation [27](c),(d)

$$\sum_{j\in Q}\{(H_{aa}+\Delta^{(i)}_a)\delta_{ij}-H_{ij}\}(C^{(1)}_{ja}+\bar{C}^{(3)}_{ja})=H_{ia}-\sum_{b\in P}^{b\neq a}H_{bi}C^{(1)}_{ia}C^{(1)}_{ib}, \quad (1.29)$$

where $\Delta^{(i)}_a$ is defined as $\Delta^{(i)}_a=H_{ai}C^{(1)}_{ia}$, and the last term on the right hand side represents coupling between the reference space through inclusion of correlation. If the solution of Eq. (1.29) is used in the effective Hamiltonian, we obtain the approximate fourth order energy. This level of approximation is now named as MRCPA(4). $\{C^{(1)}_{ja}\}, a=1,...,M$ in Eq. (1.25) or $\{C^{(1)}_{ja}+\bar{C}^{(3)}_{ja}\}, a=1,...,M$ in Eq. (1.29). By the use of the vectors thus obtained, the total energy by MRCPA(2) or MRCPA(4) will be obtained by

$$\sum_{b=1}^{M}(H_{eff})_{ab}c^{P}_{b\nu}=c^{P}_{a\nu}E_{\nu}. \quad (1.30)$$

Recently we developed a new version of MRCPA which is written by C++ and runs on a linux based computer [37](l), using data from Alchemy II [43]. The approximations introduced in this subsection, MRCPA(2) and MRCPA(4) satisfy the size consistency condition for the non-interacting two molecular problem. This is demonstrated elsewhere [27](b)(d).

1.2.2 *Comparison with other multi-reference based CEPA type methods*

In the following, we are comparing the present method with the state-selective CEPA type methods from the view point of formalism.

Since Szalay reviews state-selective methods extensively and focuses on the way of approximation on $\{\Delta\}$ and treatment of reference space by comparing various approaches, we will not go into detailed comparison, but will discuss how the MRCPA method is different from the other methods. If we approximate $\hat{P}H\hat{P}$ in H_0 by $E_0\hat{P}$ (E_0 = the first order energy for the selected state) throughout Eqs. (1.20) - (1.22) in our formalism, (or use E_0 instead of H_{aa} through Eqs. (1.25) - (1.29)), the theory falls into the category of the state-selective theory. State-selective theory uses common zeroth-order energy in equations determining correlation wavefunction by replacing $\hat{P}H\hat{P}$ by $E_0\hat{P}$. So we have to carry out a separate calculation for a higher state by changing the zeroth-order energy, E_0, to an appropriate value for the state. If we do not utilize Eq. (1.30) and take $(H_{eff})_{aa}$ as the a-th state energy, it corresponds to the state-selective theory without relaxation in the reference space due to inclusion of electron correlation. As was pointed out previously, the one-to-one mapping from the exact manifold to the reference space in the state-universal framework is not realistic if we employ full valence space or complete active space (CAS) as a reference space. In fact, solving higher state solutions, Eqs. (1.25) and (1.27) sometimes suffer from the so called "intruder state" problem. In such cases, we truncate the state and higher states in constructing the effective Hamiltonian, Eq. (1.30).

The first paper on the MRCEPA type approximation is averaged coupled-pair functional (ACPF) method by Gdanitz and Ahlrichs [25]. They proposed an energy functional which leads to state-selective equations with $\Delta = (2/n_e)E_{corr}$, where n_e is the number of correlated electrons. Szalay and Bartlett [30] further considered Δ within a framework of functional scheme and proposed a few types of multi-reference averaged quadratic coupled-cluster methods (MRAQCC) in which detailed modification on averaged values for Δ was given. The use of average value for Δ means that Δ is independent of electron pair excitation in the equation to determine correlation coefficients. It is noted that the quantity Δ has meaning of approximate pair correlation energy. As a molecule dissociates into two parts, separating electron pairs should decrease their pair correlation energy. The use of averaged value does not guarantee this condition.

Murray, Racine, and Davidson [26](b) presented a qausi-degenerated variation perturbation-averaged pair correction (QDVPT-APC) by modifying QDVPT [26](a) which is equivalent to linearized cluster expansion. The method utilizes an effective Hamiltonian formalism

$$\{\hat{P}H\hat{P}+\hat{P}H\hat{Q}[(E_0+(2/n_e)E_{corr})-H\hat{Q}]^{-1}\hat{Q}H\hat{P}\}|\Phi_0>=E|\Phi_0>, \quad (1.31)$$

where E_{corr} is defined as $E = E_0 + E_{corr}$. This could be considered as a modification of ACPF in an effective Hamiltonian formalism. MRCPA uses also effective Hamiltonian formalism. MRCPA(4) would be reduced to this scheme by solving Eq. (1.30) with the vectors obtained from similar equation as Eq. (1.29), which is modified as follows; (i) replacing H_{aa} by a constant E_0, (ii) replacing $\Delta_a^{(i)}$ by $(2/n_e)E_{corr}$, (iii) discarding the second term in the right hand side of the equation.

MCCEPA by Fink and Staemmler [29] and MRCEPA by Ruttink, van Lenthe, Zwaans and Groenenboom [28] start from the cluster expansion basing on a multi-reference function with a simple exponential ansatz and give more detailed consideration on Δ.

MR(SC)^{2}SCI by Dauday and Malrieu and co-workers [31](a,b) devise CEPA type equations via a perturbation theory within the framework of the state-selective theory.

1.3 Recent important results

In this section, some important numerical results from MRCPA calculations [37](a) - (l) are shown in accordance with the discussion in the preceding part of this note.

Although the choice of the CAS as a reference space is essential to guarantee the size-consistency, the assumption of one-to-one mapping of reference space to exact manifold is unrealistic as was pointed out in the previous subsection. The intruder state problem may occur. In application, we discard some solutions of Eq. (1.25) or Eq. (1.29), if difficulty takes place. Thus the dimension of the effective Hamiltonian is decreased in such a case.

Firstly, we present how size-consistency is realized in connection with the previous subsection. Some results of the potential surfaces including avoided crossing are shown in the next, which reflects the characteristic of the state-universal theory upon application to excited states. Further we will show our recent results in comparison with other methods and experimental results.

Dissociation energy of As_2 The dissociation energy of As_2 has been one of the most demanding problems. Since the number of electrons is moderately large and size-consistency is very important to predict the dissociation energy, the molecule was studied by MRCPA in the early stage [37](a). Using extensive basis set including two *f* functions and nine reference CSF's, we obtained 3.86 eV for the dissociation energy of As_2which is very close to experimental value of 3.99 eV. The study showed that inclusion of electron correlation between *semivalence* 3d electrons and *valence* 4s-4p electrons is very significant. Since the number of the correlating electron pair was rather large and the correlation energy contribution from 3*d* shell stabilized the molecular state effectively, the MRCPA could assess the correlation energy more appropriately and predict better dissociation energy of the molecule than MRSDCI.

Avoided crossing of adiabatic potential curves; Ga_2 Avoided crossing of the potential surfaces plays an important role to interpreting chemical reaction in the excited state. Capability of describing such situation is desirable aspect of the electronic structure theory. The state-universal approach is free from bias in calculating the total energies of the two states. Fig. 1.1 compares avoided crossing between $(1)^1\Sigma_g^+$ and $(2)^1\Sigma_g^+$ states obtained by MRSDCI with that by MRCPA [37](b).

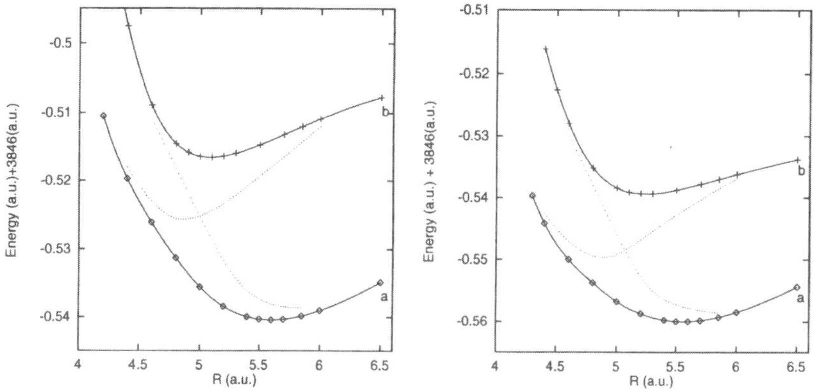

Fig. 1.1 Avoided crossing between $^1\Sigma_g^+$ and $(2)^1\Sigma_g^+$ og Ga_2: left) MRSDCI and right) MRCPA

The figure reveals obvious difference in the higher correlation energy difference between the two leading configurations.

Correlation energy difference between the high spin and low spin; FeH The lowest state of the FeH has been considered as the $^4\Delta$ state [44], but many *ab − initio* calculations showed that the $^6\Delta$ state is lower in energy than the lowest $^4\Delta$ state. We carried out very extensive calculations of MRSDCI and MRCPA with very extensive basis set $[8s6p5d2f1g]$ and $[4s2p]$ for Fe and H, respectively. Fig. 1.2 shows potential curves of the $^4\Delta$ state and $^6\Delta$ state obtained by MRSDCI, MRSDCI+Q [45], and MRCPA(4). MRSDCI calculations gave lower energy for the $^6\Delta$ state, in line with Bauschlicher's results [46](a). Davidson type quadruple correction stabilized $^4\Delta$ more than $^6\Delta$, suggesting that the higher order correlation is important in the $^4\Delta$ state in comparison with $^6\Delta$. MRCPA calculations stabilized the $^4\Delta$ state much more than the $^6\Delta$ state. The energy difference is 0.27eV which is very close to the observed value (0.25 eV) [44]. It is well recognized that the lower spin states are correlated much more than the high spin states especially in transition metal complexes. The present results are in accord with the knowledge and it is important to employ MR-CPA or similar method to studying the electronic structure of such systems.

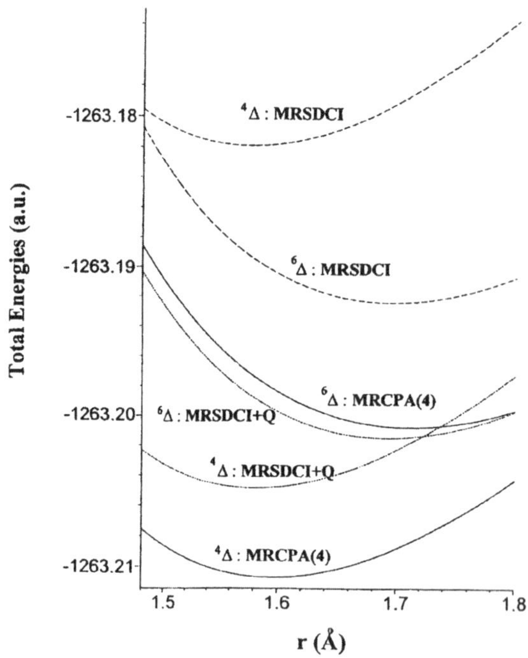

Fig. 1.2 potential curves of $^4\Delta$ and $^6\Delta$ of FeH by MRSDCI, MRSDCI+Q, and MRCPA.

Adiabatic excitation energies of lower states; FeH Fig. 1.3 shows excitation energies to the lower quartet states obtained by MRCPA(4) (using about 0.5 - 0.7 million CSF's) and compares with experimental value of $F^4\Delta$ [47] and other states obtained by MRSDCI [46](b). In the left hand side of the figure, the solid horizontal lines are the ground state level and the formaer$F^4\Delta$ state, and the broken horizontal lines are of other quarted states(MRSDCI [46](b)). The state-universal approach enabled us to obtain the lowest three solutions of the $^4\Delta$ states simultaneously. The lowest three $^4\Pi$ and two $^4\Sigma^+$ state wavefunctions are also obtained at the smae time. This situation is an advantage of the theory basing on the state-universal approach. The present $F^4\Delta$ state energy is a bit lower than experimental one by about 0.1 eV. Other excited states except for the $(2)^4\Sigma^+$ state are lower than the previous MRSDCI [46](b).

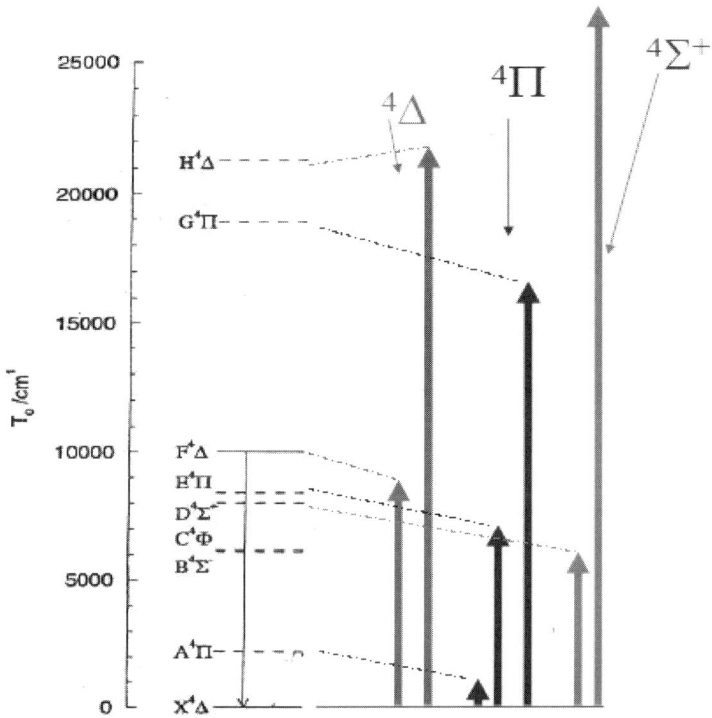

Fig. 1.3 lower adiabatic excitation energies of quartet states of FeH.

Electron affinity of FeH and NiO_2 Since Electron affinity is small in general, accurate prediction of the the affinity has been one of the most demanding problems. The quantity is the energy difference between two states in which the number of electrons is different. Calculation of the affinity is a good test of the size-extensivity of the used method. First we carried out calculation of the electron affinity of the linear O-Ni-O with extensive basis set including *f* functions. The energies of O-Ni-O and O-Ni-O^- were calculated at the respective equilibrium structures determined by MRSDCI. Table 1.1 shows the results. The worst values among used methods was given. MRSDCI+Q and ACPF gave a little smaller value

Table 1.1 Electron affinity, A_e (eV), of the linear O-Ni-O by barious methods

MRSDCI	MRSDCI+Q	ACPF	MRCPA(4)	Obs.[a]
2.44	2.80	2.76	3.04	3.05

[a] Reference [48]

Table 1.2 $R_e$and ω_e of the $^5\Delta$ state of FeH^- and electron affinity (A_e) of FeH

	MRCPA(4)	Obs.[a]
R_e ($\AA$)	1.715	1.79 ± 0.03
ω_e (cm^{-1}	1480	1300 ± 0.140
A_e (eV)	0.873	0.937 ± 0.011

[a] Reference [44]

than expeimental value and the present result gave the best approximate value.

Electron affinity of FeH was also calculated by MRCPA. The equilibrium inter-nuclear distance (R_e), the harmonic vibrational frequency (ω_e) of the $^5\Delta$ state of FeH^- and the electron affinity (A_e) of FeH. They were obtained by MRCPA(4) calculation with about 2.5 million CSF's. Calculated A_e is in very good agreement with observed value, as shown in Table -1.3. Since molecules including transition metal is highly correlated, these results of the electron affinity show how size-extensivity is inherent in the MRCPA approach.

1.4 Summary

In this review note, we mainly focus our attention on the MRCPA method which is one of non-variational variants of MRSDCI. MRCPA incorporates some significant effect inherent in the cluster expansion method into the MRSDCI framework. In contrast to the other methods which are developed as state-selective theory, the state-universal representation makes easier to applied MRCPA to the excited states as well as the lowest state of the given symmetry species. Some promising results are presented for the excited states.

Size-consistency and size-extensivity of MRCPA were demonstrated by numerical results of the systems including larger number of electrons than the systems usually considered.

As was shown in this article, we have developed a high performance program system by making use of highly developed algorithms and codes of a CI system. The method is promising in obtaining reliable results for larger systems.

Acknowledgments

This work was partially supported by a Grant-In-Aid for Scientific Research on a Priority Area (A), No. 12042201, and a fundamental research 16550001 from the Japanese Ministry of Education, Science, Sports, and Culture.

The author expresses his sicere appreciation to his colleages, Prof. E. Miyoshi, Dr. T. Sakai, Dr. Y. Mochizuki, Dr. M. Sekiya, Dr. T. K. Ghosh, Dr. M. Tomonari, Dr. T. Ishikawa, Mr. Y. Tawada, and Mr. T. Yamaki for their earnest efforts and pleasant collaboration works.

Bibliography

[1] Hirata, S. (2005) *Theor. Chem. Acc.* (an invited minireview), (published online) and references therein.

[2] Piecuch, P. and Kowalski, K. (2002) *Int. J. Mol. Sci.* **3** 676

[3] Bartlett, R. J. (1981) *Ann. Rev. Phys. Chem.* **32**, 359

[4] Urban, M., Cernusak, I., Kellö V., Noga J. in *Electron Correlation in Molecules*, ed. S. Wilson, Methods in Computational Chemistry, vol. 1, (Plenum Press, New York, 1987). (1966) *J. Chem. Phys* **45**, 4256

[5] Čížek, J. (1969) *Adv. Chem. Phys.* **14**, 35

[6] Hurley, A. C. *Electron Correlation in Small Molecules*, (Academic Press, 1976).

[7] Kutzelnigg, W. in *Methods of electronic structure theory*, vol. 3, ed. H. F. Schaefer III (Plenum Press, New York, 1977).

[8] Brandow, B. H. (1967) *Rev. Modern Phys.* **39**, 771 .

[9] (a) Lindgren, I. (1974) *J. Phys.* B **7**, 2441 ,
(b) Lindgren, I. (1978) *Intern. J. Quantum Chem. Symp.* **12**, 33 .

[10] (a) Ladig, W. O. and Bartlett, R. J. (1984) *Chem. Phys. Letters* **104**, 424 ,
(b) Ladig, W. D., Saxe, P. and Bartlett, R. J. (1987) *J. Chem. Phys* **86**, 887 .

[11] Jeziorski, B. and Monkhorst, H. J. (1981) *Phys. Rev.* A **24**, 1668 .

[12] D. Mukherjee, (1986) *Chem. Phys. Letters* **125**, 207 .

[13] Lindgren, I. and Mukherjee, D. *Phys. Rep.* **151**, 93, (1987).

[14] Meissner, L. Jankowski, K. and Wasilewski J. , (1988) *Intern. J. Quantum Chem.* **43**, 535 .

[15] Kutzelnigg W. in *Aspects of Many-Body Effects in Molecules and Extended Systems*, Lecture Notes in Chemistry vol. 50, ed. D. Mukherjee, (Springer-Verlag, 1988).

[16] (a) Banerjee, A. and Simons, J. (1981) *Intern. J. Quantum Chem.* **19**, 207,
(b) Banerjee, A. and Simons, J. (1982) *J. Chem. Phys* **76**, 4548 ,
(c) Banerjee, A. and Simons, J. (1983) *Chem. Phys.* **81**, 297 .

[17] (a) Nakatsuji, H. and Hirao, K. (1981) *Intern. J. Quantum Chem.* **20**, 1301,

(b) Hirao, K. Nakatsuji, H. (1982) *J. Comput. Phys.* **45**, 246 ,
(c) Nakatsuji, H. (1985) *J. Chem. Phys* **83**, 713 .

[18] Baker, H. and Robb M. A. (1983) *Mol. Phys.* **50**, 1077 .

[19] Tanaka, K. and Terashima H. (1984) *Chem. Phys. Letters* **106**, 558 .

[20] (a) Hoffmann, M. R. and Simons J. (1988) *J. Chem. Phys* **88**, 993 ,
(b) Hoffmann, M. R. and Simons J. (1989) *J. Chem. Phys* **90**, 3671 .

[21] Mukherjee, D. and Pal S. (1989) *Adv. Qunatum Chem.* **20**, 291

[22] Paldus J. in *Method in Computational Molecular Physics*, vol. 293 of NATO ASI, Series B: Physics, eds. S. Wilson and G. H. F. Diercksen (Plenum Press, New York, 1992)

[23] Shavitt I. in *Methods of electronic structure theory*, vol. 3, ed. H. F. Schaefer III (Plenum Press, New York, 1977).

[24] Ahlrichs R. and Scharf P., in *Ab initio Methods in Quantum Chemistry I*, ed. K. P. Lawley, Advances in Chemical Physics - vol. 67, (John Wiley and Sons, New York, 1978).

[25] Gdanitz R. J. and Ahlrichs R. (1988) *Chem. Phys. Letters* **143**, 413 .

[26] (a) Cave R. J. and Davidson E. R. (1988) *J. Chem. Phys* **89**, 6798 ,
(b) Murray C., Racine S. C. and Davidson E. R. (1992) *Intern. J. Quantum Chem.* **42**, 273 .

[27] (a) Tanaka K., Sakai T. and Terashima H., (1989) *Theoret. Chim. Acta* **76**, 213 ,
(b) Sakai T. and Tanaka K., (1993) *Theoret. Chim. Acta* **85**, 451 ,
(c) Tanaka K., Ghosh T. K. and Sakai T. (1999) *Intern. J. Quantum Chem.* **74**, 661 (d) Tanaka K., Sakai T. and Mochizuki Y. in *Recent Advances in Multireference Methods*, ed. K. Hirao (World Scientific, Singapore, 1999).

[28] Ruttink P. J. A., van Lenthe J. H., Zwaans R. and Groenenboom G. C. (1991) *J. Chem. Phys* **94**, 7212 .

[29] Fink R. and Staemmler V. (1993) *Theoret. Chim. Acta* **87**, 129 .

[30] (a) Szalay P. G. and Bartlett R. J. (1993) *Chem. Phys. Letters* **214**, 481 ,
(b) Szalay P. G. and Bartlett R. J. (1995) *J. Chem. Phys* **103**, 3600 .

[31] (a) Daudey J.-P., Heully J.-L. and Malrieu J.-P. (1993) *J. Chem. Phys* **99**, 1240 ,
(b) Malrieu J.-P., Daudey J.-P. and Caballol R. (1994) *J. Chem. Phys* **101**, 8908 .

[32] Szalay P. G. in *Recent Advances in Coupled-Cluster Methods*, ed. R. J. Bartlett,Recent Advances in Computational Chemistry - vol. 3, (World Scientific, Singapore, New Jersey, London, Hong Kong, 1997).

[33] Adamowicz L. and Malrieu J.-P. in *Recent Advances in Coupled-Cluster Methods*, ed. R. J. Bartlett, Recent Advances in Computational Chemistry - vol. 3, (World Scientific, Singapore, 1997).

[34] Mahapatra U. S., Datta B. and Mukherjee D. in *Recent Advances in Coupled-Cluster Methods*, ed. R. J. Bartlett, Recent Advances in Computational Chemistry - vol. 3, (World Scientific, Singapore, 1997).

[35] Ghosh P. , Chattopadhyay S., Jana D. and Mukherjee D. (2002) *Int. J. Mol. Sci.* **3** 733,

[36] Li X. and Paldus J. (2003) *J. Chem. Phys* **119**, 5320 .
[37] (a) K. Tanaka and Y. Mochizuki, (1998) *Theoret. Chem. Acct.* **98**, 165 .
(b) T. K. Ghosh, K. Tanaka and Y. Mochizuki,(1998) *THEOCHEM* **451**,61
(c) Y. Mochizuki and K. Tanaka, (1998) *Theoret. Chem. Acct.* **99**, 88 ,
(d) Y. Mochizuki and K. Tanaka, (1999) *Theoret. Chem. Acct.* **101**, 257 ,
(e) Y. Mochizuki and K. Tanaka, (1999) *Theoret. Chem. Acct.* **101**, 292 ,
(f) T. Sumi, E. Miyoshi, and K. Tanaka, 1999 *Phys. Rev. B* 59 6153 ,
(g) M. Tomonari, Y. Mochiduki, and K. Tanaka, (1999) *Theoret. Chem. Acct.* **101**, 332 ,
(h) K. Tanaka, M. Sekiya, and M. Yoshimine, (2001) *J. Chem. Phys* **115**, 4558 ,
(i) M. Tomonari and K. Tanaka, (2001) *Theoret. Chem. Acct.* **106**, 188 ,
(h) T. Yamaki, M. Sekiya, and K. Tanaka, (2003) *Chem. Phys. Letters* **376**, 487 ,
(i) M. Tomonari and K. Tanaka, (2003) *Mol. Phys.* **101**, 111 ,
(j) K. Tanaka and K. Nobusada, (2004) *Chem. Phys. Letters* **388**, 389 ,
(k) T. Ishikawa and K. Tanaka, (2004) *Chem. Phys. Letters* **395**, 166 ,
(l) K. Tanaka, M. Sekiya, Y. Tawada and E. Miyoshi, (2005) JCP **122**, 214315 .
[38] Davidson E. R., McMurchie L. E. and Day S. J. (1981) *J. Chem. Phys* **74**, 5491 .
[39] Gershgorn Z. and Shavitt I. (1968) *Intern. J. Quantum Chem.* **2**, 751 .
[40] Hirao K. (1991) *Chem. Phys. Letters* **190**, 374 .
[41] Roos B. O. , in *Modern Electronic Structure Theory*, ed. R. Yarkony, vol. 1, (World Scientific, New York, 1995) and references therein.
[42] (a) Meyer W. (1973) *J. Chem. Phys* **58**, 1017 ,
(b) Meyer W. (1974) *Theoret. Chim. Acta* **35**, 277 .
[43] (a) Lengthfield III B. H. (1980) *J. Chem. Phys* **73**, 382 ,
(b) Liu B. and Yoshimine M. (1981) *J. Chem. Phys* **74**, 612 ,
(c) Lengthfield III B. H. and Liu B. (1981) *J. Chem. Phys* **75**, 478 ,
(d) McLean A. D., Yoshimine M., Lengsfield B. H., Bagus P. S. and Liu B. in *Modern Techniques in Computational Chemistry: MOTECC-90*, ed. E. Clementi, (ESCOM Science Publishers B. V., Leiden, 1990) .
[44] Stevens A. E., Feigerle C. S. and Lineberger W. C. (1983) *J. Chem. Phys* **78**, 5420 .
[45] (a) Langhoff S. R. and Davidson E. R. (1974) *Intern. J. Quantum Chem.* **8**, 61 ,
(b) Davidson E. R. and Silver D. W. (1977) *Chem. Phys. Letters* **52**, 403 .
[46] (a) Bauschlicher Jr., C. W. and Langhoff S. R. (1998) *Chem. Phys. Letters* **145**, 205 ,
(b) Langhoff S. R. and Bauschlicher Jr., C. W. (1990) *J. Mol. Spectrsc.*, **141**, 243
[47] Phillips J. G., Davis S. P., Lindgren B. and Balfour W. J. (1987) *Astrophys. J.*, Supple. Ser. **65**, 721
[48] H. Wu and L.-S. Wang, (1997) *J. Chem. Phys.*, **107**, 16

CALCULATION OF NEGATIVE ION SHAPE RESOANCES USING COUPLED CLUSTER THEORY

Y. SAJEEV

Theory group, Physical Chemistry Division, National Chemical Laboratory, Pune 411 008, India

SOURAV PAL

Theory group, Physical Chemistry Division, National Chemical Laboratory, Pune 411 008, India

1. Introduction

Electronic resonance states correspond to metastable bound states coupled to continuum states, which decay by electron emission. Resonance phenomena constitute some of the most interesting features of scattering experiments [1]. Resonances which occur in electron-atom/-molecule scattering and other areas of atomic and molecular physics, are associated with quasi-bound state with a lifetime long enough to be well characterized [2]. The accurate calculation of the energies and lifetimes of resonance state is very important for describing several physical processes, such as Auger decay[3], intermolecular Coulombic decay (ICD)[4] and metastable anions [5]. Considerable progresses have been made in the development of practical methods to calculate the energy and lifetime of resonance in the framework of analytical continuation of Hamiltonian techniques [2].

2. Analytical Continuation of the Hamiltonian

In recent decades, a number of methods have been proposed to calculate the energies and width of resonances. The theoretical approaches based on the quantum theory of scattering are not fully L^2 methods because of the appearance of so called bound-free matrix elements. Much of the difficulty encountered in previous theoretical treatments of electronic resonances based either on the use of the scattering matrix [6] or on the approximate expansion of the scattering wave function in a set of square-integrable basis function [2,7] follows from two fundamental facts. First, when focusing, within conventional Hermitian quantum mechanics, on the continuous spectrum of the Hamiltonian H of the metastable system, there exist an inherent difficulty in identifying which discrete positive eigenvalue of an approximate, finite-rank representation of H corresponds most closely to the metastable state of interest. Second, as the Siegert wave function diverges asymptotically and does not belong to the Hermitian domain of the Hamiltonian, the computational difficulties in treating such a state are potentially more severe than those encountered in the treatment of bound states. The method of analytical continuation of the Hamiltonian in the complex plane eliminates many of these difficulties [2]. This approach to get the resonances from the complex eigenvalue problem, which is a direct calculation of resonance eigenvalue of the Hamiltonian, has not received enthusiasm among atomic theorists before the method of complex scaling became known. Instead of extracting the resonance parameters from the cross section, the metastable state are described by an eigenfunction belongs to a complex eigenenergy. The resonance wavefunction (The so called Gamow-Siegert wavefunction) is a solution of Schrödinger equation with purely outgoing boundary condition. The outgoing boundary condition enforces the eigenenergy to become complex, $E_{res}=E_r - i\Gamma/2$. The Simplest mathematical description of such states is that they resemble bound stationary states in that they are localized in space (at t=0), and their time evolution is given by

$$\Psi_R(r,t) = e^{-i(E-\Gamma/2)t/\hbar}\Psi_R(r) \tag{1}$$

The presence of in the energy term forces exponential decay and the probability,

$$|\Psi_R(r,t)|^2 = |\Psi_R(r)|^2 \, e^{-\Gamma t/\hbar}, \tag{2}$$

decays to zero as time passes at constant r. Therefore, the particle disappears from any given point in the coordinate space. Due to the exponential divergence, the number of particle is conserved only when the reaction coordinate, r and the time, t, approach the limit of infinity. A characteristic for these states is their exponential growth in the asymptotic region. Thus they are not square integrable and do not belong to the Hermitian domain of the Hamiltonian [8]. The Gamow-Siegert wave functions possess large amplitude in the inner molecular region resembling a bound state. The wavefunction in this region is affected by physical interactions, while the asymptotic exponentially growing part describes the decay. In fact, resonance states can be understood as a discrete state coupled to continuum. They represent a particularly challenging problem to quantum chemist because one has to treat a continuum problem and the electron correlation simultaneously. In particular, the latter effect plays a crucial role in most temporary anions. The role of correlation and relaxation in the formation and decay of metastable states has been reported [9,10].

When one solves the Schrödinger equation $H\Psi = E\Psi$ for a self-adjoint Hamiltonian subject to standard boundary condition, one obtains a spectrum $\{E\}$ of real eigenvalues E which are discrete or continuous. However H can be modified [analytically continued] in several ways such that direct access is gained to Siegert energies E_{res}. Each Siegert energy is an eigenvalue of the resulting non-Hermitian Hamiltonian, and the associated eigenfunctions are square integrable, i.e., they can be represented in Hilbert space. Analytic continuation can achieved either by complex scaling of the electronic coordinates[2,11-14] or by utilizing a complex absorbing potential (CAP) based methods[15,16]. However, even without detailing any of the discussion, it is straight forward to state that one of the major purposes of introduction of complex scaling and CAP in nonrelativistic quantum theory is to produce exactly those [non-Hermitian] operators which have the complex energies. The advantage of this general approach to investigating resonances is that the complex Siegert energy can be calculated directly within a Hilbert space of L^2 functions. Therefore, many of the methods of bound-state calculations can be readily adapted to the resonance problem in this form.

2.1. *Complex Scaling*

This method is associated with a similarity transformation in which the many-particle Hamiltonian loses its self-adjoint character. The resonance state can be described by square integrable functions associated with the eigenfunctions of a transformed Hamiltonian $\hat{U}\hat{H}\hat{U}^{-1}$, obtained from the original Hamiltonian H by an unbounded similarity transformation[17].
That is,

$$(\hat{U}\hat{H}\hat{U}^{-1})(\hat{U}\Psi_R) = (E_{res} - i\Gamma/2)(\hat{U}\Psi_R) \tag{3}$$

such that

$$\hat{U}\Psi_R \rightarrow 0 \quad \text{as } r \rightarrow \infty \tag{4}$$

and $\hat{U}\Psi_R$ in Hilbert space although Ψ_R are not. The method of complex scaling provides such an unbounded similarity transformation. The complex scaled Hamiltonian is no longer self-adjoint and the original spectrum $\{E\}$ has been changed: some energy eigenvalues are persistent, others may be lost, and new eigenvalues may occur also in the complex plane. The theory of change of spectra of a many-particle Hamiltonian associated with the unbounded similarity transformation- due to the change of boundary condition-is reviewed by Löwdin[17]. The complex-scaling operator is given by

$$\hat{U} = e^{i\theta r\partial/\partial r} \tag{5}$$

such that $\hat{U}f(r) = f(re^{i\theta})$ for any analytical function $f(r)$

By scaling the reaction coordinate, the resonance wave function becomes square integrable and, consequently, the number of particles in the coordinate space is conserved. Therefore, complex scaling has the advantage of associating the resonance phenomenon with the discrete part of the spectrum of the complex scaled Hamiltonian. Moreover, the resonance state is associated with a single square integrable function, rather than with a collection of continuum eigenstates of the unscaled Hermitian Hamiltonian. Complex scaling may be viewed as a procedure which compresses the information about the evolution of a resonance state at infinity into a small well defined part of the space.

Boundary conditions determines whether an operator has eigenvalues, and whether the corresponding eigenfunctions are L2 (bound states) or non-L2 [scattering states]. The boundary condition of square integrability is preserved during the complex scaling for a θ value of $|\theta| \geq \Pi/2$. Conversely, no new square integrable eigenfunctions with real eigenvalues suddenly appear as $r \rightarrow re^{i\theta}$. That is H and $H(\theta)$ has same real bound state eigenvalues. The continuum is rotated into the lower half complex energy plane. It is the fact that the continuous spectrum of $H(\theta)$ is different from that of H which is the key to the utility of $r \rightarrow re^{i\theta}$ transformation.

2.2. *Complex Absorbing Potentials*

The complex scaling method is consistent with the Born-Oppenheimer approximation and the application of complex scaling method to molecular potential would [applying the Born-Oppenheimer approximation after dilation], however, result in the determination of electronic spectra for unphysical complex internuclear separations[18,19]. What is done instead is to clamp the nuclei on the real axis, and scale only the electronic coordinates. The problem in this complex coordinate real axis clamped nuclei approximation arises from the fact that if the nuclear coordinates are left real, and an electronic coordinates $\vec{r}_i$ is scaled $\vec{r}_i \rightarrow \vec{r}_i e^{i\theta}$, the nuclear electron interaction

$$\sum_{\alpha} \left| \vec{r}_i e^{i\theta} - \vec{R}_\alpha \right|^{-1} \tag{6}$$

is non-analytic [20] as the argument of the absolute value can vanish for a continuous range of value such that

$$\left|\vec{r}_i e^{i\theta}\right| = \left|\vec{R}_\alpha\right|, \quad \hat{r}_i.\hat{R}_\alpha = \cos\theta, \tag{7}$$

giving rise to a continuous range line of square branch points. The existence of these continuous branch points makes the Hamiltonian non-analytic, thus making the Balslev-Combes theorem [13] inapplicable to the Born-Oppenheimer Hamiltonian. Simon [21]suggested using an exterior scaling method to avoid any interior non-analyticities by keeping the coordinates on the real axis long enough. The exterior complex-scaling operator in this case is

$$U = \begin{cases} 1 & if \quad r < r_o \\ e^{i\theta(r-r_0)\partial/\partial r} & if \quad r \geq r_o \end{cases} \tag{8}$$

and the Hamiltonian is given by

$$H(\theta) = \begin{cases} H(r) & if \quad r < r_0 \\ H\left((r-r_0)e^{i\theta}\right) & if \quad r \geq r_0. \end{cases} \tag{9}$$

In the spirit of Simon's proposition, to avoid the need of carrying out analytical continuation of the potential term in the Hamiltonian into the complex coordinate plane Rom and coworkers proposed a smooth exterior scaling path which is defined as[22]

$$f(r) = \frac{\partial F(r)}{\partial r} = 1 + (e^{i\theta} - 1)g(r) \tag{10}$$

where $F(x)$ is a path in the complex coordinate plane such that.

$$F(r) \to re^{i\theta} \quad as \quad r \to \infty \tag{11}$$

And $g(r)$ is varied from 0 to 1 value around the point $r = r^0$. If $V(r \geq r^0) = 0$, then one can use the unscaled potential $V(r)$. Note, however, that the path which defined in Eq. (11) is very general and is not necessarily limited to the case where $V\left(F(r)\right) = V(r)$ or $V\left(F(r)\right) \sim V(r)$. Consequently, the smooth exterior scaled Hamiltonian derived by Moiseyev, [23]

$$H = \frac{-\nabla^2}{2} + V\left(F(r)\right) + V_{CAP} \tag{12}$$

where V_{CAP} is a universal energy-independent complex absorbing potential (CAP). When the smooth exterior scaling is used, the CAP gets non-zero values in the region where the interaction potential vanishes. The complex smooth exterior scaling Hamiltonian is obtained by adding to the unscaled Hamiltonian matrix a matrix which represents the universal CAP.

Until fairly recently, the main theoretical method for treating the analytical continuation method for the molecular Hamiltonian relied on complex scaling based methods. Finite box methods are also particularly appealing to calculate resonance widths and energies due to the quasi bound character of resonance states. Complex absorbing potentials are best for treating the electronic resonances in molecules. [15,16] The molecular Hamiltonian is perturbed by an appropriate complex potential, which enforces an absorbing boundary condition. This artificial potential absorbs the emitted particle and consequently transforms the former continuum wavefunction into a square integrable one. CAPs were first used for this aim by Jolicard and Austin [24-27] and later by Riss and Meyer.[15,28-30] Jolicard and Austin suggested and demonstrated that the stability of the resonance eigenvalue could also be achieved by varying the strength or the location of the absorbing potential, whose job would be to absorb the perfectly outgoing Siegert state without creating any "reflection".[25] The basic concepts introduced by Jolicard and coworkers were refined and developed further by Riss and Meyer.16 However, the most serious problems with these approaches are the conditions and approximations at which the spectrum of CAP augmented Hamiltonian resembles with the spectrum of complex scaled Hamiltonian. Unlike other methods such as complex scaling method which stays on solid mathematical ground given by Balslev, Combes and Simon,[13] the use of CAP was based on the intuition and numerical experience. It has been proved that the poles of scattering matrix are also the eigenvalues of the complex scaled Hamiltonian, but it has not been proved that they are the eigenvalues of the Hamiltonian which is perturbed by a CAP. Riss and Meyer addressed the question under what condition the resonances obtained by the CAP are the poles of the scattering matrix. Their strategy and derivation is as follows.

The eigen spectrum of a perturbed Hamiltonian

$$H(\eta) = H - i\eta W(r) \tag{13}$$

has a purely discrete spectrum, where

$$H(\eta) = -\tfrac{1}{2}\frac{d^2y}{dr^2} + V(r), \quad \eta > 0 \tag{14}$$

and $W(r)$ is a piecewise continuous coulomb potential which satisfies the properties. The exact prerequisites that $W(r)$ must satisfy are derived in Ref. [16]. For $\eta \to 0$, the eigenvalues of $H(\eta)$ converge towards the poles of the Green function on its physical sheet provided $\pi \geq 0 \geq \arg(E) \geq 0$. Applying a CAP is not equivalent to complex scaling: it becomes equivalent only in the limit $\eta \to 0^+$. Since this limit cannot be easily carried out unless there is a complete basis set, there is always [at finite η] nonanalytic perturbation. These large values of η causes artificial reflections. There have been many correction schemes devised for removing the artificial reflections. [16,24] The correction process would not be necessary if reflections were avoided. Assume a $W_0(r)$ which is twice the differentiable and satisfies

$$W_0(r) = \begin{cases} = 0 & \text{for} \quad r \leq r_c \\ > 0 & \text{for} \quad r > r_c \end{cases} \tag{15}$$

$W_0'(r) \geq 0$ and $W_0(r) \to \infty$ for $r \to \infty$.

The parameter r_c is the cut off parameter. For a reflection free CAP, the regular solution is identical to the exact Siegert resonance wavefunction up to r_c, and E_0 is the exact Siegert resonance energy. This perturbation is particularly undesirable on the target, so the CAP is set to zero in the vicinity of the target. There may, however, be a non-negligible artificial perturbation of the projectile if η cannot be chosen small enough [if the basis set is too small]. Riss and Meyer proposed another method called "transformative CAP" [TCAP]. [28,29] In exchange part, it implies a modified kinetic energy instead of adding a local complex potential.

There is a connection here with the smooth exterior scaling [SES] with the "transformative CAP" [TCAP] [29-31]. The TCAP and SES in fact become identical for cut-off potentials. For the TCAP method, Riss and Meyer started from the Hamiltonian perturbed by a CAP and ended up with a complex-scaled operator. Instead, For the SES method, Moiseyev started with the complex coordinate method and ended up with a non-scaled Hamiltonian perturbed by a perfectly absorbing [universal] "potential" which is energy and problem independent. For a detail discussion of comparisons between the smooth exterior CAP, which is derived from a complex scaling method and the other forms artificial perturbations applied to trap the projectile, we refer to the review article by N. Moiseyev, and more references therein. The order of presentation just outlined is by no means the order in which the idea of CAP was originally developed. Historically, the subject developed in terms of the artificial potential generated by several workers [16,24-31]. Only later was the CAP potential derived from the first principles of Quantum mechanics by N. Moiseyev [23] to provide a proper justification of the results already derived.

The method of using a CAP is closely related to complex scaling. The major advantage of the former is its simplicity. The CAP procedure is minimally invasive in the sense that neither the internal structure of the physical Hamiltonian is affected nor is there any need to use other basis sets than usual real Gaussians[32]. Thus, this method offers great promise for the determination of accurate Siegert energies in a computationally viable form by a relatively straightforward modification of existing electronic structure codes for bound states.[10,32,33]In addition to the calculations and methods mentioned above, the application of complex scaling and CAP methods in the framework of post Hartree -Fock method have been documented.

3. Negative Ion Resonances

The negative ion resonances in electron-molecule collision appear as sharp changes in scattering cross section at low incident electron energies (typically 1-10 eV). At some incident energies, the electron wavefunction has large amplitude within the target. This happens only when the incident energy falls in one of the discrete bands, where the incident electron finds a comfortable quasi-stationary orbit in the field of target molecule. The quasi stationary nature of the compound state is usually guaranteed by either one of the two following mechanisms. The first possibility and the most common situation that causes the appearance of resonance is an effective potential made up of attractive potential [attractive polarization force at small distances] combined with a repulsive potential (repulsive centrifugal force at long distances) produces a barrier in the potential. For energies below the maximum in the barrier, there would be bound states inside the attractive part of the potential if tunneling could be ignored. However, the quantum mechanical tunneling permits particle 'trapped' inside the attractive part of the potential to escape to infinity, and the tunneling rate depends on the height and thickness of the barrier. Conversely, particles incident on the potential at energy close to the virtual states are able to penetrate inside the attractive barrier. Once the electron has entered the region inside the barrier, it will take some time before the electron leaks out to the outer region again by a tunnel effect. This type of resonance is called *shape resonance* or *potential resonance* since the resonance state is produced by an appropriate shape of the effective interaction potential between the electron and the molecule.

The second possibility arises when the inelastic channels are introduced. By exciting the target molecule, the electron loses its energy. Suppose that the incident electron energy is not large that after the excitation the electron energy becomes negative, and furthermore, its value coincides with one of the bound-state energies allowed in the field produced by the excited target molecule. Then it will take some time before the electron gets its energy back from the target and escape to outside. Thus one has a new type of resonance process which is called *core-excited type I* resonance or the resonance of *Feshbach resonance.*

There exists also shape resonances associated with the effective potential in the inelastic channel. They are called *core-excited type II* resonance or *core excited shape resonances*.

All these resonance states can be considered as quasistationary states of negative ion M^- formed by the electron and the target molecule M. In the elastic collision experiments, the electron-atom or electron-molecule shape resonance can be thought of as metastable anionic state which decay by electron emission and the temporary anion formation in this case can be represented like

$$e + A \rightarrow \left(A^-\right)^* \rightarrow e + A\,. \tag{16}$$

The resonance energy of the metastable state relative to the target can be defined as

$$E_{res} = E_{\left(A^-\right)^*} - E_A \tag{17}$$

which is nothing but the electron affinity of the target in the complex plane. The configuration of a Feshbach resonance is that of an excited state of M, usually referred to a parent state, with an additional electron in an excited orbital. Alternatively the resonance configuration may be viewed as that of a state [often ground state] of M^+, referred to as grandparent state, with the addition of two electrons in excited orbitals. Feshbach resonances are generally long lived compared to typical vibrational period [10^{-14} s], as their decay involves the rearrangement of the orbitals of both excited electrons. Thus the temporary negative ions make many vibrations before decaying. Since the lifetimes of the Feshbach resonances tend to be long, resonance width are correspondingly small [typically less than 20 meV], and so the resonances appear as narrow structures in the cross sections of elastic and inelastic processes.

The structural and spectroscopic properties of electronic resonance states are similar to those of bound states and their study reveals deep physical insight into the complex many-body effects governing molecular physics and this is the reason why this states are of particular importance to physical chemists. Structurally, resonances provide information on metastable negative ions, negative electron affinities, orbital energies of unbound orbitals, and doubly excited electronic states.

4. CAP method in the Fock space Coupled Cluster theoretical framework

Recently CAP techniques at the multireference configuration interaction level [34], multireference coupled cluster level [35,36], and in the context of electron propagator theories [38,39] have been applied to the calculation of shape resonances in molecules. In these calculations the resonance energies and width are calculated as a difference between the ground state total energies of the (N+1) and the neutral target. In the negative ion shape resonances in elastic scattering, where the trapping of the projectile electron occurs in the potential well created by low lying electronic states, target electrons are not affected dynamically, except through a minor polarization. The vertical energy difference approximation, i.e., by assuming the same geometry for (N+1) and N electron system, adopted in the above mentioned methods, determines the resonance energy, is valid for this kind of short-lived resonances, where the time delay is too small for the nuclei to relax. The electron affinity studies provide the simultaneous calculation of both energy (real part) and width [twice the imaginary part] of electron attachment shape resonances (E_s^{N+1}-$E_0^{\,N}$), where s labels a stationary state and $E_0^{\,N}$ is the ground state total energy of the neutral N-electron target. However, it is well known that the CI method is not a very desirable electronic structure technique mainly because of size-inextensive manner in which the dynamic correlation is included. Coupled cluster (CC) [40,41] based methods, with exponential wave-operator, includes this dynamic correlation efficiently and in a size-extensive manner[42]. Further, the Fock space multireference coupled cluster (FSMRCC) [42-45] treats the non-dynamic correlation inherent in the ionized states and thus this class of method represents method of choice in the electronic structure calculation. In addition, the evaluation of energy differences can be done in a direct manner in the MRCC method

Since the complex absorbing potential serves to render the wave function of the projectile square-integrable, while it must leave the target unaffected, the CAP must be introduced, in principle, only into the description of the (N+1)/(N-1)-electron state. The N-electron ground state may be described by the Hamiltonian without CAP. In order for the CIP method to have a CAP unperturbed target state, we eliminate the effect of the CAP on the Hartree-Fock ground state by the replacement

$$\hat{W} \rightarrow \hat{P}\hat{W}\hat{P} \ , \tag{18}$$

where

$$\hat{P} = \sum_i |\phi_i\rangle\langle\phi_i| \tag{19}$$

projects onto the subspace of unoccupied orbitals. The redefinition is easily accomplished by setting

$$\langle\phi_p|\hat{W}|\phi_q\rangle = 0 \tag{20}$$

if either p or q is an occupied orbital.

As our objective is to introduce the CAP potential—a one-body potential— into the FSMRCC theory, we would need to introduce it in both the (0,0) and (1,0) sectors to completely define the valence universal wave operator for the CAP perturbed Hamiltonian,

$$H(\eta) = H - i\eta W. \tag{21}$$

The restricted Hartree-Fock determinant $|\phi_0\rangle$ corresponding to the unperturbed N-electron target is chosen as the vacuum. With respect to this vacuum, holes and particles are defined. Depending on the energies of interest, these are further divided into active and inactive sets such that each determinant ϕ_i of the model space has m active particles and n active holes. For example, when studying an (N+1)/(N-1) electron state, the model space consists of determinants representing one active particle/hole. These are called one-particle or one-hole model spaces. The active particles or holes can be so defined as to make the model space complete. For each sector (m,n) of Fock space, our zeroth-order wave function consists of a linear combination of model-space functions:

$$\Psi_\mu^{0(m,n)} = \sum_i C_{\mu i}\phi_i^{(m,n)} \ . \tag{22}$$

$C_{\mu i}$ are the model space coefficients and (m,n) denotes the particle-hole rank of the Fock space sector. The exact (1,0) sector eigenfunctions of $H(\eta)$ can be written in the Fock space method using Lindgren's normal-ordered ansatz [44] as,

$$\Psi_\mu^{(1,0)}(\eta) = \Omega(\eta)\Psi_\mu^{0(1,0)} \tag{23}$$

$$\Omega(\eta) = \{e^{\tilde{T}^{(1,0)}(\eta)}\} \ , \tag{24}$$

where $\Omega(\eta)$ is the wave operator, and the curly brackets indicate normal ordering of the operator within them The wave operator in FSMRCC theory is known to be valence-universal [42-44] and this ensures connectivity and size-extensivity of the Fock space equations. In general, the wave operator for the Fock space of m active particles and n active holes correlates all lower active particle-active hole Fock space sectors. The cluster operator $\tilde{T}^{(m,n)}(\eta)$ corresponding to the wave operator for the (m,n) sector is defined to consist of cluster operators for the lower Fock space sectors[42]:

$$\tilde{T}^{(m,n)}(\eta) = \sum_{a=0}^{m}\sum_{b=0}^{n} T^{(a,b)}(\eta) \ . \tag{25}$$

Hence, the operator used is capable of describing the problem of lower valence sectors. In particular,

$$\tilde{T}^{(1,0)}(\eta) = T^{(0,0)}(\eta) + T^{(1,0)}(\eta) \, , \tag{26}$$

where $T^{(0,0)}(\eta) = T(\eta)$ is the cluster operator for the single reference coupled cluster case, containing only hole-particle excitation operators

The normal ordering of the wave operator prevents the different T operators to contract among themselves and leads to a decoupling of the Bloch equations [46] for different sectors. Equations for cluster amplitudes for different sectors can be solved using the subsystem-embedding-condition (SEC), [42,44] i.e., first the equations for the lowest sectors are solved. With the T operators of the lower sectors as constants, equations for higher Fock space sectors are solved progressively upwards. The Bloch equations [46] for general (m,n) sector can be written as

$$\begin{aligned} &P^{(k,l)}H(\eta)\Omega(\eta)P^{(k,l)} = P^{(k,l)}\Omega(\eta)H_{eff}(\eta)P^{(k,l)} \\ &k = 0,1,...m \\ &l = 0,1,...n \end{aligned} \tag{27}$$

$$\begin{aligned} &Q^{(k,l)}H(\eta)\Omega(\eta)P^{(k,l)} = Q^{(k,l)}\Omega(\eta)H_{eff}(\eta)P^{(k,l)} \\ &k = 0,1,...m \\ &l = 0,1,...n \end{aligned} \tag{28}$$

where

$$H_{eff}(\eta) = P^{(m,n)}\Omega(\eta)^{-1}H(\eta)\Omega(\eta)P^{(m,n)} \tag{29}$$

is an effective Hamiltonian[42] whose eigenvalues determine the roots of the Fock space sector (m,n):

$$\mathbf{H}_{eff}(\eta)\mathbf{C} = \mathbf{CE} \, . \tag{30}$$

The symbols $P^{(k,l)}$ and $Q^{(k,l)}$ used in the above equations denote the projection operator for the (k,l) sector of model space and its complementary space, respectively.

The vacuum expectation value of the effective Hamiltonian is the energy of the closed-shell N-electron state (E^{SRCC}) or zero-valence problem. Diagrammatically, this can be dropped easily and the eigenvalues of the resultant effective Hamiltonian is equivalent

$$\bar{H}(\eta)_{N,eff} = H(\eta)_{eff} - E^{SRCC}(\eta). \tag{31}$$

where $\bar{H}_N = (H_N e^{T^{(0,0)}})_{connected,open}$.

Diagonalizing the effective Hamiltonian in the above equation we directly obtain the energy difference between our state defined by (m, n) model space and the couple cluster state defined by (0, 0) model space function. For an (1,0)model spaces these diagonalization of the effective Hamiltonian yield the correlated electron.

$$\bar{H}^{(1,0)}_{N,eff} = (C^{(1,0)}_{PP})\varepsilon^{EA}_p (C^{(1,0)}_{PP})^{-1}; \quad \bar{H}^{(0,1)}_{N,eff} = (C^{(0,1)}_{PP})\varepsilon^{IP}_p (C^{(0,1)}_{PP})^{-1} \tag{32}$$

The $^2\Pi_g$ shape resonance is identified by plotting the complex electron affinity spectra using this method is shown in Figure 1. The eigenvalues that move quickly into the complex plane as η is increased demonstrate strong interaction with the complex absorbing potential. [37] The associated eigenfunctions are, in contrast to the wave function of a resonance in the presence of a CAP, not localized in the vicinity of the target. Thus, the behavior of their η trajectories identifies them as continuum states (discretized by the absorbing potential and the finite basis set). In Figure 1, the second eigenvalue from the right shows a pronounced stabilization in the complex plane. This is the root associated with the resonance. The resonance parameters are determined by analyzing the stabilization behavior of the η trajectory of this eigenvalue.

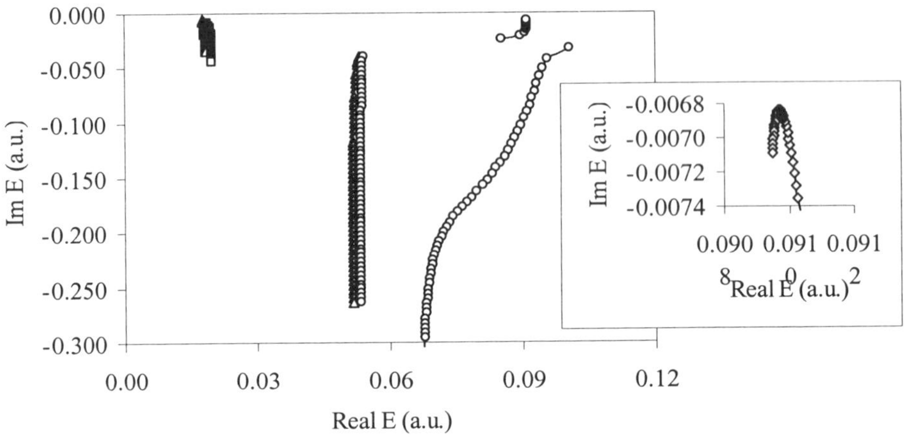

Figure 1 The η trajectories of the complex eigenvalues from CAP-FSMRCC calculations. The η values vary from 0.0010 to 0.0065. is incremented linearly in steps of 0.0001. The region of stabilization of the trajectory corresponding to the resonance (second from the right) is magnified in the lower panel. (This figure has been taken from Ref. 37)

5. Conclusion

In this article, we describe the introduction of complex absorbing potential in highly correlated Fock space multi-reference coupled-cluster theory to describe electron resonances in molecules. It is important to introduce electron correlation and our results reflect the importance of correlation in description of resonances.

References

1. J. R. Taylor, Scattering Theory: The quantum theory on nonrelativistic collisions (John Wiley & Sons, New York ,1972).
2. W. P. Reinhardt, *Ann. Rev. Phys. Chem.* **33**, 223 (1982).
3. *The Auger effect and other radiationless transition*, edited by E. H. Burhop, (Cambridge monographs on physics, Krieger, Melbourne, 1980).
4. L. S. Cederbaum, J. Zobeley and F. Tarantelli, *Phys. Rev. Lett.* **79**, 4778 (1997); J. Zobeley, L. S. Cederbaum and F. Tarantelli, *J. Chem. Phys.* **108**, 9737 (1998); R. Santra and L. S. Cederbaum, *Phys. Rep.* **368**, 1 (2002).
5. K. D. Jordan and P. D. Burrow, *Chem. Rev.* **87**,557 (1987).
6. D. G. Truhlar, in *Modern Theoretical Chemistry,* vol. **8**, edited by G. A. Segal (Plenum, New York, 1977).
7. A. U. Hazi, H. S. Taylor, *Phys. Rev. A* **1**, 1109 (1970).
8. A.I. Baz', Ya.B. Zel'dovich, and A.M. Perelomov, *Scattering, Reactions, and Decays in Nonrelativistic Quantum Mechanics* (Nauka, Moscow, 1971); N. Moiseyev and J. O. Hirschfelder, J. Chem. Phys., **88**, 1063 (1988).
9. N. Moiseyev and F. Weinhold, *Phys. Rev. A* **20**, 27 (1979); M. N. Medikeri, and M. K. Mishra, *Adv. Quantum Chem.* **27**, 223 (1996).
10. T. Sommerfeld, U. V. Riss, H.-D. Meyer, L. S. Cederbaum, B. Engels, and H. U. Suter, *J. Phys. B* **31**, 4107 (1998).
11. N. Moiseyev, *Phys. Rep.* **302**, 211 (1998).
12. B. R. Junker, *Adv. At. Mol. Phys.* **18**, 207 (1982); *Int. J. Quantum. Chem.* **14,** No. 4 (1978) is devoted entirely to complex scaling and more references therein
13. E. Balslev and J .M. Combes, *Commun. Math. Phys.* **22,** 280 (1971); J. Aguilar and J. M. Combes, *ibid.* **22,** 269 (1971);B Simon, *ibid .* **27**, 1 (1972); B. Simon, Ann. *Math.* **97,** 247(1973)
14. C.W. McCurdy, *Autoionization: Recent Developments and Applications,* edited by A. Temkin (New York: Plenum,1985) pp153-70
15. J. G. Muga, J. P. Palao, B. Navarro, I. L. Egusquiza, *Phys, Rep*, **395**, 357(2004).
16. U.V. Riss and H.-D Meyer, *J. Phys. B.* **26**, 4503 (1993).
17. P. O. Löwdin *Adv. Quantum. Chem.* **20**, 87 (1989).
18. B. R. Junker, *Phys. Rev. Lett.* **44**, 1487 (1980).
19. J. N. Bardsley, *Int J. Quantum. Chem*, **14**, 343 (1978).
20. B. Simon, *Int J. Quantum. Chem*, **14**, 529 (1978).
21. B. Simon, *Phys Lett. A*, 71, 211, (1979).
22. N. Rom, E. Engdahl, N. Moiseyev, *J. Chem. Phys.* **93,** 3413 (1990).
23. N. Moiseyev, *J. Phys. B.* **31**,1431 (1998).
24. G. Jolicard, J. Humbert, *Chem. Phys.* **118,** 397 (1987).
25. G. Jolicard, E.J. Austin, *Chem. Phys. Lett.* **121**, 106 (1985).
26. G. Jolicard, E.J. Austin, *Chem. Phys.* 103, 295 (1986).
27. G. Jolicard, C. Leforestier, E. J. Austin, *J. Chem. Phys.* **88,** 1026 (1988).
28. U.V. Riss, H.D. Meyer, *J. Phys. B* **28**, 1475 (1995).
29. U.V. Riss, H.D. Meyer, *J. Phys. B* **31**, 2279 (1998).
30. U.V. Riss, H.D. Meyer, *J. Chem. Phys.* **105,** 1409, (1996).
31. N. Rom, N. Moiseyev, *J. Chem. Phys.* 99, 7703 (1993).
32. R. Santra and L. S. Cederbaum, *J. Chem. Phys.* **115**, 6853 (2001).
33. R. A. Donnelly, *J. Chem. Phys.* **84**, 6200 (1986).
34. T. Sommerfeld, U. V. Riss, H.-D. Meyer, L. S. Cederbaum, B. Engels, and H. U. Suter, J. Phys. B **31**, 4107 (1998).
35. Y. Sajeev and S. Pal, *Mol. Phys.* **103**, 2267,(2005).
36. Y. Sajeev, R. Santra, and S. Pal, *J. Chem. Phys.* **122**, 234320 (2005).
37. Y. Sajeev, R. Santra, and S. Pal, *J. Chem. Phys*,**123**,204110 (2005).
38. R. Santra and L. S. Cederbaum, *J. Chem. Phys.* **117**, 5511 (2002).
39. S. Feuerbacher, T. Sommerfeld, R. Santra, and L. S. Cederbaum, *J. Chem. Phys.* **118**, 6188 (2003).

40. J. Cizek, *J. Chem. Phys.* **45**,4256 (1966); *Adv. Chem. Phys.* **14**, 35 (1969).
41. J. Paldus, J. Cizek, I. Shavitt, *Phys. Rev. A.* **5**, 50 (1972), J. Paldus, *J. Chem. Phys.***77**, 303 (1977), R. Offermann, W. Ey, and H. Kümmel, *Nucl. Phys.* A **273**, 349 (1976); R. Offermann, *Nucl. Phys.* A **273**, 368 (1976); W. Ey, *Nucl. Phys. A* **296**, 189 (1978).
42. D. Mukherjee and S. Pal, *Adv. Quantum. Chem.* **20**, 291 (1989).
43. U. Kaldor and M. A. Haque, *Chem. Phys. Lett.* **128**, 45 (1986); U. Kaldor and M. A. Haque, *J. Comp. Chem.* **8**, 448 (1987); D. Mukherjee, *Pramana* **12**, 1 (1979); M. A. Haque and D. Mukherjee, *J. Chem. Phys.* **80**, 5058 (1984).
44. I. Lindgren and D. Mukherjee, *Phys. Rep.* **151**, 93 (1987).
45. J. Paldus, L. Pylypow and B. Jezioroski, in *Many-body methods in Quantum Chemistry, Lecture notes in Chemistry,* edited by U. Kaldor (Springer-Verlag, 1989)Vol.**52**, p151, J. Paldus, P. Piecuch, B. Jezioroski, and L. Pylypow, in *Recent Advances in Many-Body Theories*, edited by T. L. Ainsworthy, C. E. Campbell, B. E. Clements, and E. Krotschek (Plenum, New York, 1992), Vol.**3**, p287; L. Meissner, K. Jankowski and J. Wasilewski, *Int. J. Quant. Chem.* **34**, 535 (1888).
46. C. Bloch, *Nucl. Phys.* **6**, 329 (1958).

Optical frequency standard with Sr^+: A theoretical many-body approach

Chiranjib Sur , K. V. P. Latha, Rajat K. Chaudhuri, B. P. Das
NAPP Theory Group, Indian Institute of Astrophysics, Bangalore 560 034, India
D. Mukherjee
Indian Association for the Cultivation of Science, Kolkata - 700 032, India

Abstract

Demands from several areas of science and technology have lead to a worldwide search for accurate optical clocks with an uncertainty of 1 part in 10^{18}, which is 10^3 times more accurate than the present day cesium atomic clocks based on microwave frequency regime. In this article we discuss the electric quadrupole and the hyperfine shifts in the $5s\,^2S_{1/2} \longrightarrow 4d\,^2D_{5/2}$ clock transition in Sr^+, one of the most promising candidates for next generation optical clocks. We have applied relativistic coupled cluster theory for determining the electric quadrupole moment of the $4d\,^2D_{5/2}$ state of $^{88}Sr^+$ and the magnetic dipole (A) and electric quadrupole (B) hyperfine constants for the $5s\,^2S_{1/2}$ and $4d\,^2D_{5/2}$ states which are important in the study of frequency standards with Sr^+. The effects of electron correlation which are very crucial for the accurate determination of these quantities have been discussed.

PACS number(s). : 31.15.Ar, 31.15.Dv, 32.30.Jc, 31.25.Jf, 32.10.Fn

1 Introduction

The frequencies at which atoms emit or absorb electro-magnetic radiation during a transition can be used for defining the basic unit of time [1, 2, 3]. The transitions that are extremely stable, accurately measurable and reproducible can serve as excellent frequency standards [1, 2]. The current frequency standard is based on the ground state hyperfine transition in ^{133}Cs which is in the microwave regime and has an uncertainty of one part in 10^{15} [4]. However, there is a search for even more accurate clocks in the optical regime. The uncertainty of these clocks is expected to be about 1 part in 10^{17} or 10^{18} [5]. Some of the prominent candidates that belong to this category are $^{88}Sr^+$ [6, 7], $^{199}Hg^+$ [8], $^{171}Yb^+$ [9], $^{43}Ca^+$ [10], $^{138}Ba^+$ etc. Indeed detailed studies on these ions will have to be carried out in order to determine their suitability for optical frequency standards. In a recent article [11] Gill and Margolis have discussed the merits of choosing $^{88}Sr^+$ as a candidate for an optical clock. Till recently, the most accurate measurement of an optical frequency was for the clock transition in $^{88}Sr^+$ which has an uncertainty of 3.4 parts in 10^{15} [12]. However, recently, Oskay et al. [13] have measured the optical frequency of $^{199}Hg^+$ to an accuracy of 1.5 parts in 10^{15} and further improvements are expected [14].

In this article we concentrate on strontium ion (Sr^+) which is considered to be one of the leading candidates for an ultra high precision optical clock [11]. The clock transition in this case is $5s\,^2S_{1/2} \longrightarrow 4d\,^2D_{5/2}$ and is observed by using the quantum jump technique in single trapped strontium ion. When an atom interacts with an external field, the standard frequency may be shifted from the resonant frequency. The quality of the frequency standard depends upon the accurate and precise measurement of this shift. To minimize or maintain any shift of the clock frequency, the interaction of the atom with it's surroundings must be controlled. Hence, it is important to have a good knowledge of these shifts so as to minimize them while setting up the frequency standard. Some of these shifts are the linear Zeeman shift, quadratic Zeeman shift, second-order Stark shift, hyperfine shift and electric quadrupole shift. The largest source of uncertainty in frequency shift arises from the electric quadrupole shift of the clock transition because of the interaction of atomic electric quadrupole moment with the gradient of external electric field. In this article we have applied relativistic coupled-cluster (RCC) theory, one of the most accurate atomic many-body theories to calculate the electric quadrupole moment (EQM) and the hyperfine constants for the energy levels involved in the clock transition.

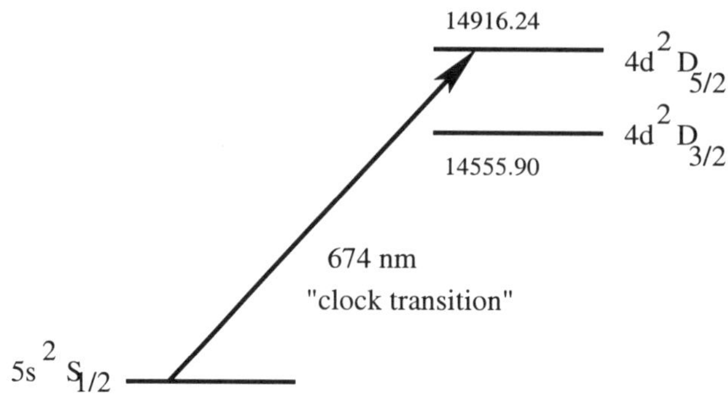

Figure 1: Clock transition in $^{88}Sr^+$. Excitation energies of the $4d\ ^2D_{3/2}$ and $4d\ ^2D_{5/2}$ levels are given in cm^{-1}.

2 Relativistic coupled-cluster theory

The relativistic and dynamical electron correlation effects can be incorporated in many-electron systems through a variety of many-body methods [15, 16, 17, 18]. The relativistic coupled cluster (RCC) method has emerged as one of the most powerful and effective tools for high precision description of electron correlations in many-electron systems [17, 18]. The RCC is an all-order non-perturbative scheme, and therefore, the higher order electron correlation effects can be incorporated more efficiently than using the order-by-order diagrammatic many-body perturbation theory (MBPT). RCC is equivalent to all order relativistic MBPT (RMBPT). The RCC results can therefore be improved by adding the important omitted diagrams with the aid of low order RMBPT. We have applied RCC theory to calculate atomic properties for several systems and more details can be obtained from Ref. [19].

Here we present a brief outline of RCC theory. We begin with N-electron Dirac-Coulomb Hamiltonian (H) which is expressed as

$$H = \sum_{i=1}^{N} \left[c\vec{\alpha}_i \cdot \vec{p}_i + \beta mc^2 + V_N(r_i) \right] + \sum_{i<j}^{N} \frac{e^2}{r_{ij}}, \tag{1}$$

with the Fermi vacuum described by the four component Dirac-Fock (DF) state $|\Phi\rangle$. The normal ordered form of the above Hamiltonian is given by

$$H_N = H - \langle\Phi|H|\Phi\rangle = \sum_{ij} \langle i|f|j\rangle \left\{a_i^\dagger a_j\right\} + \frac{1}{4} \sum_{i,j,k,l} \langle ij||kl\rangle \left\{a_i^\dagger a_j^\dagger a_l a_k\right\}, \tag{2}$$

where

$$\langle ij||kl\rangle = \langle ij|\frac{1}{r_{12}}|kl\rangle - \langle ij|\frac{1}{r_{12}}|lk\rangle. \tag{3}$$

Following Lindgren's formulation of open-shell CC [20], we express the valence universal wave operator Ω as

$$\Omega = \{\exp(\sigma)\}, \tag{4}$$

and σ being the excitation operator and curly brackets denote the normal ordering. The wave operator Ω acting on the DF reference state gives the exact correlated state. The operator σ has two parts, one corresponding to the core and the other to the valence sector denoted by T and S respectively. In the singles and double (SD) excitation approximation the excitation operator for the core sector is given by

$$T = T_1 + T_2 = \sum_{ap} \left\{a_p^\dagger a_a\right\} t_a^p + \frac{1}{4} \sum_{abpq} \left\{a_p^\dagger a_q^\dagger a_b a_a\right\} t_{ab}^{pq}, \tag{5}$$

t_a^p and t_{ab}^{pq} being the amplitudes corresponding to single and double excitations respectively. For a single valence system like Sr^+, the excitation operator for the valence sector turns out to be $\exp(S) = \{1 + S\}$ and

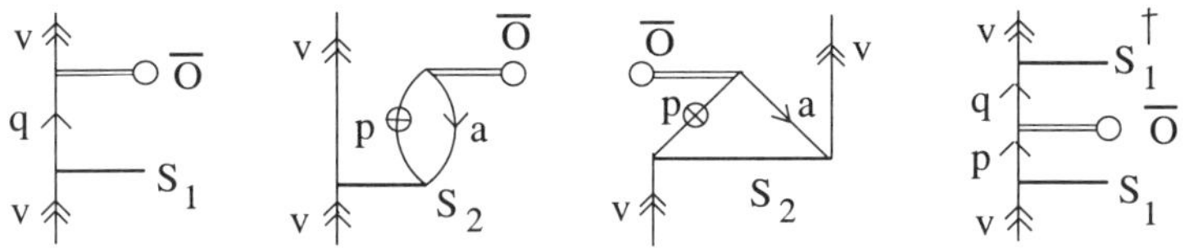

Figure 2: Some many-body diagrams representing the electric quadrupole/hyperfine interaction. Holes (occupied orbitals, labeled by a) and particles are denoted by the lines directed downward and upward respectively. The double line represents the interaction vertex. The valence (labeled by v) and virtual orbitals (labeled by $p, q, r..$) are depicted by double arrow and single arrow respectively whereas the orbitals denoted by $\oplus$ can either be valance or virtual.

$$S = S_1 + S_2 = \sum_{k \neq p} \{a_p^\dagger a_k\}\, s_k^p + \sum_{bpq} \{a_p^\dagger a_q^\dagger a_b a_k\}\, s_{kb}^{pq}\,, \tag{6}$$

where s_k^p and s_{kb}^{pq} denotes the single and double excitation amplitudes for the valence sectors respectively. In Eqs. (5) and (6) we denote the core (virtual) orbitals by $a, b, c...$ $(p, q, r...)$ respectively and k corresponds to the valence orbital.
The corresponding correlated closed shell state is then

$$|\Psi\rangle = \exp(T)\,|\Phi\rangle\,. \tag{7}$$

The exact open shell reference state is achieved by using the techniques of electron attachment. In order to add an electron to the kth virtual orbital of the N electron DF reference state we define

$$|\Phi_k^{N+1}\rangle \equiv a_k^\dagger\,|\Phi\rangle \tag{8}$$

with the particle creation operator $a_k^\dagger$. Then by using the excitation operators for both the core and valence electron the $(N+1)$ electron exact state is defined as [20]:

$$|\Psi_k^{N+1}\rangle = \exp(T)\,\{1+S\}\,|\Phi_k^{N+1}\rangle\,. \tag{9}$$

We write the expectation value of any operator O in a normalized form with respect to the exact state $|\Psi^{N+1}\rangle$ as

$$\langle O\rangle = \frac{\langle\Psi^{N+1}|\,O\,|\Psi^{N+1}\rangle}{\langle\Psi^{N+1}|\,\Psi^{N+1}\rangle} = \frac{\langle\Phi^{N+1}|\,\{1+S^\dagger\}\exp(T^\dagger)O\exp(T)\,\{1+S\}\,|\Phi^{N+1}\rangle}{\langle\Phi^{N+1}|\,\{1+S^\dagger\}\exp(T^\dagger)\exp(T)\,\{1+S\}\,|\Phi^{N+1}\rangle}\,. \tag{10}$$

For computational simplicity we store only the one-body matrix element of $\overline{O} = \exp(T^\dagger)O\exp(T)$. $\overline{O}$ may be expressed in terms of uncontracted single-particle lines [21]. The fully contracted part of $\overline{O}$ will not contribute as it cannot be *linked* with the remaining part of the numerator of the above equation. First few terms of Eq.(10) can be identified as $\overline{O}$, $\overline{O}S_1$ and $\overline{O}S_2$ which we identify as dressed Dirac-Fock (DDF), dressed pair-correlation (DPC) and dressed core-polarization (DCP) respectively. The other terms are being identified as , $S_1^\dagger\overline{O}S_1$, $S_1^\dagger\overline{O}S_2$, $S_2^\dagger\overline{O}S_1$ and $S_2^\dagger\overline{O}S_2$ and are classified as higher order effects. We use the term 'dressed' to describe the modification of the operator O due to core-core correlation effects. In Fig. 2 we replace the operator O by the dressed operator $\overline{O}$ (the dressed quadrupole/hyperfine interaction operator) which includes the core excitation effects and the respective figures are termed as 2(a) $\overline{O}S_1$, 2(b+c) $\overline{O}S_2$ and 2(d) $S_1^\dagger\overline{O}S_1$. Here 2(a) and 2(b+c) represent the DPC and DCP effects respectively. 2(b) is known as the direct and 2(c) is the exchange DPC diagram. Figure 2(d) refers to one of the higher order pair correlation effect which belongs to the set, termed as 'others'.

3 Electric quadrupole shift

The largest source of systematic frequency shift for the clock transition in Sr^+ arises from the electric quadrupole shift of the $4d\,^2D_{5/2}$ state caused by its electric quadrupole moment of that state and the interaction of the external electric field gradient present at the position of the ion. The electric quadrupole moment in the state $4d\ ^2D_{5/2}$ was measured experimentally by Barwood *et al.* at NPL [22]. Since the ground state $5s\ ^2S_{1/2}$ does not possess any electric quadrupole moment, the contribution to the quadrupole shift for the clock frequency comes only from the $4d\ ^2D_{5/2}$ state.

The interaction of the atomic quadrupole moment with the external electric-field gradient is analogous to the interaction of a nuclear quadrupole moment with the electric fields generated by the atomic electrons inside the nucleus. In the presence of the electric field, this gives rise to an energy shift by coupling with the gradient of the electric field. Thus the treatment of electric quadrupole moment is analogous to its nuclear counterpart. The quadrupole moment $\mathbf{\Theta}$ of an atomic state $|\Psi(\gamma, J, M)\rangle$ is defined as the diagonal matrix element of the quadrupole operator with the maximum value M_J, given by

$$\mathbf{\Theta} = \langle\Psi(\gamma JJ)|\,\Theta_{zz}\,|\Psi(\gamma JJ)\rangle\ . \tag{11}$$

Here γ is an additional quantum number which distinguishes the initial and final states. The electric quadrupole operator in terms of the electronic coordinates is given by

$$\Theta_{zz} = -\frac{e}{2}\sum_j \left(3z_j^2 - r_j^2\right), \tag{12}$$

where the sum is over all the electrons and z is the coordinate of the jth electron. To calculate the quantity we express the quadrupole operator in its single particle form as

$$\Theta_m^{(2)} = \sum_m q_m^{(2)}. \tag{13}$$

More details about evaluation of electric quadrupole moment using RCC theory is described in our recent paper [23]. The electric quadrupole shift is evaluated using the relation

$$\langle\Psi(\gamma JFM_F)|\,\Theta\,|\Psi(\gamma JFM_F)\rangle = \frac{-2A\left[3M_F^2 - F(F+1)\right]\langle\Psi(\gamma JF)\|\,\Theta^{(2)}\,\|\Psi(\gamma JF)\rangle}{\left[(2F+3)(2F+2)(2F+1)2F(2F-1)\right]^{1/2}} \times \mathcal{O}(\alpha,\beta) \tag{14}$$

and

$$\mathcal{O}(\alpha,\beta) = \left[(3\cos^2\beta - 1) - \epsilon(\cos^2\alpha - \sin^2\alpha)\right]. \tag{15}$$

Here γ specifies the electronic configuration of the atoms and F and M_F are the total atomic angular momentum (nuclear + electronic) and its projection; α and β are the two of the three Euler angles that take the principal-axis frame of the electric field gradient to the quantization axis and ϵ is an asymmetry parameter of the electric potential function [24].

4 Hyperfine shift

The frequency standard is based on the ^{88}Sr isotope. In addition to $^{88}Sr^+$, the odd isotope $^{87}Sr^+$ has also been proposed as a possible candidate for an optical frequency standard [25]. An experiment has been performed in NPL to measure the hyperfine structure of the $4d\ ^2D_{5/2}$ state in $^{87}Sr^+$[25]. Theoretical determination of hyperfine constants is one of most stringent tests of accuracy of the atomic wave functions near the nucleus. Also accurate predictions of hyperfine coupling constants require a precise incorporation of relativistic and correlation effects.

Unlike the even isotope ($^{88}Sr^+$) of strontium ion, $^{87}Sr^+$ has a non zero nuclear spin ($I = \frac{9}{2}$) and the $m_F = 0$ levels for both the $^2S_{1/2}$ and $^2D_{5/2}$ states are independent of the first order Zeeman shift. Here, in table 2 we present the results of our calculation of the magnetic dipole (A) hyperfine constant for the $5s\,^2S_{1/2}$ and $4d\ ^2D_{5/2}$ states and the electric quadrupole hyperfine constant (B) for the $4d\ ^2D_{5/2}$ state of $^{87}Sr^+$ and compare with the measured values. More details of our calculation can be found in Ref [26].

The hyperfine interaction is given by

$$H_{hfs} = \sum_k M^{(k)} \cdot T^{(k)}, \tag{16}$$

where $M^{(k)}$ and $T^{(k)}$ are spherical tensors of rank k, which corresponds to nuclear and electronic parts of the interaction respectively. The lowest $k = 0$ order represents the interaction of the electron with the spherical part of the nuclear charge distribution.

In the first order perturbation theory, the energy corresponding to the hyperfine interaction of the fine structure state $|JM_J\rangle$ are the expectation values of H_{hfs} such that

$$\begin{aligned} W(J) &= \langle IJFM_F | \sum_k M^{(k)} \cdot T^{(k)} | IJFM_F \rangle \\ &= \sum_k (-1)^{I+J+F} \left\{ \begin{array}{ccc} I & J & F \\ J & I & k \end{array} \right\} \langle I \| M^{(k)} \| I \rangle \langle J \| T^{(k)} \| J \rangle . \end{aligned} \tag{17}$$

Here **I** and **J** are the total angular angular momentum for the nucleus and the electron state, respectively, and $\mathbf{F} = \mathbf{I} + \mathbf{J}$ with the projection M_F.

4.1 Magnetic dipole hyperfine constant

Hyperfine effects arise due to the interaction between the various moments of the nucleus and the electrons of an atom. Nuclear spin gives rise to a nuclear magnetic dipole moment which interacts with the electrons and thus gives rise to magnetic dipole hyperfine interaction defined by the magnetic dipole hyperfine constant A. For an eigen state $|IJ\rangle$ of the Dirac-Coulomb Hamiltonian, A is defined as

$$A = \mu_N \left(\frac{\mu_I}{I} \right) \frac{\langle J \| T^{(1)} \| J \rangle}{\sqrt{J(J+1)(2J+1)}}, \tag{18}$$

where μ_I is the nuclear dipole moment defined in units of Bohr magneton μ_N. The magnetic dipole hyperfine operator $T_q^{(1)}$ which is a tensor of rank 1 can be expressed in terms of single particle rank 1 tensor operators and is given by the first order term of Eq. (17)

$$T_q^{(1)} = \sum_q t_q^{(1)} = \sum_j -ie\sqrt{\frac{8\pi}{3}} r_j^{-2} \overrightarrow{\alpha_j} \cdot \mathbf{Y}_{1q}^{(0)}(\widehat{r_j}) . \tag{19}$$

Here $\overrightarrow{\alpha}$ is the Dirac matrix and $\mathbf{Y}_{kq}^{\lambda}$ is the vector spherical harmonics. The index j refers to the j-th electron of the atom with r_j its radial distance and e is the magnitude of the electronic charge.

4.2 Electric quadrupole hyperfine constant

The second order term in the hyperfine interaction is the electric quadrupole part. The electric quadrupole hyperfine constant is defined by putting $k = 2$ in Eq. (17). The nuclear quadrupole moment is defined as

$$T_q^{(2)} = \sum_q t_q^{(2)} = \sum_j -er_j^{-3} C_q^{(2)}(\widehat{r_j}), \tag{20}$$

Here, $C_q^{(k)} = \sqrt{\frac{4\pi}{(2k+1)}} Y_{kq}$, with Y_{kq} being the spherical harmonic. Hence the electric quadrupole hyperfine constant B in terms of the nuclear quadrupole moment Q_N is

$$B = 2eQ_N \left[\frac{2J(2J-1)}{(2J+1)(2J+2)(2J+3)} \right]^{1/2} \langle J \| T^{(2)} \| J \rangle . \tag{21}$$

The corresponding shift in the energy levels are known an hyperfine shift which is expressed as

$$W_{hyp} = W_{M1} + W_{E2} = A\frac{K}{2} + \frac{B}{2} \frac{3K(K+1) - 4I(I+1)J(J+1)}{2I(2I-1)2J(2J-1)}, \tag{22}$$

where $K = F(F+1) - I(I+1) - J(J+1)$.

Table 1: Electric quadrupole moment for the $4d\ ^2D_{5/2}$ state of $^{88}Sr^+$ in units of ea_0^2. PW corresponds to our present CCSD(T) calculation and MCDF for Multi-configuration Dirac-Fock.

	PW	MCDF [27]	Experiment [22]
$4d_{5/2}$	2.94	3.02	2.6±0.3

5 Results and discussions

The occupied and the virtual orbitals used in the calculation are obtained by solving the Dirac-Fock (DF) equation for Sr^{++} for a finite Fermi nuclear distribution. These orbitals are linear combinations of Gaussian type functions on a grid [28]. The open shell coupled cluster (OSCC) method is used to construct different single valence states whose reference states correspond to adding a particle to the closed shell reference state. We use the singles-doubles and partial triples approximation, abbreviated as CCSD(T) and excitations from all the core orbitals have been considered. We have estimated the error incurred in our present work, by taking the difference between our RCC calculations with singles, doubles as well as the most important triple excitations (CCSD(T)) and only single and double excitations (CCSD).

5.1 Electric quadrupole moment

We present our results of the electric quadrupole moment for the $4d\ ^2D_{5/2}$ state of $^{88}Sr^+$ in table 1. The value of Θ in the $4d\ ^2D_{5/2}$ state measured experimentally is $(2.6 \pm 0.3)ea_0^2$ [22]. Our calculated value for the $4d\ ^2D_{5/2}$ stretched state is $(2.94 \pm 0.07)ea_0^2$, where e is the electronic charge and a_0 is the Bohr radius. We analysed our results and have found that the DDF contribution is the largest. The leading contribution to electron correlation comes from the DPC effects and the DCP effects are an order of magnitude smaller. This can be understood from the DPC diagram (Fig.2(a)) which has a valence electron in the $4d_{5/2}$ state. Hence the dominant contribution to the electric quadrupole moment arises from the overlap between virtual $d_{5/2}$ orbitals and the valence, owing to the fact that S_1 is an operator of rank 0 and the electric quadrupole matrix elements for the valence $4d_{5/2}$ and the diffuse virtual $d_{5/2}$ orbitals are substantial. On the other hand, in the DCP diagram (Fig.2(b+c)), the matrix element of the same operator could also involve the less diffuse s or p orbitals. Hence, for a property like the electric quadrupole moment, whose magnitude depends on the square of the radial distance from the nucleus, this trend seems reasonable, whereas for properties like hyperfine interaction which is sensitive to the near nuclear region, the trend is just the opposite for the d states [29]. As expected, the contribution of the DHOPC effect i.e., $S_1^\dagger \bar{O} S_1$ (Fig.2(d)) is relatively important as it involves an electric quadrupole matrix element between the valence $4d_{5/2}$ and a virtual $d_{5/2}$ orbital.

5.2 Hyperfine constants

The magnetic dipole (A) and electric quadrupole (B) hyperfine constants for the of $^{87}Sr^+$ are given in table 2 along with the calculated and experimental results. The $g_i = \mu_N \left(\frac{\mu_I}{I}\right)$ value used for the calculation is from Ref. [30]. PW corresponds to the 'present work' using CCSD(T). CC stands for coupled-cluster calculation by Martensson-Pendrill [31] and Nayak and Chaudhuri [32], DF-AO for Dirac-Fock with all order core polarization effect by Yu *et al.* and the column 'others' refers to the calculation by Yuan *et al.* using relativistic linked cluster many-body perturbation theory (RLCMBPT) [33] and by one of the authors using relativistic effective valence shell Hamiltonian method [34]. Information on A for the $4d\ ^2D_{5/2}$ state is very important in connection with optical frequency standards [12, 25]. The measured value of this quantity is 2.1743 ± 0.0014 MHz [25], whereas the previously calculated values vary from 1.07 MHz [31] and 2.507 MHz [35]. Our calculated value of A is 2.16 ± 0.02 MHz; this is the most accurate theoretical determination of A for the $4d\ ^2D_{5/2}$ state to date. For the $5s\ ^2S_{1/2}$state our calculated value of A is 997.26 ± 0.03 MHz. To analyze the result we focus on the various many-body effects contributing to the calculation of A. The most important many-body diagrams are presented in Fig. 2. We have noticed that for $5s\ ^2S_{1/2}$ state the dominant contribution is at the DDF level $\sim 72\%$. However, for the $4d\ ^2D_{5/2}$ state the DCP effect is larger and its sign is opposite that of the

Table 2: Magnetic dipole (A) hyperfine constant for the $5s\,^2S_{1/2}$ and $4d\,^2D_{5/2}$ states and the electric quadrupole hyperfine constant (B) for $4d\,^2D_{5/2}$ state of $^{87}Sr^+$ in MHz.

		PW	CC	DF-AO [35]	Others	Experiment
$5s_{1/2}$	A	-997.26	-1000 [31]	-1003.18	-987 [33]	-1000.5±1.0 [36]
			-999.89 [32]		-1005.74 [34]	-990 [37]
						-993.5 [38]
						-1000.473 673 [39]
$4d_{5/2}$	A	2.16	1.07 [31]	2.51		2.1743±0.0014 [25]
			1.87 [32]			
	B	47.8	54.4 [31]			49.11±0.06 [36]
			51.12 [32]			

DDF contribution. In their calculations Martensson [31] and Yu *et al.* [35] have found similar trends. We have seen that the higher order correlation effects contribute significantly in determining A for this state - collectively they are 60% of the total value but opposite in sign. In the earlier calculations the determination of the higher order effects was not as accurate as ours.

The calculated value for the electric quadrupole hyperfine constant (B) for the $4d\,^2D_{5/2}$ state is 47.8 ± 0.2 MHz which deviates $\sim 2.7\%$ from the central experimental value. The earlier determination of B was off by $\sim 11\%$ from the experiment. Since the other state involved in the clock transition is spherically symmetric the electric quadrupole hyperfine constant is zero and there will be no hyperfine shift due to B for the $5s\,^2S_{1/2}$state.

From the figure presented in Fig. 2 it is clear that the DPC effect involves the hyperfine interaction of a valence electron and the residual Coulomb interaction, i.e for $5s\,^2S_{1/2}$ state the hyperfine matrix element becomes $\langle [4p^6]5s_{1/2}|h_{hfs}|\,[4p^6]q\rangle$, where q can be any virtual orbital and h_{hfs} is the single-particle hyperfine operator. Since only $s_{1/2}$ and $p_{1/2}$ electrons have a sizable density in the nuclear region, the DPC effect is dominant for $5s_{1/2}$ state but not for the $4d_{5/2}$ state. On the other hand, the DCP effects represent the hyperfine interaction of a polarized core electron with any virtual electron (see Fig. 2b,c). For $4d_{5/2}$ state it is clear that the hyperfine matrix element for the DPC effect is much smaller than the DCP effect, which plays the most important role in determining the value of A. The core polarization contribution is so large for this state that it even dominates over the DDF contribution.

6 Conclusion

In conclusion, we have performed an *ab initio* calculation of the electric quadrupole moment for the $4d\,^2D_{5/2}$ state of $^{88}Sr^+$ to an accuracy of less than 2.5% using the RCC theory. This is the first application of RCC theory to determine the electric quadrupole moment (EQM) of any atom and is currently the most accurate determination of EQM for the $4d\,^2D_{5/2}$ state in Sr^+. We have also calculated the magnetic dipole (A) hyperfine constant for the $5s\,^2S_{1/2}$ and $4d\,^2D_{5/2}$ states and electric quadrupole hyperfine constant (B) for the $4d\,^2D_{5/2}$ state of $^{87}Sr^+$. Evaluation of correlation effects to all orders as well as the inclusion of the dominant triple excitations in our calculation was crucial in achieving this accuracy. The magnitude of electric quadrupole moment depends on the square of the radial distance from the nucleus, whereas properties like hyperfine interaction are sensitive to the near nuclear region. The accurate determination of quantities like electric quadrupole moment and hyperfine constants establish the fact that RCC is very powerful and efficient method for determining atomic properties near the nuclear region as well as at large distances from the nucleus. Our result will lead to a better quantitative understanding of the electric quadrupole shift of the resonance frequency of the clock transition in Sr^+.

Acknowledgments : Financial support from the BRNS, DAE for project no. 2002/37/12/BRNS is gratefully acknowledged. The computations are carried out in our group's Xeon and the Opteron computing cluster at IIA.

References

[1] J.C. Bergquist, S. R. Jefferts, and D. J. Wineland, *Physics Today*, **54**, No.3, 37, (2001).

[2] W. M. Itano, Proc. *IEEE*, **79**, 936-942 (1991).

[3] W. M. Itano and N.F. Ramsey, *Sci. Am.*, **269**, 56-65 (1993).

[4] `http://tf.nist.gov/cesium/atomichistory.htm`

[5] L. Hollberg *et al.*, *J. Phys. B*, **38**, S469-S495 (2005).

[6] J. E. Bernard *et al.*, *Phys. Rev. Lett.*, **82**, 3228 (1999).

[7] H. S. Margolis *et al.*, *Phys. Rev. A*, **67**, 032501 (2003).

[8] R. Rafac *et al.*, *Phys. Rev. Lett.*, **85**, 2462 (2000).

[9] J. Stenger *et al.*, *Opt. Lett.*, **26**, 1589 (2001).

[10] C. Champenois *et al.*, *Phys. Lett. A*, **331**, 298 (2004).

[11] P. Gill and H. Margolis, *Physics World*, 35, May (2005).

[12] H. S. Margolis *et al.*, *Science*, **306**, 1355 (2004).

[13] W. H. Oskay, W. M. Itano and J. C. Bergquist, *Phys. Rev. Lett.*, **94**, 163001 (2005).

[14] W. M. Itano, NIST, *Private Communications*.

[15] I. P Grant, *Phys. Scr.*, **21**, 443 (1980).

[16] S. A. Blundell, W. R. Johnson and J. Sapirstein, *Phys. Rev. A*, **43**, 3407 (1991).

[17] D. Mukherjee and S. Pal, *Adv. Quant. Chem.*, **20**, 281 (1989) and the references therein.

[18] U. Kaldor, *Lecture Notes in Physics, Microscopic Quantum many-body theories and their applications*, p.71, Eds. J. Navarro and A. Polls, Springer-Verlag-Berlin, Heidelberg and New York (1998) and references therein.

[19] B. P. Das *et al.*, *J. Theor. and Comp. Chem.*, **4**, 1 (2005) and references therein.

[20] I. Lindgren and J. Morrison, *Atomic Many-Body Theory* (Springer, Berlin) 1985.

[21] G. Gopakumar *et al.*, *Phys. Rev. A.*, **64**, 032502 (2001).

[22] G. P. Barwood *et al.*, *Phys. Rev. Lett.* **93**, 133001 (2004).

[23] C. Sur *et al.*, **physics/0507199** (To be appeared in *Phys. Rev. Lett.* 2006).

[24] P. Dube *et al.*, *Phys. Rev. Lett.* **95**, 033001 (2005).

[25] G. P. Barwood, K. Gao, P. Gill, G. Huang and H. A. Klein, *Phys. Rev. A*, **67**, 013402 (2003).

[26] C. Sur *et al.*, (Submitted to *J. Phys. B* 2006).

[27] W. Itano, *Phys. Rev. A*, **73**, 022510 (2006).

[28] R. K. Chaudhuri, P. K. Panda and B. P. Das, *Phys. Rev. A*, **59**, 1187 (1999).

[29] C. Sur, B. K. Sahoo, R. K. Chaudhuri, B. P. Das and D. Mukherjee, *Eur. Phys. J. D*, **32**, 25 (2005).

[30] `http://www.webelements.com`

[31] A. M. Martensson-Pendrill, *J. Phys. B*, **35**, 917 (2002).

[32] M. K. Nayak and R. K. Chaudhuri, *Eur. Phys. J. D*, **37**, 171 (2006).

[33] X Yuan *et al.*, *Phys. Rev. A.*, **52**, 197 (1995).

[34] R. K. Chaudhuri, K. Freed, *J. Chem Phys.*, **122**, 204111 (2005).

[35] K. Yu, L. Wu, B. Gou and T. Shi, *Phys. Rev. A*, **70**, 012506 (2004).

[36] F. Buchinger *et al.*, *Phys. Rev. A*, **41**, 2883 (1990).

[37] R. Beignang, W. Makat, A. Timmermann and P. J. West, *Phys. Rev. Lett.*, **51**, 771 (1983).

[38] K. T. Lu, J-Q Sun and R. Beignang, *Phys. Rev. A*, **37**, 2220 (1988).

[39] H. Sunaoshi *et al.*, *Hyperfine Interact.*, **78**, 241(1993).

Fast Heavy Ion Collisions with H_2 Molecules And Young Type Interference

Lokesh C. Tribedi
Tata Institute of Fundamental Research, Homi Bhabha Road, Colaba, Mumbai 400005

Deepankar Misra
Tata Institute of Fundamental Research, Homi Bhabha Road, Colaba, Mumbai 400005

We have investigated the Young type interference effect in electron emission spectrum from molecular H_2 in Coulomb ionization induced by fast heavy ions. The details of the derivation of the oscillations due to this process from the electron double differential distributions are discussed. In addition, we have explored the effect of Compton profile, on such interference oscillations, which gives rise a double-peak structure in case of low energy collisions for which the binary encounter peak starts overlapping with soft collision electrons. The measured DDCS as well as the interference patterns are compared with molecular CDW-EIS models.

1. Introduction

The low energy electrons emitted in heavy ion atomic collisions provide crucial information on the various mechanisms responsible for the ion-atom or ion-molecule ionization [1-4]. It has been demonstrated that double differential cross sections (DDCS) of electron emission in terms of emission angle (θ) and electron energies (ε) (i.e $d^2\sigma / d\Omega\, d\varepsilon$) can be very useful tool to investigate some of the salient features of the ion-atom collisions and can provide a stringent test to the theoretical methods. Several experimental investigations are carried out to explore this aspect in more details. The important features of such electron DDCS spectrum are the soft collision electrons (SE), electron capture into continuum (ECC) cusp and binary encounter (BE) peak. The ECC cusp is observed at zero degree and is composed of electrons emitted with a velocity (v_e) equal to the projectile velocity (v_p) and arises due to the projectile-electron interaction. The first Born (B1) calculations can not reproduce this peak since in this model such interaction is treated only as a perturbation which produces the transition of the active electron from a bound state to a continuum state. The soft collision electrons are the lowest energy electrons ($\varepsilon = v_e^2/2 \rightarrow 0$) which are mainly produced in large impact parameter collisions. Although the low energy electrons compose the bulk of the cross sections, it is very difficult to detect these low energy electrons as their trajectories get very much affected by the residual electric and magnetic fields present at the interaction region. The ionized electron moves in the presence of two moving sources of Coulomb potentials, which due to their long range nature, can distort the initial and final state wave functions even when the two centres are far away. Such two center effect leaves its signature in the strong forward focussing of the electron emitted in fast ion atom collisions [4-14]. The BE electrons are produced due to the headon collisions between the projectile and the target electrons in case of zero degree electron emission [4]. In the projectile frame the BE electrons can be viewed as due to 180° Rutherford scattering of the electrons from the bare projectile nucleus and, therefore, show up in the DDCS spectrum as a broad peak centred around electron velocity $v_e \cong 2v\cos\theta$. The width of the BE peak arises due to the Compton profile of the electrons in the target initial state. The separation of the soft and hard collision branches in the recoil-ion longitudinal momentum spectrum corresponding to the soft collision and binary encounter mechanisms was investigated in Refs. [12,13]. In case of the soft electrons although, the projectile velocity considered here is much larger than the velocity of the electron in the atom, ($v_p/v_e \sim 11.5$ a.u.) it is seen that the first Born calculation (B1) fails to explain the energy and angular distributions of the ionized electrons. This is a consequence of the fact that, B1 accounts only for the target center effects and doesn't consider the effect of the receding projectile after the electron has been ionized. In order to account for the two center effects, the CDW-EIS (the continuum distorted wave eikonal initial state) model has been developed by Crothers et al [15], extended and fine tuned by Fainstein and coworkers [16-20] over the period of time and has been tested by experimental data [5,7,8,10,11,12,14].

Low energy electron emission spectrum from the simplest diatomic molecule H_2 in fast ion collisions manifests yet another important aspect of ion-atom ionization with a fast projectile. Since the two indistinguishable H-atoms may be considered as coherent sources of electrons, their contributions to electron emission add coherently thereby giving an interference effect. The electron emission from H_2 may be closely related to well known Young's two-slit experiment which provided the crucial input to the development of the quantum mechanics. How the indistinguishability of the two atoms play a role in the particle induced ionization or other processes remains to be explored. Although this effect has been known, for many years, to exist in the case of photo-ionization [21,22,23] and in energy loss of H_2 in solids [24]. It is only very recently, that the interference effect has been observed in the electron spectra in heavy ion induced ionization of H_2 using 60 MeV/u Kr 34+ [25,26]. It was predicted that high velocity ions, for which the dipole term contributes largely to the DDCS, were necessarily required in order to observe such effect [25]. Following the recent observations of such interference in electron emission from H_2 in fast heavy ion collisions [25,26,27] the interest in the Young type interference has been renewed. The various aspects of the interference effect in fast [28,29,30,31,32] and slow ion [33] collisions are being investigated both theoretically and experimentally.

The amplitude of oscillation (due to interference), being quite small, is difficult to be observed in such DDCS spectrum owing to its steep dependence on the electron energy. To enhance the visibility of the oscillatory structure in the DDCS spectrum of H_2 target, it is necessary to divide it by twice the DDCS of atomic H . The oscillatory variation of the derived ratio R around 1.0 then signifies the interference effect. The interference effect manifests itself as oscillations in the ratio of the doubly differential cross sections (DDCS) for molecular and atomic hydrogen targets. However, experiments using atomic H are rare due to experimental difficulties. To circumvent this problem theoretical DDCS for atomic, or effective atomic H have been employed [25,27]. In such case the derived oscillations depend very much on how accurately the theoretical model reproduces the DDCS for atomic H itself. Also, the shapes of the oscillations are very sensitive to the parameters chosen for the calculations, such as, the effective atomic number Z_{eff} of H. It was therefore suggested by Misra et al. [26] that for the determination of fully experimental interference effects complementary measurements of the DDCS for H and H_2 are required. Oscillations in the DDCS ratios were then observed by Tribedi and coworkers in their early works [10] and also recently [26] in such experiment. In the theoretical front, Galassi *et al.* [30] have recently modified the CDW-EIS model for H_2 molecule to explain the interference effect. The main feature of this model is to represent the initial bound state by a two-center molecular wavefunction. Within the impact parameter approximation the transition amplitude reduces to a coherent sum of atomic transition amplitudes for each molecular center.

In this article we give some examples of the DDCS measured for the forward and backward angles in order to study the two center effect. Then we show examples of interference effect for various backward and forward electron emission angles. In low energy collisions since the binary encounter electrons have a substantial overlap with the soft collision electrons for large forward angles (such as for 60^0 or 75^0), oscillatory structures due to interference get influenced by the difference in the Compton profiles of H and H_2. We show that, a double peak structure with a valley in between arises in the DDCS-ratio due to this effect.

2. Experimental Details

The experimental set up and measurement techniques are standard and can be found in several publications. We will describe, in brief, the set up used in recent electron spectroscopic studies in TIFR. Bare C ions at different velociries (v=10-15 a.u.) were obtained from the BARC-TIFR Pelletron accelerator at TIFR, Mumbai. The energy and charge state selected fast ion beam was collimated by three sets of four-jaw-slits arrangements and was made to pass through another aperture of diameter 3 mm before it collides with the target gas. This second aperture was also used to prevent the scattered beam and the secondary electrons from entering the chamber. The current on the aperture was read separately and made negligible by reducing the beam dimension by the four-jaw-slits. This was necessary in order to reduce the background electron counts arising from slit scattering. Two layers of μ-metal shield (of thickness about 0.3 mm) inside the chamber were enough to reduce the stray magnetic field below 10 mG in the region where the electrons travel before entering the analyzer.

The electrons emitted in ionization of target were energy analyzed with the help of a hemispherical electrostatic analyzer before they were detected by a channel electron multiplier (CEM). The analyzer is made of two hemispheres of diameter 50 and 70 mm made from oxygen free high conductivity copper. The spherical surfaces were coated with carbon soot to reduce the secondary electron production from the copper surface due to the electron bombardment. Before entering the analyzer the electrons had to pass through a collimator made of a copper tube with two rectangular grounded apertures one on each end. These two apertures of widths 4 and 3 mm mainly define the effective path-length solid-angle integral. Additional apertures at entrance and the exit of the analyzer were biased with a small voltage V_0 in order to pre-accelerate the electrons entering the analyzer. It was found that V_0 = +5 V was enough to improve the collection efficiency of the low energy electrons. The resolution of the spectrometer was about 5%. The energy-analyzed electrons were detected by a channel electron multiplier mounted on the exit of the analyzer. The cone of the CEM was biased at +100V to help the low energy electrons reach the detector.

Earlier measurements have shown that with this bias the efficiency of the CEM is constant within 4% in the present energy range. The spectrometer could be rotated between 20° and 160° and the electrons were detected at 10 to 12 different angles at an interval of about 15°. The data was collected in fine energy steps between 1 to 500 eV and in some cases up to a few keV. . Hydrogen gas was flooded inside the chamber through a 6 mm hole from one of the side ports. The gas pressure was kept very low i.e., ~ 0.1 mTorr for $E_e < 50$ eV and 0.3 mTorr for higher energy (E_e) electrons. The chamber base pressure was typically ~ 10^{-8} mTorr. A PC based data acquisition system was developed in house which was equipped with the automation and control system for driving the HV power supplies to bias the inner and outer hemispheres with suitable voltage for detecting definite energy electrons.

3. Theoretical models

Under the frame work of Born approximation, the cross section for ion induced electron emission from two centers separated by a distance d (d >> a, atomic dimensions) is proportional to the square of the Born matrix element:

$$\frac{d^3\sigma_{H_2}}{dq d\Omega_e d\varepsilon_e} \sim \left|\langle\varphi_k\right|e^{iq.r}\left|\varphi_0\rangle\right|^2, \tag{1}$$

where r is the coordinate of the active electron and q is the momentum transferred to the system; $d\Omega_e$ and ε_e being the solid angle and the energy of the outgoing electron, respectively. The initial wave function, φ_o, describes the ground state wave function of the two center system, in this case it refers to the ground state of H_2 molecule. The final state wave function φ_k represents the outgoing electron with momentum k. If the initial wave function, $\varphi_0 = \frac{1}{N}\left[\varphi_{1s}(\vec{r}) + \varphi_{1s}(\vec{r} - \vec{d})\right]$ is approximated by the normalized linear combination of the 1s orbital of two H atoms separated by a distance d, following Messiah's formalism [35] for scattering from two identical centers, we can write the cross section for electron emission from two centers as:

$$\frac{d^3\sigma_{H_2}}{dqd\Omega_e d\varepsilon_e} = \frac{d^3\sigma_{2H}}{dqd\Omega_e d\varepsilon_e}\{1+\cos[(\vec{k}-\vec{q}).\vec{d})]\}, \qquad (2)$$

The cross section term in the right side of Eq.2 corresponds to the electron emission from two independent H atom centers, denoted by the label 2H. The bracketed term represents the interference caused by the coherent emission of electrons from the two centers separated by a distance, d (the average inter-nuclear distance of H_2, taken to be 1.41 a.u.).

Since in most of the experiments one observes the electron emission form a randomly oriented sample of molecules, it is necessary to compare the experimental results with the cross sections which are averaged over the random orientation of the inter-nuclear axis. The above expression can be easily averaged over all molecular orientations leading to the expression;

$$\frac{d^3\sigma_{H_2}}{dqd\Omega_e d\varepsilon_e} = \frac{d^3\sigma_{2H}}{dqd\Omega_e d\varepsilon_e}\left[1+\frac{\sin\left(\left|\vec{k}-\vec{q}\right|d\right)}{\left|\vec{k}-\vec{q}\right|d}\right]. \qquad (3)$$

From the above equation, it is evident that the effect of interference is still preserved even after the averaging over the random orientations. Once the averaging has been done, in order to compare the results with the experimental double differential cross sections (DDCS), integration over the momentum transfer, q has to be performed. This can be done by following the pioneering work by Bethe. The cross section is split into two parts, namely the dipole part which arise from soft ("gentle") collisions and the binary part, which is a result of violent collision of projectile ion with the target electrons. The dipole part of the spectrum has maximum contribution near the minimum momentum transfer $q_{min} = \Delta\tilde{\varepsilon} v_p$ and for relatively faster collisions, v_p ~10-12 a.u. , $q \approx 0$. Hence a peaking approximation can be applied to the dipole part of the cross section. A similar peaking approximation can also be applied to the binary part by setting $k\text{-}q = p_i$, the mean initial momentum of the bound electron. Under the above two assumptions, Eq. 3 reduces to [25];

$$\frac{d^2\sigma_{H_2}}{d\Omega_e d\varepsilon_e} = \frac{d^2\sigma_{2H}^{dip}}{d\Omega_e d\varepsilon_e}\left[1+\frac{\sin(kd)}{kd}\right]+\frac{d^2\sigma_{2H}^{bin}}{d\Omega_e d\varepsilon_e}\left[1+\frac{\sin(p_i d)}{p_i d}\right] \tag{4}$$

The low energy region of spectrum is mainly dominated by dipole part of the cross section, and the higher energy side is dominated by the binary part.

Assuming p_i to be a constant, it is evident from Eq. 4, that one full oscillation in the cross section takes place for the product $kd = \pi$, giving $0 < k < 4.3$ a.u., and since $k = \sqrt{2\varepsilon}$, this range of k corresponds to a range in ε between 0 and 250 eV. Hence the interference structure is mainly seen in the dipole dominated part of the electron emission cross section, for which it was believed that [25] higher velocity collisions (v_p~ 50 a.u.) are preferable. However, it has now been shown [26] that the interference has a signature even at relatively low collision velocities (v_p~6 a.u.). A careful inspection of Eq. 4 reveals the fact that interference oscillations don't have an angular dependence. This is due to the fact that the information is lost when the integration over q is performed under peaking approximation. However, an angular distribution study of the effect shows a difference in the oscillation frequency between the angles. It was later noted by Nagy et. al. [29], that the integration over the momentum transfer q leaves behind the angular dependence through the parallel component of the momentum transfer, which in turn is related to the parallel component of the outgoing electron's momentum $k_{||} = k \cos\theta$. The model proposed in [29] could only explain the data for the forward electron emission processes. However, the model completely fails to explain the unusual angular dependence of the backward electron emission. Recently, Galassi et.al. [30,34] have developed an *ab initio* theoretical calculation based on the Molecular Continuum Distorted Wave Eikonal Initial State (MCDW-EIS) approximation. This model calculation is an extension to the earlier atomic CDW-EIS calculation in which the initial wave function is constructed by the normalized linear superposition of the 1s atomic orbital of H and the experimental binding energy for single ionization of H_2 has been used. The calculation of effective atomic no Z_{eff} has been found out from the minimization of the ground state energy of H_2. Under the impact parameter approximation, the resultant electron emission amplitude reduces to the coherent sum of two amplitudes corresponding to the independent atomic amplitudes. The model calculation also shows an angular frequency variation that is not proportional to the parallel component of the momentum; rather it shows a higher oscillation frequency for the backward emission processes showing a better agreement with the experimental results.

4. Results and Discussions

4.1 *Electron DDCS*

Fig. 1 shows the measured energy distribution of DDCS for electron emission for 6 MeV/u C^{6+} + H_2 collision along with the CDW-EIS predictions. The data presented for the four forward angles shows a decrease of cross section over several orders of magnitude as the electron ejection energy is increased. The agreement with the model calculation is excellent. A small disagreement can be noticed above 200-300 eV for 45^0 and 60^0. At low energies, for a few eV, the cross section reaches a maximum due to the contribution of the soft electron emission process. The structure at the higher energy side of the DDCS plot for

60^0 emission angle is due to the binary encounter (BE) electrons. The peak position

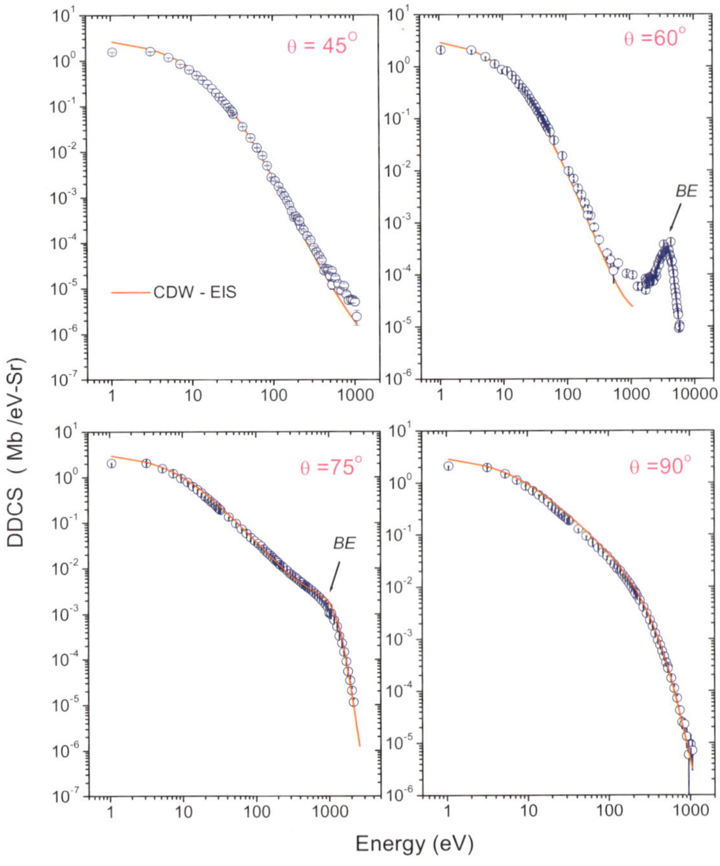

Figure 1. Electron DDCS (double differential cross sections) as a function of electron energy. The electrons are emitted in ionization of H_2 in collisions with of 72 MeV bare Carbon ions. The data in four panels are shown for four different emission angles (θ) : 45^0, 60^0,75^0 and 90^0. The BE denotes the position of the binary encounter peaks.

(E_{BE}) of the BE peak is proportional to the impact energy and varies as $\cos^2\theta$ for a given impact energy (E_p), θ being the angle of emission [4]:.

$$E_{BE} = 4t\cos^2\theta - I,$$

where

$$t = 548.4\left(E_p / M_p\right)(MeV/u),$$

where, t is the cusp energy in eV and I is the ionization potential of the 1s electron, E_p, M_p being the energy (in MeV) and the mass (in a.m.u.) of the projectile ions. As can be seen from Fig.1 and from above equation that the BE peak shifts towards the lower energy side as the angle of emission is increased. It can also be noted from Fig.1 that, as we move from 60^0 to 90^0 through 75^0, the BE peak starts to merge with the low energy continuum part of the spectrum and finally at 90^0 the peak completely disappears contributing to the whole energy range of the spectrum. Therefore, the binary encounter electrons have a substantial contribution, at the lower energy end of the spectrum, for forward emission angles close to 90^0 and at lower impact energies. The width of the BE peak is a manifestation of the Compton profile (CP) of the 1s electron of the target atom. The BE process can be regarded as a Rutherford back scattering of the target electrons when viewed from rest frame of the projectile. For bare projectiles, DDCS for the BE electron production is given by [4]:

$$\left(\frac{d^2\sigma_{H_2}}{d\Omega_e d\varepsilon_e}\right)_{BE} = \left(\frac{J(Q)}{V_p + Q/m_e}\right)\left(\frac{d\sigma}{d\Omega_e}\right)_{Ruth}$$

where $[ds/dE]_{Ruth}$ is the Rutherford scattering cross section in the rest frame of the projectile, and V_p is the lab frame projectile velocity. J(Q) represents the

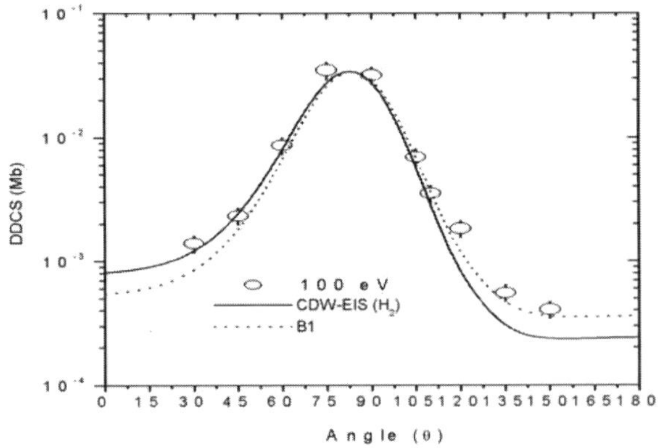

Figure 2. Angular distribution of DDCS of electrons with energy 100 eV emitted in collision of 72 MeV bare carbon ions with H_2. The solid and the dashed lines are the predictions of molecular CDW-EIS and the first Born (B1) calculations.

Compton profile of the target electrons where the component of the target electron's momentum projected along the beam axis is given by $Q=[2m_e(E\_\{proj\}+I)]^{1/2}- m_eV_p,$ with E_{proj} is the outgoing electron's energy in the projectile frame.

A typical angular distribution is shown in Fig. 2 for emitted electron energy ε=100 eV . The distribution peaks at around 80^0 and is well reproduced by the molecular CDW-EIS calculations. However one can see a disagreement for the large backward angles. The discrepancy with the B1 calculation is obvious as far as the angular distribution and the magnitudes are concerned.

A large forward backward asymmetry can be noticed which is a signature of the so called two center effect (TCE) e.g., $\sigma(30^0) / \sigma(150^0) \sim 3.3$. The CDW-EIS predicts even larger value (~5.0) for this ratio indicating even larger asymmetry.

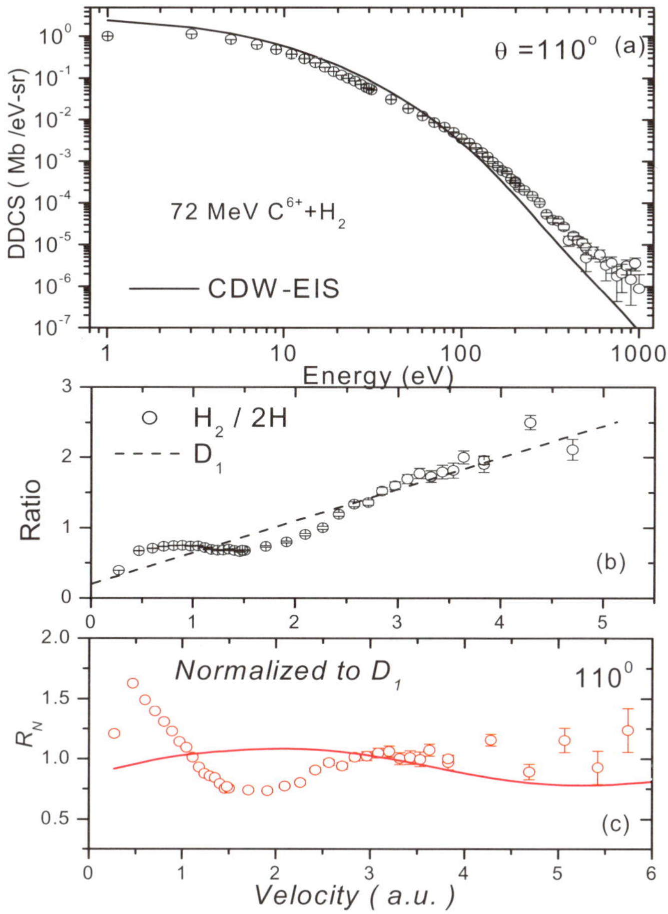

Figure 3. (a) DDCS for the energy distribution of electrons emitted in collision of 72 MeV bare Carbon ions with H_2 at 110^0. (b) The DDCS ratio i.e. DDCS (H_2)/DDCS(2H). The DDCS (2H) is taken from CDW-EIS calculations. The dashed line D_1 is fitted straightline through the ratio data. (c)

The normalized ratios derived by dividing the ratios (in (b)) by the line D_1. The solid line in (c) is the predictions of the molecular distorted wave calculation (CDWEIS).

The figure 3 (a) shows the electron double differential cross section, (DDCS) for the ionization of H_2 at an emission angle of 110^0 along with the molecular CDW-EIS calculations by Galassi *et al.,* [29,34]. The DDCS varies over several orders of magnitude in an energy interval of few hundred eV and reaches a maximum value for very low energy electrons. The overall agreement with the CDWEIS model can be termed excellent considering the wide variation of the cross sections in the given energy range. A very good agreement with the model is achieved for the low energy electrons whereas a noticeable discrepancy can be seen for large emission energies i.e. above 200 eV. This indicates the failure of the model in explaining the mechanism of the backward emission process especially at higher energy side of the spectrum.

This conclusion gets even more support from the spectrum for largest backward angle (i.e. 150^0) as shown in Fig 4(a). for which the model deviates from the data below and above 50 eV.

4.2. *DDCS ratio and interference*

Since the energy dependence of the DDCS varies over several orders of magnitude over an energy interval of 200 eV, it becomes very difficult to notice very small effects like interference in the DDCS spectrum hence to increase the visibility of the effect; the molecular cross sections are divided by the corresponding atomic cross sections.

In Fig. 3 (b) and (c), we show the ratio of the experimentally measured DDCS of H_2 to the theoretically calculated DDCS of H for 110^0 emission angle (open circles). The dashed line (D_1) in Fig. 3 is a straight line fitted through the data points. It is evident from this figure that, the cross section ratio goes through an oscillation around the fitted straight line. This oscillation can be attributed to the Young type interference effect. This oscillation can be shown to be around a horizontal line by normalizing the data points by the fitted straight line as explained in [25,26] i.e. by dividing the ratio data by the fitted straight line. The model predicts an oscillation which has the same period s observed but differs in the magnitude and the phase.

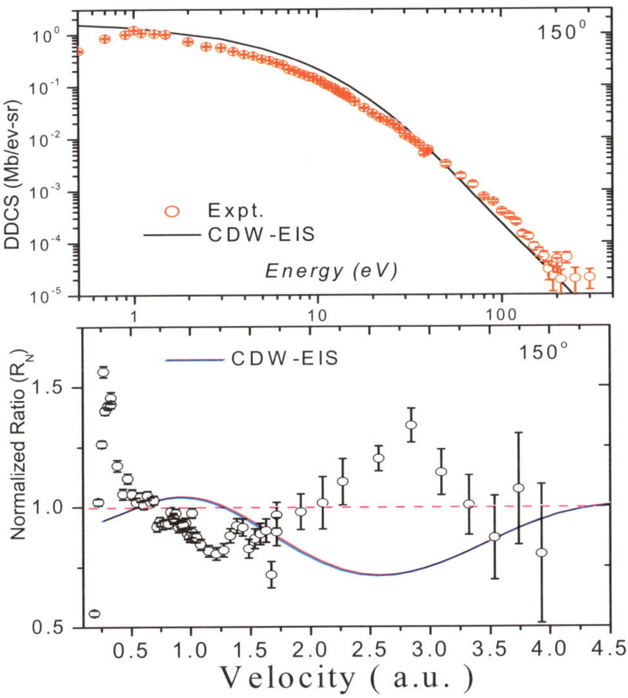

Figure 4. Electron DDCS measured for 72 MeV C^{6+} $+H_2$ (upper panel) and interference oscillation (lower panel) at backward emission angle 150^0 [26]. The solid line in both the panels are the molecular CDW-EIS calculations.

4.3. *Effect of Compton profiles*

The open circles in fig.4 correspond to the cross section ratios for 75^0 emission angle for which the BE peak is around 8.6 a.u. and has an overlap with the soft electrons. The solid curve represents the full molecular CDW-EIS calculation including interference effect and difference in Compton profiles between H_2 and H. The dashed curve is CDW-EIS calculation using an effective Z model i.e. replacing H_2 by an atomic target with equivalent binding energy. Therefore it does not include the interference effect, but includes the mismatch in the

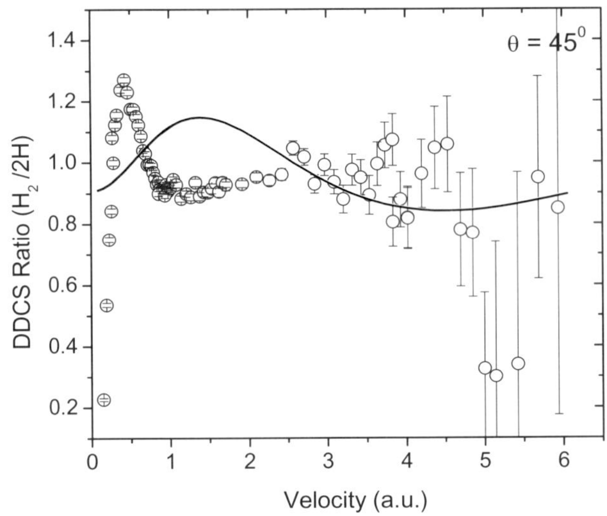

Figure 5. The DDCS ratio (H_2/2H) showing interference effect at forward angle 45°, along with CDW-EIS prediction (solid line).

Compton profiles of H_2 and H alone. It can be seen from fig. 4, that the calculation including the interference effect explains qualitatively the data having a two peak structure with a valley in between, whereas the atomic type effective-Z calculation involving only the mismatch of the Compton profiles does not reproduce this qualitative behaviour. It has been shown explicitly that the inclusion of interference effect as well as the Compto profile is required to reproduce the experimental structures [36,37].

In the later case a double peak structure with a valley in between has been observed in the DDCS-ratio spectrum which could be explained qualitatively in terms the interplay between interference process and the difference in the Compton profiles of H_2 and H [36,37].

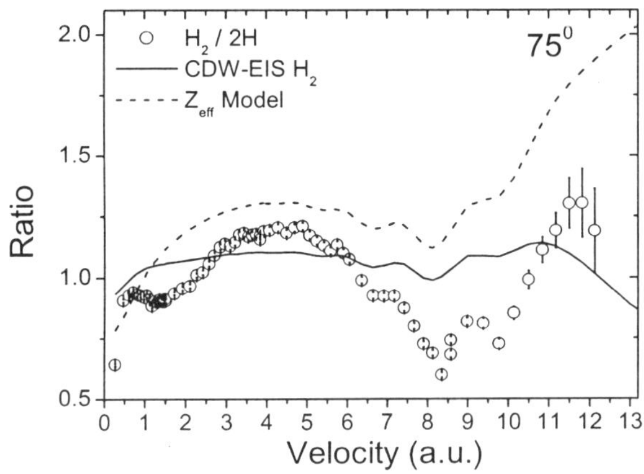

Figure 6. DDCS ratios of H_2 and H for 75^0 [26] in collisions with 72 MeV C^{6+} ; Solid line: full molecular CDW-EIS calculation [30] including interference effect; Dashed line: an effective atomic type CDW-EIS (Z_{eff}=1.19) calculation (see text).

5 Conclusions

We have studied the interference effect in case of H_2 for backward electron emission as well as for large forward angles where the binary encounter electrons interfere with the soft collision electrons. It has been shown that the DDCS for all the angles are reproduced very well by the molecular CDW-EIS calculations with some small deviations for the large back ward angles. The ratios of the electron DDCS of H_2 and 2H show an oscillation having a period of about 4 a.u. of electron velocity and are due to Young type interference. The interference patterns are more clearly visible for backward angles and have higher frequency than that for forward angles. Also the effect of Compton profile has been shown to influence the shape of the forward angle providing an oscillatory structure with a double hump. In general, the molecular CDW-EIS calculations reproduce the oscillatory structure, but with a different phase and magnitude.

6 Acknowledgement

The authors thank all the members of the group Umesh Kadhane, Yeshpal Singh, Aditya Kelkar, K.V. Thulasiram aand W. A. Fernandes who have contributed substantially to the experimental work as well as P. D. Fainstein for theoretical support. The BARC-TIFR Pelletron accelerator facility staff for the smooth running of the machine during the experiments. Funding from TIFR and the DST Swarnajayanti fellowships are gratefully acknowledged.

7 References

1. G.B. Crooks and M.E. Rudd, Phys. Rev. Lett. **25**, 1599 (1970).
2. Electron emission in heavy ion-atom collision, N. Stolterfoht, R.D. DuBois, R.D. Riverolla, Springer series on Atoms and Plasmas, 1997.
3. K.G. Harrison and M. Lucas, Phys. Lett. A **33**, 149 (1970).
4. D.H. Lee, P. Richard, T.J.M. Zouros, J.M. Sanders, J.L. Shinpaugh, and H. Hidmi, Phys. Rev. A 41, 4816 (1990).
5. S. Suarez, C. Garibotti, W. Meckbach, and G. Bernardi, Phys. Rev. Lett. 70, 418 (1993).
6. M. W. Gealy, G.W. Kerby III, Y.-Y. Hsu, and M.E. Rudd, Phys. Rev. A51, 2247 (1995).

7. J.O.P. Pedersen, P. Hvelplund, A. Petersen and P. Fainstein, J. Phys. B 24, 4001 (1991).

8. N. Stolterfoht, H. Platten, G. Schiwietz, D. Schneider, L. Guly\'as, P.D. Fainstein and A. Salin, Phys. Rev. A 52, 3796 (1995).

9. Lokesh C. Tribedi, P. Richard, D. Ling, Y.D. Wang, C.D. Lin, R. Moshammer, G.W. Kerby III, M.W. Gealy and M.E. Rudd, Phys. Rev. A 54, 2154 (1996).

10. L. C. Tribedi et al., J. Phys. B 31, L369 (1998).

11. Lokesh C. Tribedi, P. Richard, Y. D. Wang, C. D. Lin, R. E. Olson and L. Gulyas, Phys. Rev. A 58, 3626 (1998).

12. Lokesh C. Tribedi, P. Richard, Y. D. Wang, C. D. Lin and R. E. Olson, Phys. Rev. Lett. 77 3767 (1996).

13. Y.D. Wang, Lokesh C. Tribedi, P. Richard, C.L. Cocke, V.D. Rodriguez, and C.D. Lin, J. Phys. B 29, L203 (1996).

14. L.C. Tribedi, P. Richard, Y. D. Wang, C. D. Lin and L. Gulyas and M. E. Rudd, Phys. Rev. A 58, 3619 (1998).

15. D.S.F. Crothers and J.F. McCann, J. Phys. B16, 3229 (1983).

16. P.D. Fainstein, V.H. Ponce and R.D. Rivarola, J. Phys. B21, 287 (1988).

17. P.D. Fainstein, V.H. Ponce and R.D. Rivarola, J. Phys. B24 3091 (1991)

18. P.D. Fainstein, L. Gulyas, and A. Salin, Phys. Rev. A53, 3243 (1996).

19. P. D. Fainstein, L. Gulyás, F. Martin, and A. Salin, Phys. Rev. A 53, 3243 (1996).

20. L. Gulyas, P. D. Fainstein, and A. Salin, J. Phys. B 28, 245. (1995)

21. H. D. Cohen and U. Fano, Phys. Rev 150, 30 (1966).

22. M. Walter and J. S. Briggs, J. Phys. B 32, 2487 (1999).

23. R. Dorner et al., Phys. Rev. Lett., 81 5776 (1998).

24. N.R. Arista Nucl. Instr. Methods Phys. Res. B, 164, 108 (2000).

25. N. Stolterfoht et al., Phys. Rev. Lett., 87 023201 (2001).

26. Deepankar Misra, U. Kadhane, Y.P. Singh, L. C. Tribedi, P.D. Fainstein and P. Richard Phys. Rev. Lett. 92, 153201 (2004).

27. N. Stolterfoht et al., Phys. Rev. A. 67, 030702(R) (2003).

28. L. Sarkadi, J. Phys B 36, 2153 (2003)

29. L. Nagy, L. Kocbach, K. P\'ora, and J.P. Hansen J. Phys B 36, 2153 (2003)

30. M. E. Galassi, R. D. Rivarola, P. D. Fainstein and N. Stolterfoht, Phys. Rev A 66, 052705 (2002).

31. S. Hossain, A. L. Landers, N. Stolterfoht and J. A. Tanis, Phys. Rev. A 72, 010701(R) (2005).

32. A. Landers et al. Phys. Rev A 70, 042702 (2004)

33. F. Fremont et al. Phys. Rev. A 72, 050704 (2005)

34. M. E. Galassi, R. D. Rivarola and P. D. Fainstein, Phys. Rev. A 70, 032721 (2004).

35. Messiah, Quantum Mechanics (North-Holland, Amsterdam, 1970), Vol. II , pp. 848-852

36. Deepankar Misra, U. Kadhane, Y.P. Singh, L. C. Tribedi, P.D. Fainstein and P. Richard, Phys. Rev. Lett., 95 , 079302 (2005).

37. J. A. Tanis, S. Hossain, B. Sulik, and N. Stolterfoht, Phys. Rev. Lett., 95 , 079301 (2005).

Estimation of ion kinetic energies from time-of-flight and momentum spectra

B Bapat

Physical Research Laboratory
Navrangpura, Ahmedabad 380009
E-mail:bapat@prl.res.in

Keywords: fragment ions, ion kinetic energy, time-of-flight spectrometry, momentum spectrometry

1. Introduction

Conventional study of photoprocesses and charged-particle interactions with atoms and molecules has developed along three main streams: electron spectroscopy, photon spectroscopy and ion mass spectrometry. While the first two have blossomed into major fields with applications to non-atomic physics as well, the ion channel has received attention only in the last decade. In early studies recoil-ion analysis remained restricted to identification of the species. Very little information about their angular or energy distributions was available.

The situation has changed with the advent of Recoil Ion Momentum Spectroscopy, in which, besides identification of the recoil-ion species, the recoil momentum vector is completely determined. The idea behind this technique is simple. Suppose there is a well-localised and nearly stationary ensemble of target atoms or molecules, and one of these undergoes a collision with an ion or absorbs a photon. Several reactions may occur which result in the formation of ions and electrons. The motion of these fragments in a known, uniform electric field is completely determined by their initial position, and the momentum balance in the reaction. We detect the ion at some position (x, y) in a fixed plane (the plane of a position-sensitive detector) perpendicular to the applied electric field at a time t after the instant of the reaction. For decades together, studies of ion kinematics have been largely based on measurement of the time-of-flight.

2. Information from time-of-flight spectra

The spectrum of time-of-flight (TOF) of ions propagating in a uniform electric field primarily gives information about m/q distribution of the ions. In addition, the shapes of the mass peaks carry information about the kinetic energy (KE) of the ions.

In the commonly used Wiley-McLaren design TOF spectrometers [2], ions are formed in a small collision volume and are extracted by a uniform extraction field $\mathcal{E}_s$ over length s, followed by a field-free drift of length $2s$, at the end of which is placed a detector. The TOF of an ion of mass m and charge q is approximately given by

$$t = 2\left(\frac{2s}{\mathcal{E}_s}\right)^{1/2}\left(\frac{m}{q}\right)^{1/2} - \frac{p_z}{q\mathcal{E}_s}\,, \tag{1}$$

where p_z is the momentum of the ion along the spectrometer axis at the instant of formation. p_z has a distribution $f(p_z)$, leading to a TOF distribution $l(t)$, whose of FWHM we denote by Δt. A frequently used estimate of the KE of ions is based on the FWHM of the TOF peak:

$$\langle K^{\Delta}\rangle = 3\,(q\mathcal{E}_s)^2\Delta t^2/8m. \tag{2}$$

This value is an underestimate of the mean KE of the ions, especially when Δt is determined by a gaussian fit to $l(t)$. The situation is remedied by the proposition of Busch [1], that the mean KE is exactly related to σ_t^2, the variance of the TOF distribution:

$$\langle K^{\sigma}\rangle = 3(q\mathcal{E}_s)^2\sigma_t^2/2m. \tag{3}$$

This equality holds subject to the condition that there is no velocity discrimination in the spectrometer. Furthermore, the estimate needs correction for instrumental broadening, which is not an easy task.

A method to unambiguously extract a mean value of the KE has been proposed in the literature [1]. We have tested this proposition using an instrument that employs time-of-flight as well as spatial dispersion of ions, thereby enabling the measurement of each momentum component independently.

In terms of the momentum components, K^{σ} is identical (within a constant factor) to the mean-squared of p_z, we can conclude conclude that

$$\langle K^{\sigma}\rangle = 3\,\bar{p_z^2}/2m = (\bar{p_x^2} + \bar{p_y^2} + \bar{p_z^2})/2m.$$

For an estimate to be acceptable, it should be tested against the above value. It is against this definition that the other estimates will be tested.

3. Application to dissociative ionisation

In the case of dissociative ionisation, $f(p)$ may be a multi-modal distribution, due to a multitude of pre-cursor states leading to the same final ion species. Under these circumstances, estimation of $\langle K\rangle$ purely on the basis of TOF widths can lead to grossly erroneous results. The estimates suggested above can be validated only if TOF measurements and direct energy dispersive measurements are made concomitantly.

Our momentum spectrometer permits us to do precisely this. The momentum spectrometer is described in detail in [3]. It is a combination of a single field time-of-flight spectrometer and a position resolving detector. The ion acceleration region is 11 cm long followed by drift region of 22 cm, in conformity with the Wiley–McLaren condition $d = 2s$ [2]. The ion detector is a channelplate of 76 mm open diameter, with a position-encoding delay-line anode [4]. Position resolution of 250 μm and timing resolution of 500 ps are specified. Ion detection involves simultaneous derivation of a fast timing signal of the particle hit and its position on the detector. Besides the ions, ejected electrons are also detected, after acceleration over 3 cm in the opposite direction. The electron detector is a 40 mm diameter channelplate. The TOF start is obtained from the ion-hit, while the stop is obtained by a delayed coincidence with the ejected electron. The ejected electron is solely used for the purpose of TOF coincidence, its kinematics are not analysed. Ion TOF (t) and position (x, y) information is stored event-by-event as (t, x, y) triplets, and the event list can be sorted at will. The spectrum of any one of the three variables can be obtained by simply ignoring the other two in the event triplet. The (t, x, y) information can be mapped one-to-one onto the momentum components (p_z, p_x, p_y) respectively.

We have analysed the KE of ions using a momentum spectrometer, based on combined time-of-flight and position dispersion of energetic ions [5]. The process studied is electron-impact dissociative ionisation of CO_2. A TOF spectrum of the fragment ions formed in the process is shown in Figure 1. We have ascertained that the transmission losses are not significant for the collision system under investigation. The distribution of the three momentum components of O^+ ions is also shown in Figure 1. It is evident from the

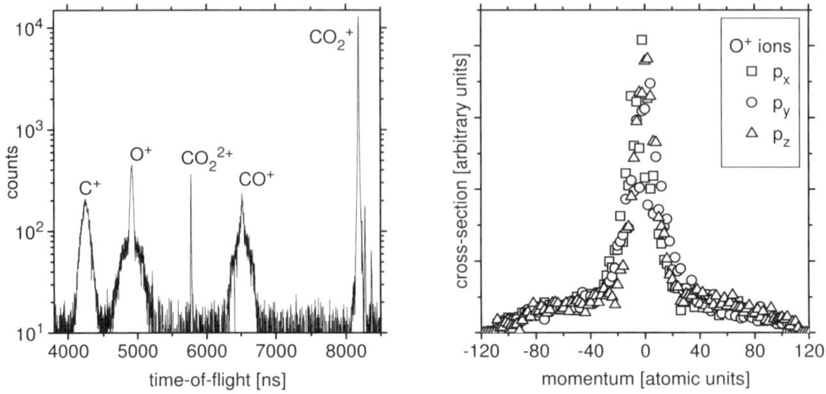

Fig. 1. The first panel shows a time-of-flight spectrum of ions formed by dissociative ionisation of CO_2 by electron impact. Note that the CO_2^+ and CO_2^{++} peaks are narrow, since these have only thermal energies. The fragment ions have broad peaks, indicating a broad kinetic energy distribution. The second panel shows the momentum distribution of the three components of momentum of O^+ ions, determined from simultaneously recorded TOFand position spectra.

figure, that within the limits of experimental errors, the three momentum components are identically distributed. This is true for other ions as well. The results of our analysis are summarised in the table below. The FWHM-based estimate is consistently lower than the variance-based estimate, the latter being in agreement with the direct momentum-based measurement. Even for those ions, whose momentum (scalar value) distribution $f(p)$ is bimodal, the estimate based on the TOF variance is found to be in excellent agreement with the momentum-based value.

Table 1. Estimates of the mean kinetic energy of various fragment ions, based on a direct momentum measurement, on the variance of the TOF distribution and the FWHM of the TOF distribution [p^2 in atomic units, K in eV].

Ion	$\bar{p}_x^2$	$\bar{p}_y^2$	$\bar{p}_z^2$	$\langle K^{\mathbf{p}} \rangle$	$\langle K^{\sigma} \rangle$	$\langle K^{\Delta} \rangle$
CO^+	2281	2231	2237	1.78	1.77	1.01
O^+	1559	1588	1557	2.17	2.16	0.70
C^+	1024	1072	1078	1.96	1.99	1.78

Another common, but erroneously used estimate of the KE of ions is based on the so-called forward–backward difference. Recent references to this practice are [6,7]. The original paper of Wiley and McLaren merely offers an equation connecting the kinetic energy of ions emitted isotropically to the flight time, provided the kinetic energy is single valued. For a single valued distribution, this is an exact result. Over the course of time, the value given by the above equation has come to be used for even those cases where there is a distribution of kinetic energies. The temptation to do this arises from a peculiar experimental artefact. If the kinetic energy distribution is broad, then the high energy ions are generally not transmitted to the detector, if they are not emitted more or less along the spectrometer axis. The ones that are, may be "forward" or "backward" emitted, resulting in a double-peaked TOF distribution. The time difference of the two peaks *does* represent the kinetic energy of the high energy ions that are ejected axially, but it by no means represents the entire distribution. Intriguingly, this seems not to have been realised so far. We offer here systematic analysis of one such case, involving dissociative ionisation of CO by 1300 eV electrons. For a glimpse of the difficulty in and the importance of correctly obtaining the kinetic energies of ion in a fragmentation by TOF and related methods, the reader is referred to [8,9]

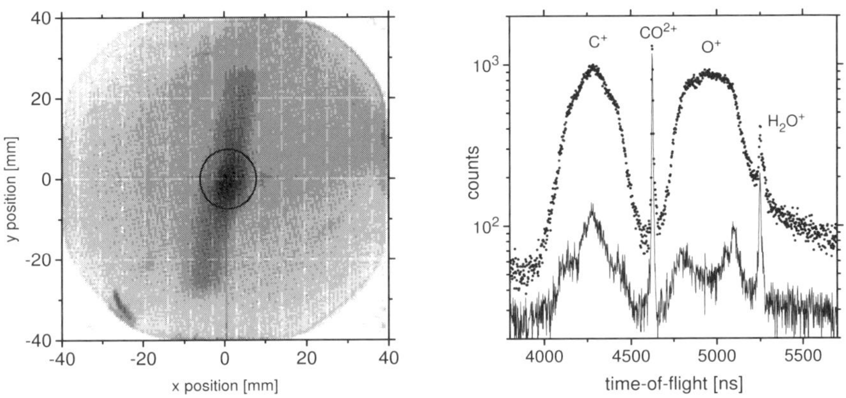

Fig. 2. On the left, position distribution of ions on the detector. The circle indicates the acceptance limit of the detector imposed artificially to mimic a typical TOF configuration. In the right panel, the time-of-flight distribution corresponding to full acceptance is shown by a dotted line and that corresponding to restricted acceptance is shown as a continuous line. Note that the restriction leads to a double-peak structure for O^+.

Consider the distribution of ions on our detector, arising from dissociative ionisation of CO. To mimic a typical TOF instrument, we mask out most of the aea of the detector, except a central circle of diameter 10 mm.

This masking is done by software, as the entire event list data is acccessible in a post-facto analysis. The position distribution and the time-of-flight distributions, with and without the mask are shown in Figure 2. It is clear, that for energetic ions, restricting the acceptance gives rise to the "forward" peak and the "backward" peak. We compare the values of the kinetic energies based on the time-of-flight in case of restricted acceptance as well as complete acceptance. This value is compared with the momentum-based value for the complete measurement. The comparison is shown in Table 2. It is clear the the KE value based on forward–backward difference is an overestimate. We also find that even in the case of restricted acceptance, the variance-based estimate is closer to the true value of mean kinetic energy.

Table 2. Comparison of the kinetic energy of C^+ and O^+ ions resulting from electron impact on CO, obtained by TOF variance, TOF forward–backward differences and momentum measurement.

Ion	$\langle K^{\sigma} \rangle$	$\langle K^{\mathrm{f-b}} \rangle$	$\langle K^{\mathrm{p}} \rangle$
C^+	6.18	9.21	4.10
O^+	4.93	7.23	3.71

4. Conclusions

In conclusion, we have shown that the mean kinetic energy of fragment ions analysed by a TOF mass spectrometer is most reliably determined by the variance of the TOF line. The commonly used FWHM-based estimate is invariably and underestimate. Another common practice, of estimating the mean kinetic energy of ions by the so-called forward–backward difference is not only based on an experimental artefact, but also leads to an overestimate of the kinetic energy.

References

1. F von Busch *J. Phys. B* **34** 431–438 (2001).
2. W C Wiley and I H McLaren *Rev. Sci. Instr.* **26** 1150 (1955).
3. V Sharma and B Bapat *European Physical Journal D* **37** 223–229 (2006).
4. O. Jagutzki, V. Mergel, K. Ullmann-Pfleger, L. Spielberger, U. Meyer, R. Dorner, H. Schmidt-Bocking in *Imaging Spectroscopy IV, Proceedings of International Symposium on Optical Science Engineering and Instrumentation* 3438, 322 (1998); see also `http://www.roentdek.com`.

5. B Bapat and V Sharma *Int. Journal of Mass Spec.* **251** 10–15 (2006).
6. S. De, P.N. Ghosh, A. Roy, C.P. Safvan *Nucl. Instru. and Meth. B* **243** 435–44 (2006).
7. A Rentenier, D Bordenave-Montesquieu, P Moretto-Capelle and A Bordenave-Montesquieu *J. Phys. B* **36** 1585–1602 (2003).
8. M Tarisien, L Adoui, F Frémont, D Lelièvre, L Guillaume, J-Y Chesnel, H Zhang, A Dubois, D Mathur, S Kumar, M Krishnamurthy and A Cassimi *J. Phys. B* **33** L11–L20 (2000).
9. B. Siegmann, U. Werner, R. Mann, N. M. Kabachnik, and H. O. Lutz *Phys. Rev. A* **62** 022718 (2000).

Third-order optical susceptibility of metal nanocluster-glass composites

Binita Ghosh and Purushottam Chakraborty
Saha Institute of Nuclear Physics
1/AF Bidhannagar, Kolkata 700 064, India

1. General Introduction

Metal-nanoclusters in dielectric materials having centers of symmetry, such as glasses, have recently attracted immense interest in materials science research for their potential applications in nonlinear optics. Various types of glasses are very attractive materials for these applications, because of their relatively fast response time. Furthermore, glass-based nanocomposites are, in general, expected to play an important role as materials for various nanotechnology applications, due to low cost, ease of processing, high durability, resistance and high transparency, as well as the possibility of tailoring the behavior of glass-based structures. Since the first attempt of Faraday [1] to explain the nature of the color induced in glasses by small metallic precipitates, many studies have been dedicated to the properties of metal nanocluster composite glass. In general, the physical properties of these systems change dramatically in the transition from atom to molecule to cluster to solids, where the cluster regime is characterized by the confinement effects, causing thereby these systems particularly interesting.

Although electronics technologies have made great advances in device speed, optical devices can function in the time domain inaccessible to electronics. In the time domain less than 1 ps, optical devices have no competition. Photonic or optical devices are designed to switch and process light signals without converting them to electronic form. The major advantages that these devices offer are speed and preservation of bandwidth. The switching is accomplished through changes in refractive index of the material that are proportional to the light intensity. The third-order optical susceptibility, $\chi^{(3)}$, known as 'optical Kerr susceptibility' which is related to the non-linear portion of the total refractive index, is the non-linearity which provides this particular feature. Future opportunities in photonic switching and information processing will depend critically on the development of improved photonic materials with enhanced Kerr susceptibilities, as these materials are still in a relatively early stage of development.

Glasses possess macroscopic inversion symmetry, thereby allowing only odd-order optical non-linearities. The third-order nonlinear optical phenomena prevail inherently in glasses because of the deficiency of long-range periodicity unlike crystals. Introducing metal nanoclusters in glass matrices can further enhance the third-order nonlinear optical response of the host matrices. Nonlinear optical susceptibilities seem to become dominant in a material when the driving optical field, E is much greater than or at least comparable to the intra-atomic field, which is of the order of 10^{10} V/m. For an optical field intense enough to drive the electron beyond the quadratic minimum of the inter-atomic potential, the response becomes increasingly nonlinear.

Melt-quenching and heat treatment processes are the most simple and long-used methods for preparing glasses that contain metal colloids. The properties of the composites, prepared by melt-glass fabrication method, can only be partially tailored, because the thermodynamics and the chemistry of glass formation by melting put several constraints to the possibility of controlling the formation and size distribution of the clusters. Sputter-deposition methods have found applications in the preparation of thin glass films containing metal clusters The main drawbacks of the method are again the lack of control of the structure of the composite and size distribution of the clusters. Ion implantation has been shown to produce high-density metal colloids in glasses and in other materials. The high-precipitate volume fraction and small size of nanoclusters in glasses lead to the generation of third-order susceptibility much greater than those for metal-doped solids. This has stimulated interest in the use of ion implantation to make nonlinear optical materials. An attractive feature of ion implantation to form these nanoclusters, compared to the classical technique of melt-glass fabrication process, i.e. mixing selected metal powder with molten glass, is that the linear and nonlinear properties occur in a well defined space in an optical device, and by using focused ion beams, point quantum confinement may be accomplished. The third-order optical nonlinear responses of the metal nanocluster-glass composites can be understood in the framework of classical (dielectric) and quantum confinement effects.

The formation of colloidal metal particles in glass matrices is evidenced by an absorption band, revealed in the linear optical absorption measurements in the UV-visible range, at the characteristic surface plasmon resonance (SPR) frequency of the metal. Heat treatments on the metal-implanted silica glasses give pronounced effects on the formation of the metal colloids of larger sizes through aggregation of the implanted metal atoms, arising out of greater mobility of the metal atoms under heating or due to the formation of metal oxide clusters under beam induced mixing. Either of the two effects can be manifested by a shift of the SPR band towards higher wavelength. It has been demonstrated that metal nanoclusters of various sizes can be safely created in 'as implanted' as well as in annealed glass samples, as evidenced by the pronounced optical absorption bands in both the cases. Rutherford Backscattering spectrometry (RBS), a non-destructive ion beam analytical technique, has been used to measure the concentrations of these nanoclusters as a function of depth of the materials. The nonlinear optics experiments like Z-scan technique, Degenerate Four-Wave Mixing (DFWM), etc. can measure the nonlinear optical properties (nonlinear absorption coefficient and nonlinear refractive index, simultaneously) of these systems. The present work reports a brief overview on the synthesis of various metal nanocluster-glass composites and their nonlinear optical responses in the nano to picosecond time domains..

2. Basics of Nonlinear Optics

Nonlinear optics is the branch of optics that describes the behavior of light in a nonlinear medium, i. e, medium in which the polarization **P** responds nonlinearly to the electric field **E** of the light. Nonlinearity in optics is manifested by changes in the optical properties of a medium as the intensity of the applied optical field is increased or when one or more fields are introduced. Typically, only a laser light is sufficiently intense to modify the optical properties of a material system. In fact, the beginning of the field of nonlinear optics is taken to be the discovery of second-harmonic generation by Fraken *et al.* in 1961 [2], shortly after the demonstration of the first working laser by Maiman in 1960 [3].

When a neutral atom is placed in an electric field **E**, the positively charged core and the negatively charged electron cloud surrounding it are both influenced by the electric field. The nucleus is pushed in the direction of the field and the electrons in the opposite direction. In principle, if the field is large enough it can pull the atom apart completely, 'ionizing' it. With less extreme fields, however, equilibrium is soon established, for if the center of the electron cloud does not coincide with the nucleus, leaving the atom 'polarized' with positive charge shifted slightly one way, and negative the other. The atom now has a tiny dipole moment which points in the same direction as **E**. Typically, the polarization **P** (dipole moment per unit volume) is directly proportional to the field, provided E is not too strong. That is

$$P = \varepsilon_0 \chi E \tag{1}$$

where, χ is the dielectric susceptibility and ε_0 is the dielectri constant of the medium. The value of χ depends on the microscopic structure of the substance in question and also on the external electric field. Materials obeying the above equation are known as linear dielectrics. Linearity means that the displacement of the electrons is proportional to the acting force. Only under this condition the electrons oscillate sinusoidally, with amplitude proportional to the acting force (apart from a phase shift).

The above stated model of the elastically bound electron is too simple and is only an approximation valid for small amplitudes. The deviations become significant at high oscillation energy. If this energy exceeds the value of the binding energy, the electron is set free. At this level the electron oscillates non-harmonically. Higher frequencies appear in the displacement and in the polarization P. In this case, the linear equation between polarization and the field amplitude is no longer valid. The polarization is a complicated function of the field amplitude. It contains not only the linear term E, but also higher order terms. This is traditionally introduced by noting that the dielectric susceptibility can be expanded in a Taylor series in terms of the oscillating optical field. In an isotropic medium the general relation between the polarization **P** (dipole moment per unit volume), induced by the applied electric field **E**, is expressible as a Taylor expansion involving only the magnitude, because the direction of the polarization coincides with that of the field, namely

$$P = \varepsilon_0 \left(\chi^{(1)} E + \chi^{(2)} E^2 + \chi^{(3)} E^3 + \ldots\ldots\ldots\ldots\ldots\ldots \right. \tag{2}$$

In this expansion $\chi^{(1)}$ is the normal or linear dielectric susceptibility of the medium and is related to the refractive index η by $\eta^2 = 1 + \chi^{(1)}$. It contributes to all the phenomena associated with linear optics, such as reflection, refraction, interference etc. It is generally much larger than the nonlinear coefficients $\chi^{(2)}$, $\chi^{(3)}$ and so forth. In general, $\chi^{(n)}$, the complex dielectric susceptibilities, are tensors of rank (n+1) and are related to the microscopic (electronic and nuclear) structure of the material.

It immediately becomes clear what is meant by the terms in the equation (2) if we direct a (laser) light beam with frequency ω on to the sample. The second term then reads

$$\chi^{(2)} E_0^2 \sin^2 \omega t = \frac{1}{2}\chi^{(2)} E_0^2 (1 - \cos 2\omega t) \tag{3}$$

The cos2ωt term in Eq.(2a) tells us that a contribution to the polarization is created, which oscillates at a frequency of 2ω, and is radiated by the sample, the so called ' second harmonic generation'. If two fields with different frequencies ω_1 and ω_2 interact in the sample via $\chi^{(2)}$ effects, a similar approach as in Eq.(2a) gives contribution to the polarization which oscillate with frequencies $\omega_1 \pm \omega_2$ which are also radiated. $\chi^{(2)}$ and all other even terms vanish for crystals whose symmetry elements contain the inversion. This is immediately clear by letting $\mathbf{E} \rightarrow -\mathbf{E}$ and consequently $\mathbf{P} \rightarrow -\mathbf{P}$. The even-order nonlinear optical interactions can occur only in non-centrosymmetric crystals, that is, in crystals that do not display inversion symmetry [4]. Since liquids, gases, amorphous solids (such as glass) have inversion symmetry, $\chi^{(2)}$ vanishes identically for such media and consequently they cannot produce second-order nonlinear optical interactions. On the other hand, odd-order nonlinear optical interactions can occur both for centrosymmetric and non-centrosymmetric media.

3. Physical Origin of Nonlinear Optical Phenomena

The equation for the polarization **P** links the material properties with Maxwell's equations for the propagation of electromagnetic waves. It does not give any indication as to the physical mechanisms which give rise to the optical nonlinearities. The origin of the nonlinear susceptibility can be viewed classically as the response of an electron driven by an electromagnetic field in an anharmonic potential well, resulting from the inter-atomic electric field **E** in the crystalline solid. When a light wave propagates through an optical medium, the oscillating electromagnetic field exerts a polarizing force on all the electrons comprising the medium. Because the inner electrons of the atoms are tightly bound to the nuclei, the major polarizing effect is exerted on the outer or valence electrons. The inter-atomic field is of the order of 10^{10} V/m. For driving optical fields **E** much weaker than that, the polarization response is essentially linear. For an intense optical field (comparable to or greater than the inter-atomic field) enough to drive the electron beyond the quadratic minimum of the interatomic potential, the response becomes increasingly nonlinear.

The most obvious way of approaching the origin of optical nonlinearities and to gain some appreciation of the scales involved is simply to consider that the electric field obtainable by focusing a laser beam of 20 W average power down to a spot size of 20 μm results in an electric field which is of the order of 10^7 V/m, which can significantly perturb the interatomic field in a typical solid. We thus expect that under conditions of nonresonant excitation the second-order susceptibility $\chi^{(2)}$ will be of the order of $\chi^{(1)}/E_{at}$ [5]. For condensed matter $\chi^{(1)}$ is of the order of unity, and we hence expect that $\chi^{(2)}$ will be of the order of $1/E_{at}$, or that

$$\chi^{(2)} \cong 5\times10^{-8}\, cm/statvolt = 5\times10^{-8}\left(cm^3/erg\right)^{1/2} = 5\times10^{-8}\, esu$$

Similarly, we expect $\chi^{(3)}$ to be of the order of $\chi^{(1)}/E_{at}^{2}$, which for condensed matter is of the order of

$$\chi^{(3)} \cong 3\times10^{-15}\, cm^2/statvolt^2 = 3\times10^{-15}\, cm^3/erg = 3\times10^{-15}\, esu$$

These predictions are in fact quite accurate, as one can see by comparing these values with the actual measured values of $\chi^{(2)}$ and $\chi^{(3)}$. For certain purposes, it is useful to express the second- and third-order susceptibilities in terms of fundamental physical constants. Noting that the number density N of the condensed matter is of the order $(a_0)^{-3}$, we find that

$$\chi^{(2)} \cong h^4 / 16\pi^4 m^2 e^5 \text{ and } \chi^{(3)} \cong h^8 / 2^8 \pi^8 m^4 e^{10}.$$

The physical origin of nonlinear optical phenomena can be categorized as either structural or compositional [6]. Here 'structural' refers to light induced structure changes, such as a change of electronic density, average interatomic distances, molecular orientation, phase transition, etc. These phenomena belong to the intrinsic category. The 'compositional' refers to light induced chemical composition changes such as molecular dissociation, polymerization etc. Intrinsic nonlinearity violates the principle of superposition arising from a nonlinear response of the individual molecule unit cell to the fields of two or more light waves, while extrinsic nonlinearity is related to changes in the composition of the medium that results from the absorption or emission of light. In either type of nonlinearity, the optical properties of the medium depend on the intensity of light and the order of nonlinearity can be classified according to the power of the intensity involved. For example, in the case of harmonic generation, the intensity of the nth harmonic is proportional to the nth power of the intensity of the fundamental, and in case of optical Kerr effect the intensity of the Kerr signal is proportional to the product of probe beam intensity and the square of the intensity of the pump beam.

The optical Kerr effect can be defined as either light induced double refraction or an intensity dependence of refractive index. The intensity dependent refractive index η is usually expressed as

$$\eta = \eta_0 + \eta_2 I \tag{4}$$

where η_0 is the linear refractive index of the material and is a function of $\chi^{(1)}$ [7]. I is the intensity of light averaged over a period and η_2 is the nonlinear refractive index of the material. The total absorption α under saturation can be expressed as [8]:

$$\alpha = \frac{\alpha_0}{\left(I + I/I_S\right)} \tag{5}$$

where α_0 is the linear absorption coefficient in the absence of saturation, I_S is the saturation intensity. In the weak saturation limit, i.e, $I/I_S << 1$, Eq.(5) can be reduced to

$$\alpha = \alpha_0\left(1 - I/I_S\right) \tag{6}$$

$$= \alpha_0 - \alpha_0 I/I_S$$

$$= \alpha_0 + \beta I \tag{7}$$

β is the coefficient characterizing the nonlinear absorption and is known as the ' two-photon absorption coefficient'. For all-optical devices, the nonlinear refractive index and the two-photon absorption play critical roles. The nonlinear refractive index and the two-photon absorption coefficient are related to the real and imaginary parts of $\chi^{(3)}$ according to the following equations:

$$\eta_2 = 12\frac{\pi \operatorname{Re}\left[\chi^{(3)}\right]}{\eta_0} \tag{8}$$

$$\beta = 96\frac{\pi^2 \omega \operatorname{Im}\left[\chi^{(3)}\right]}{\eta_0^2 c^2} \tag{9}$$

Eqs. (8) and (9) are valid in the case of negligible linear absoption, i.e. when $\alpha_0/2kn_0 << 1$, where $k=2\pi/\lambda$. Strong linear absorption provides a coupling between $\operatorname{Re}[\chi^{(3)}]$ and β, and between $\operatorname{Im}[\chi^{(3)}]$ and n_2 when transforming from experimental quantities to susceptibilities [9]. The coupling arises from the complex nature of the linear absorption index, defined as $n_0 - i(\alpha_0/2k)$:

$$\operatorname{Re}\chi^{(3)}(e.s.u) = \frac{cn_0^2}{120\pi^2}\left(n_2 - \frac{\alpha_0\beta}{4k^2 n_0}\right) \tag{10}$$

$$\operatorname{Im}\chi^{(3)}(e.s.u)=\frac{cn_0^2}{120\pi^2}\left(\frac{\beta}{2k}+\frac{\alpha_0 n_2}{2kn_0}\right) \tag{11}$$

Quantities on the right hand sides of Eqs. (10) and (11) are in S.I. units.

4. Materials for Nonlinear Optics

Nonlinear optical materials play a pivotal role in the future evolution of nonlinear optics in general and its impact in technology and industrial applications in particular. Our goal is to find and develop materials having large nonlinearities and satisfying at the same time all the technological requirements for applications such as wide transparency range, fast response, high damage threshold but also processibility, adaptability and interfacing with other materials. The nonlinear response of a material is determined by the nature of the electronic environment of the medium, its symmetry for even-order coefficients, and the nature of the nonlinear interaction. Molecules and molecular solids with π-delocalized systems exhibit much larger nonlinear responses than covalently bonded molecules and solids. Molecules with electron donor and acceptor groups connected by the π-orbitals give the largest nonlinear responses. Nonlinearity can show up in the following materials.

Semiconductors:. One nonlinear phenomenon in semiconductors is the photorefractive effect in the bulks and in the films. Carriers are photo-excited from deep traps to the conduction band, and then retrapped in non-illuminated regions, including a periodic charge distribution that gives rise to a spatially varying refractive index. Other phenomena in semiconductors are electro-absorption and electro-refraction.

Quantum dots: These are small devices that contain tiny droplets of atoms of metals or semiconductors. They are usually not self-supported, rather are synthesized in semiconductor materials, polymers or insulators. They have typical dimensions ranging from nanometers to a few microns. The size and shape of these structures and therefore, the number of electrons they contain, can be precisely controlled. Like in an atom, the energy levels in a quantum dot become quantised due to the confinement of electrons. Quantum dots are zero-dimensional semiconductor or metal clusters with diameters smaller than twice the excitonic Bohr radius. Bohr radii are much smaller than the wavelength of light. Electrons in quantum dots are confined between infinite potential barriers. It has been demonstrated that the inherent nonlinear optical response of a dielectric host material such as glass, polymers, etc may be enhanced by several orders of magnitude by introducing small semiconductor or metal clusters within the host matrices.

Organic compounds: Organic compounds with π-electrons delocalization are very important nonlinear materials. Their nonlinear response can be extremely high under resonant conditions. The basic molecular structure required for the construction of organic single crystals is asymmetric with electron donating and accepting groups connected by a conjugated structure. There is a trade off between the nonlinearity and the wavelength range of optical transparency. The hyperpolarizibility of an organic molecule is proportional to the fifth power of the πconjugation length. In the one-dimensional system, it is expected that the optical band gap becomes narrower as more aromatic groups are introduced into the molecular backbone to lengthen its π-conjugation. The band gap depends on the degree to which the π-electrons are delocalized along the molecular backbone. The second hyperpolarizibilities of organic molecules may be substantially improved if proper functional groups are chosen for incorporation into the building blocks for a particular molecule.

Inorganic glasses: Glasses are optically isotropic media due to lack of long-range periodicity. Although they are structurally ordered on the scale of atomic dimensions, atomic sites are physically inequivalent. They can be highly transparent, giving rise to non-resonant third order nonlinearities. The electronic nonlinear response of glasses is generated by the valence metal cations or by the *p-d* orbital overlap. The *d* orbital contributions become dominant when the bond length is less than 2 Å. The absolute value of the third-order nonlinear optical susceptibility gets enhanced when metal nanoclusters are formed within the glass matrices by suitable methods.

4.1 Choice of Nonlinear Materials:

Selection of Nonlinear Optical (NLO) materials for practical applications is, however, not easy because there is no ideal material for all applications and as a result there is no universal definition of the 'figure of merit' of materials. NLO effects have important applications in optical communications where optical switching and optical signal processing devices are essential elements. Each of these devices essentially uses some material with inherent $\chi^{(3)}$ or high-intensity optical field induced $\chi^{(3)}$. A 'figure of merit' is helpful in comparing materials for all optical devices but it must be formulated with a specific application in mind. A proposed 'figure of merit' for NLO materials compares the refractive index change created by the nonlinear coefficient [10]. The 'figure of merit' is defined as

$$F = \text{Fast index change required for switching / resultant thermal index change.}$$

$$= \frac{\Delta \eta_{sat}}{\tau \alpha_0}$$

where τ is the response time or relaxation time of the nonlinearity. $\Delta\eta_{sat}$ is the change in refractive index at saturation and α_0 is the linear absorption coefficient. It is clear that an ideal nonlinear material should have larger nonlinearity, fast response time and minimal absorption in the wavelength range of interest. In addition these materials must have thermal relaxation times that are sufficiently rapid so that there is no heat build-up, because this would produce an effective refractive index just due to the change in temperature.

Glass is widely used as the base material within which nanoclusters are grown, mainly due to three reasons:

(i) Inherent nonlinear susceptibility of glasses is, in general, quite small and can be enhanced by several orders of magnitude by introducing small metal clusters in them. Furthermore, glass materials have high 'figure-of-merit' [11].

(ii) Glass materials have a more open structure compared to crystalline quartz s, which provides an added advantage for the growth of nanoclusters. Metal nanoclusters can be easily accommodated into the open spaces of the glass material.

(iii) A distinction has to be considered as to whether the frequency of the applied optical field is resonant or non-resonant with an electronic transition in the material. In the non-resonant case, which is applicable to most of the glasses for wavelengths around 1 μm, the optical non-linearity is of electronic origin. The electronic non-linear response of glasses is generated by the electronic structure change due to the distortion of the electron orbits about the nuclei (polarizability) under the influence of a high intensity optical field. However, the non-linearities in glasses are small and require high optical intensities for switching. This drawback can, however, be compensated by the advantages due to an extremely low absorption coeffcient and the high optical damage threshold of glasses for practical applications. As the origin of nonlinearity is electronic and off-resonance, they have the fastest response times (~femtoseconds).

4.2. Glass Structure:

The meaning of the term 'glass' is often taken to be that proposed by the American Society for Testing Materials: ' Glass is an inorganic product of fusion which has been cooled to a rigid condition without crystallizing'. This definition was modified in 1976 by the U.S National Research Council as: 'Glass is an X-ray amorphous material which exhibits the glass transition, this being defined as the phenomenon in which a solid amorphous phase, with changing temperature, exhibits a more or less sudden change in the derivative thermodynamic properties, such as heat capacity and expansion coefficient, from crystal-like to liquid-like values'.

The oxygen polyhedra found in oxide crystals must also be present in glasses, but in case of the crystalline forms the same geometrical pattern is repeated over large distances throughout the material; angles between bonds of the same type are constant and distances between pairs of atoms are constant. In the glass the structure is non-periodic with varying bond angles and separations. Figure 1 shows the hypothetical two-dimensional compound X_2O_3 for the crystalline and glassy form of the material. One feature of the glassy structures is that the structures are relatively open – as seen in figure and contain large voids.

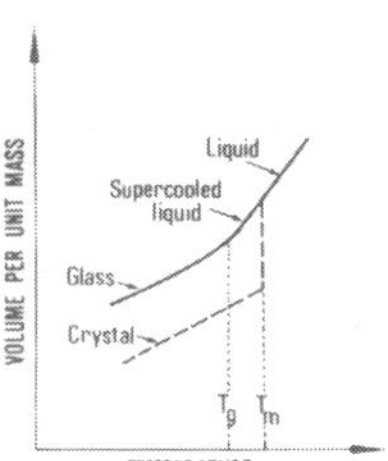

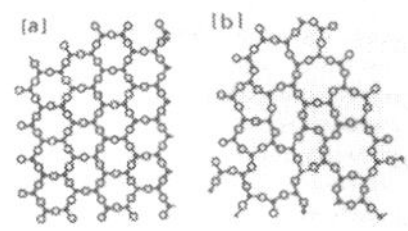

Figure 1:
Hypothetical two-dimensional compound X_2O_3 for (a)crystalline and (b) glassy form of the material.

In most multicomponent oxide glasses, there are both bridging and non-bridging oxygens in the glass network (i.e. for a silicate glass, $Si\text{-}O^{+}Na^{-}$). These non-bridging ionic bonds possess larger n_2 than the bridging oxygen of the more covalent Si-O-Si bonds. A correlation between the non-bridging oxygen content and the non-linear indices was reported by Adair *et al.* [12]. In silicates modified with transition metal cations, optical properties are mostly determined by the concentration of metal cations rather than by the number of non-bridging oxygens, i.e. non-linearity is dominated by the bond-polarizability of the metal oxygen bonds.

5. Optical absorption of metal nanocluster-glass composites

In general, when an electromagnetic wave propagates in the composite medium, it does not resolve the individual scattering centers, and the medium appears as a homogeneous medium characterized by an effective complex dielectric function. If we consider a composite consisting of small particles occupying a relative volume fraction $p \ll 1$ in a dielectric host, the complex dielectric constant of the metal cluster, which is a function of the frequency of the incident electromagnetic field by $\varepsilon_m(\omega) = \varepsilon_1(\omega) + i\varepsilon_2(\omega)$. The optical response of the metal nanocluster glass composite systems may be approached following two classes of models, namely, (i) discrete island model, in which each scattering center is considered, and (ii) effective medium theory, in which one scattering center is considered, the rest of the surrounding medium being averaged into a homogeneous medium. Considering the first kind of approach, Maxwell-Garnet [13] considered that the metal clusters embedded in a host matrix of dielectric constant ε_h are polarized by an optical field. The Lorentz local-field relationships show that the effective dielectric constant ε_{eff} of such a composite medium is given by

$$\frac{\varepsilon_{eff} - \varepsilon_h}{\varepsilon_{eff} + 2\varepsilon_h} = p\frac{\varepsilon_m - \varepsilon_h}{\varepsilon_m + 2\varepsilon_h} \tag{12}$$

For small volume fractions, Eq.(12) may be expanded to the first order in the volume fraction. The optical absorption coefficient is related to the imaginary part of ε_{eff}. Mie was the first to derive an exact description of the optical absorption and elastic scattering by a collection of metal spherical clusters suspended in a transparent medium. By adding the contribution of single clusters, the theory accounts correctly for many experimental cases, provided that interaction effects among the clusters as well as multiple scattering effects are missing. This is concerned, however, with a large class of inhomogeneous composite glasses, where the cluster density is anyhow sufficiently low to allow a description within this regime. The optical absorption coefficient α of a collection of uniform spheres, very small compared to the wavelength, λ, of light and embedded in a medium of refractive index n_d, is well described by the Mie scattering theory. The general expression for the extinction coefficient, which includes both absorption and scattering, is given by [14]

$$\alpha = \frac{6\pi p n_d}{\lambda}\left\{-\operatorname{Im}\sum_{i=1}^{\infty}\left[(-1)^i\left(\frac{a_i - b_i}{2\gamma^3}\right)\right]\right\} \tag{13}$$

where $\gamma = 2\pi n_d R/\lambda$ (R = particle radius). The quantities a_i and b_i are the partial wave contributions to the scattered wave by electric (a_i) and magnetic (b_i) multipoles. The number of terms used in Eq.(13) in order to represent satisfactorily expected extinction of an assembly of particles depends on the particle size R. For very small particles (R <100 Å), Eq.(13) is often approximated by using only the electric dipole term, for which the relation reduces to [15]

$$\alpha = p\frac{18\pi n_d^{\ 3}}{\lambda}\frac{\varepsilon_2}{\left(\varepsilon_1 + 2n_d^{\ 2}\right)^2 + \varepsilon_2^2} \tag{14}$$

This expression has a maximum at λ for which the condition $(\varepsilon_1 + 2n_d^{\ 2}) = 0$ is satisfied. This is due to the absorption resonance known as the surface plasmon resonance (SPR). Plasmons are the quanta associated with longitudinal waves propagating in matter through the collective motion of conduction band electrons. Surface plasmons are a subset of these 'eigen-modes' of the electrons, which are bound to regions in the material where the optical properties reverse, i.e. the interface between a dielectric and conducting medium. The position, width and shape of the SPR band are determined by the metal dielectric function, also on the size, shape and concentration of the particles and on the surrounding dielectric medium. The position of the resonant absorption band changes with the cluster size, but remains more or less unchanged for the clusters with less than 10 nm diameter, as was predicted by Mie in his scattering theory. Moreover, quantum corrections to the classical absorption become significant as the cluster dimension approaches 1 nm diameter [16].

If the complete expression for the dielectric constant of a metal, when it is determined by free electrons, is substituted in Equation 14, the result is as follows [17]

$$\alpha = \frac{9\pi p n_d^3 c}{\sigma}\frac{\lambda^2}{\left(\lambda_m^2 - \lambda^2\right)^2 + \lambda^2\lambda_m^4 / \lambda_a^2} \tag{15}$$

where $\lambda_m = \lambda_c\left(\varepsilon_0 + 2n_d^2\right)^{1/2}$ is the wavelength at which the maximum absorption takes place. $\lambda_c = (2\pi c)^2 m / 4\pi N_e \varepsilon_m^{\ 2}$ and $\lambda_a = \frac{2\lambda_c^{\ 2}\sigma}{c}$. Equation (15) gives a band of Lorentzian shape: if the band is narrow, its width *w* at half maximum absorption (FWHM), i.e. bandwidth, is λ_m^2 / λ_a and is given by

$$w = \frac{\lambda_m^2}{\lambda_a} = \left(\varepsilon_0 + 2n_d^{\,2}\right)c/2\sigma \tag{16}$$

The d.c. conductivity, σ, is given by [17]

$$\sigma = \frac{N_e e^2 R}{m u_F} \tag{17}$$

where R is the particle radius, $u_F = \left(\frac{2E_F}{m}\right)^{1/2}$ is the electron velocity at the Fermi energy $E_F = \frac{h^2}{2m}(3n8\pi)^{2/3}$ and N_e is the number of electrons per unit volume. Therefore, the experimental result of constant band position is reasonable, because λ_m does not depend upon the conductivity, σ, which is determined by the mean free path of the electrons and therefore, upon the particle size. Alternatively, Equation (16) shows that the dependence of bandwidth upon particle size, i.e. $w \propto 1/R$, as obtained from Equations (16) and (17), results from the changing mean free path of the electrons.

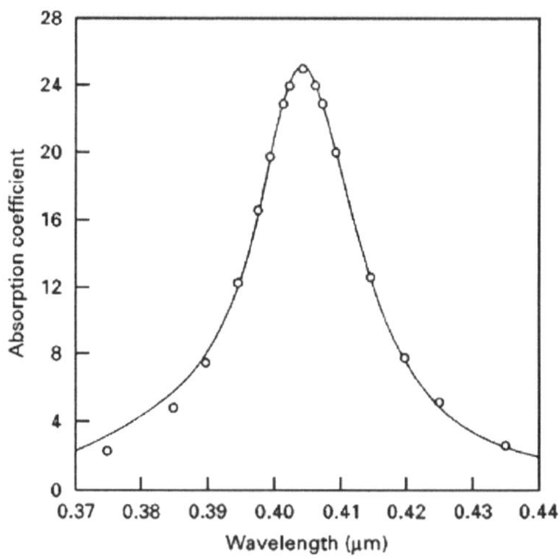

Figure 2: (o) Measured and (___) calculated absorption band for silver particles of about 10 nm diameter, suspended in glass.

Figure 2 gives a comparison between the experimentally measured absorption band for silver particles of about 10 nm diameter, suspended in glass, and the calculated one from Equation (15), using the measured λ_m and bandwidth [18]. The agreement between these curves confirms that the absorption results from the free electrons in the silver particles. The slight increase in absorption at lower wavelengths probably results from the ultraviolet irradiation of the cerium present in the glass matrix [17]. Eq.(13) predicts a shift of the resonance to longer wavelengths and increasing half-widths for the particles of radius $R \geq 10$ nm. Figure 3 shows the optical extinction per unit Ag concentration as a function of wavelength for various values of colloidal radius.

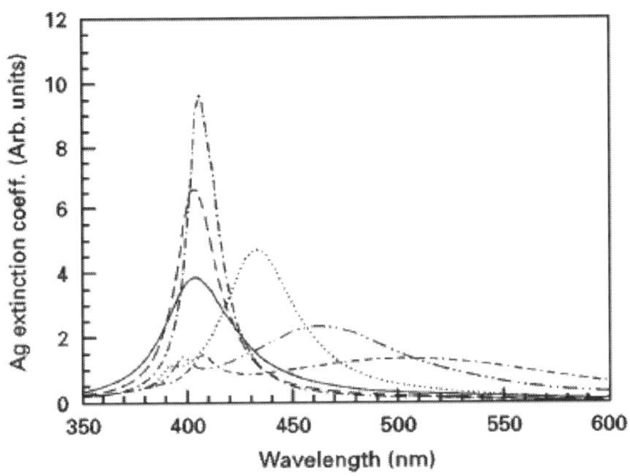

Figure 3: Mie extinction coefficient of silver colloids of various radii as a function of wavelength of the incident light. R (nm): (__) 2.5, (_ _) 5, (_ .) 10, (… ...) 25,(_ .. _) 35, (_ _ . _ _) 45.

The average radius of metal spheres, small compared to the wavelength of light, can be approximately estimated from the resonance optical absorption spectrum as per the Mie scattering formula.

$$r_{metal} = \frac{v_F}{\Delta\omega_{1/2}} = \frac{v_F}{2\pi c_0\left(\frac{\Delta\lambda}{\lambda_p^2}\right)} \tag{18}$$

Where v_F = Fermi velocity (1.57×10^6m/s for Cu) and $\Delta\omega_{1/2}$ is the full width at half maximum (FWHM) of the SPR absorption when plotted as a function of angular frequency ω, c_0 is the speed of light in vacuum, λ_p is the wavelength where absorption peak appears, and $\Delta\lambda$ gives the bandwidth. This expression, however, provides a very crude estimate of the metal particle dimension as it is reliable only when the metal colloids are strictly spherical. It is also difficult to estimate the particle dimensions from the optical absorption spectra as the main features of the SPR bands (i.e wavelength, position and FWHM) depend on size, size distribution, shape, lattice parameters and filling factors of the metal particles.

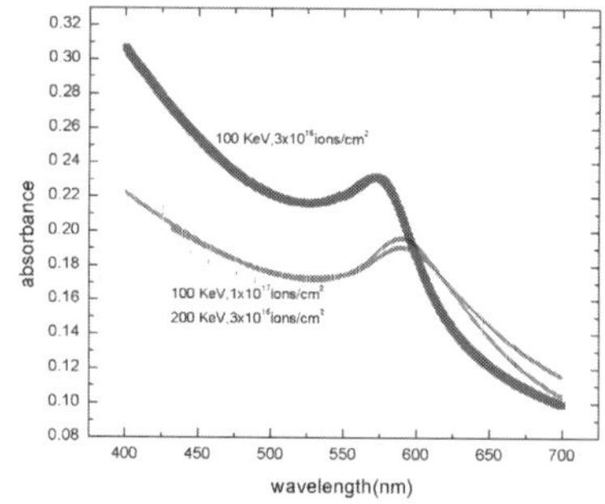

Figure 4: Optical absorption spectra for as-implanted Cu^{+} ion-implanted glass

Figure 4 shows the linear optical absorption bands for different Cu–implanted samples (as-implanted) for various doses. As evidenced from the figures, the absorption bands appeared at around 589 nm for the samples (100 keV, 1×10^{17} ions/cm^2 and 200 keV, 3×10^{16} ions/cm^2) and at around 580 nm for the sample (100 keV, 3×10^{16} ions/cm^2). These absorption bands give the evidence of the formation of copper nanoclusters in glasses even without annealing. This can be attributed to the relatively greater mobility of copper atoms even at ambient conditions. The peak position at 580 nm matches considerably well with the expected SPR band for Cu nanoclusters in glass. The shift in the peak position towards higher wavelength (589 nm) for the other two samples due to the change in the cluster size can be ruled out, as according to the Mie theory SPR wavelengths seem to remain unaltered for all the cluster dimensions less than 10 nm. The estimated cluster sizes in the present cases have been found to be much below 10 nm. The shift in the peak position could be due to the beam-induced effects resulting in a possible formation of copper oxide clusters. This perception is probably justified in the present case, as there was a similar observation in the shifting of SPR band towards lower photon energy due to formation of copper oxide nanoclusters under low temperature oxygen annealing [19]. Another similar observation was reported for the case of a preferential oxidation of copper in a study of chemical or radiation-assisted selective dealloying in Cu-Au nanoclusters [20]. With respect to single element case, binary metal nanoclusters offer higher degrees of freedom for the control of the material features, namely, cluster compositions and crystal structures. The sequential ion implantation of two different metal species may give rise to various different nanoclusters, with the presence of separated family of pure metal clusters, crystalline alloy structures, or core-shell structures. Figure 5 shows a planer bright field cross-sectional TEM micrograph of the Au_3Ag_6 alloy clusters formed by double implantation of Au+Ag in silica after thermal treatment at 800°C performed for 1 hour [21].

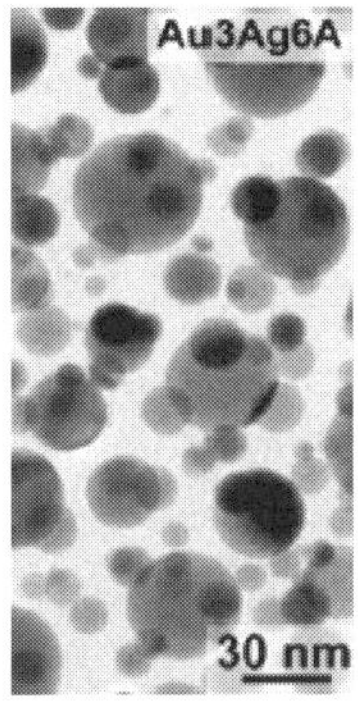

Figure 5: TEM micrograph of Au3Ag6 alloy nanoclusters prepared by sequential implantation of gold and silver in silica.

When particle volume fraction is greater than a few percent, the interaction between the particles causes a red shift of the absorption band, an effect for which the Maxwell-Garnett picture can give an account. On the other hand, increasing the concentration of the particles also leads to optical broadening due to statistical effects. The plasmon response in semiconductor nanoclusters tends to be much less significant than in metals; the real part of the dielectric constant for most semiconductors varies too weakly with frequency to reach a zero in the denominator of Eq (12).

Generally speaking, the dielectric function of nanoclusters can be calculated in a classical way, changing the relaxation time in the Drude free-electron model and taking into account the confinement into clusters smaller than the mean free-path of the electrons in the bulk material, as well as in the quantum picture of electric dipole transitions in a spherical box. The optical absorption of the glass matrix containing the metal clusters is usually expressed starting from either the Mie theory or from the Maxwell-Garnett effective dielectric function model. In the first case, the contribution of each cluster to the absorption is calculated, where as in the second case, an effective dielectric function is calculated for the assembly of clusters, and the absorption coefficient is obtained from the imaginary part of this function. Two confinement effects essentially govern the nonlinear optical behavior of the metal nanocluster-glass composite: one is the dielectric or classical confinement and the other is the quantum confinement.

5.1 Classical Confinement:

The classical or dielectric confinement effect can be illustrated with the help of Figure 6. For visible and infrared wavelengths, the optical field has nearly constant magnitude over the entire volume of the cluster. When subjected to an optical field, the conduction band electrons oscillate against the ionic charge background, and the surface provides a resonant restoring force; the process has an interesting analog in the giant dipole resonance. The motion of the electronic charge under the influence of the rapidly oscillating optical field alters the field in the vicinity of the cluster, as described by Maxwell-Garnett and Mie.

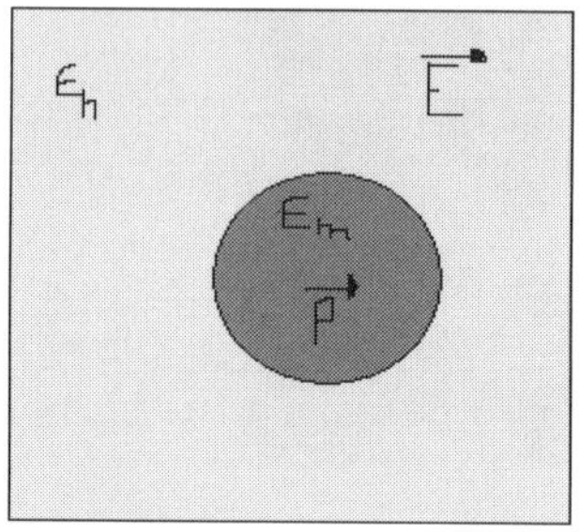

Figure 6: Spherical nanoparticle of metal of radius a, with dielectric constant ε_m, embedded in a glass matrix of dielectric constant ε_h, with dipole **P** induced by the field **E**.

The effects of classical confinement on third-order susceptibility of the composite medium can be derived by applying Maxwell's equations for the case in which the susceptibility is nonlinear; the calculation is carried out to the first order in the electric field. For a charge-free medium, it is $\nabla.D = \nabla.\left[\varepsilon E + \chi^{(3)}.|E|^2.E\right] = 0$ and, if the radius of the nanoparticles a satisfies the inequality $(\omega n_0 a / c << 1)$, and if $a << \delta$, where δ is the skin-depth for the metal, and then it is also approximately true that $\nabla \times E = 0$. This is consistent with a quasi-static regime approximation. For example, for visible frequencies, the approximation is accurate for particles with diameters less than 50 nm. Under these circumstances, it can be shown that the nonlinear optical susceptibility of the composite assumes a form analogous to the nonlinear conductivity of granular materials [22],

$$\chi^{(3)}_{eff} = p.\chi^{(3)}_{QD} \frac{\left\langle |E|^2 E^2 \right\rangle}{E_0^4} \tag{19}$$

where the brackets denote a spatial average over the fields, $\chi^{(3)}{}_{QD}$ denotes the susceptibility of the metal quantum dot (QD) and E_0 is the electric field far from the nanoclusters. For spherical metal nanoclusters, the field E inside the particle is given by [23],

$$E = E_0 \frac{3\varepsilon_2}{\varepsilon_1 + 2\varepsilon_2} \tag{20}$$

Substituting Eq. (20) in Eq. (19), we have for the effective third-order nonlinear optical susceptibility of the composite,

$$\chi_{eff}^{(3)} = p.\chi_{QD}^{(3)}\left|\frac{3\varepsilon_2}{\varepsilon_1+2\varepsilon_2}\right|^2\left[\frac{3\varepsilon_2}{\varepsilon_1+2\varepsilon_2}\right]^2 \equiv p.\chi_{QD}^{(3)}|f_c|^2(f_c)^2 \tag{21}$$

where the quantity f_c may be considered as a measure of the local-field enhancement of the polarization. More rigorous, self-consistent treatments using a jellium model [24] for the metal particle yield the same result for this special case.

5.2 Quantum Confinement:

The mean-field theory of classical confinement effect gives an expression relating the effective susceptibilities of the composite to the susceptibilities of the metal cluster itself. The susceptibilities of the metal clusters, on the other hand, have to be calculated from a microscopic picture of the electron gas. In the case of metal nanoclusters embedded in silicate glasses, confinement is strong. The electronic distribution rapidly decreases to zero at distances of the order of $r_F = 1/k_F$, which is about a few angstroms. The electrons are viewed, at a first approximation, as free particles moving inside metal nanocrystallites, with a spherical potential well that can be well approximated as of infinite height. Indeed, the band gap of silica is about 9 eV, and the energy barriers for a transition between the filled metal valence states and the conduction band of the fused silica are about 3.8 eV for Cu and 4.1eV for Au, respectively [25]. For spherical quantum dots (QD), the electronic structure is described by the well-known functions for a 'particle in a box', which depend on principal quantum number (n), spin and angular momentum quantum numbers (l,m), and a band index b [26],

$$\psi_{bnml}(r) = Y_{lm}(\theta,\phi)\frac{1}{R}\left(\frac{2}{r}\right)^{1/2}\frac{J_{l+1/2}(k_{nl}r)}{J_{l+3/2}(k_{nl}R)}U_b(r) \tag{22}$$

where the functions J's are the usual Bessel functions of half-integer argument and the $Y_{lm}(\theta,\phi)$ are the standard spherical harmonics. If one assumes another shape for the quantum dot, a cube, for example, the form of the wave function changes accordingly. The energy levels of a spherical particle with bandgap energy (E_{gap}) are given by

$$E_{bnl} = \pm\frac{E_{gap}}{2} + \frac{\hbar^2 k_{nl}^2}{2m_b} \tag{23}$$

For metals, of course, the bandgap vanishes, and the effective masses of electrons and holes are roughly equal to each other. As a rule of thumb, one can set the density of states equal to the number of valence electrons divided by the Fermi energy. A more exact expression for the density of states at Fermi level is [27]

$$D(E_F) = \frac{V.m_e k_F}{2\hbar^2\pi^2} = \frac{2\pi.a^3(3\pi^2 n_e)^{1/3}}{2\hbar^2\pi^2} \tag{24}$$

where a is the particle radius and n_e is the electron density. In noble metals, typical Fermi energies are of the order of a few eV; for a 1 nm diameter quantum dot, the level spacings are of the order of tens of meV. Because this is substantially greater than the bandwidth of picosecond lasers, this implies that quantum dots in this size range may be viewed as having distinct, well-separated energy levels. Quantum mechanically, there are three possible optical transitions in a metal nanoclusters, each with its own characteristic size dependence. These three cases are: (1) *intraband transitions*, (2) *interband transitions*, and (3) *hot-electron transitions*.

The *intraband transitions* originate in the filled conduction-band states near the Fermi level and terminate in other conduction-band states which satisfy the selection rules for electric-dipole transitions. Because both the initial and final states are free-electron-like, these transitions show the strongest quantum-confinement effects, because the initial and final states both 'feel' the effects of the boundary surface of the quantum dot.

The *interband transitions* are from the *d*-like orbitals of the valence band to the empty conduction-band states, and these transitions are the ones, which produce the characteristic colors of metals. These states are only weakly dependent on quantum size - effects because the initial state is already localized in space. The electric-dipole transitions between the d-band states and the quantum-confined conduction-band states gives rise to a third-order nonlinear optical susceptibility [28]. The transition probabilities are proportional to the density of states at the levels consistent with the photon excitation energies – that is, to the oscillator strengths of the allowed transitions. These oscillator strengths increase for specific transitions as the level separation increases, and so become more size-dependent at smaller quantum-dot radii.

Finally, the *hot-electron transitions* are those in which an electron in the conduction band absorbs a photon and is heated, losing its energy by electron-phonon scattering or collisions with the walls. Indeed, when light of frequency near that of plasmon resonance is incident upon a metal cluster-doped glass, part of the energy is transferred to the metal particles. This energy partly promotes d-electrons to the conduction band, and the rest is absorbed by the conduction electrons, which have relatively weak specific heat, and thus can be raised to high temperatures. During the electron-lattice thermalization time, the Fermi-Dirac distribution is modified, part of the one-electron levels below the Fermi one being emptied and part of the levels above becoming occupied. The hot-electron transitions can be particularly strong and can, under some circumstances, be the dominant transitions for a metallic nanoclusters in an optical field, giving positive imaginary (absorptive) contribution to the third-order susceptibility. While these transitions produce a third-order nonlinear susceptibility with magnitudes as 10-10^4 times as large as the intraband transitions at a given wavelength, the excited states tend to have relaxation times of the order of hundreds of picoseconds.

6. Ion Implantation: Synthesis of Metal Nanoclusters

It has been demonstrated that the inherent nonlinear response of a dielectric host material such as glass, polymers etc, can be enhanced by several orders of magnitude by introducing small metal clusters within the host matrices. There are both chemical and physical routes to synthesize nanoclusters. Amongst the chemical routes, Sol-Gel has been a well-known technique to grow metal nanoclusters in glasses [29]. Although there are several physical and chemical techniques to grow nanoclusters, it would be desirable to use a method in which the nonlinearity could be confined to specific patterned regions in order to provide effective designs of integrated optical devices. This clearly calls for the use of ion implantation. Although several other methods exist for introducing metal into the insulating substrate, the ion implantation has the advantage of being a generally applicable process; it can be performed at an ambient temperature, it has no side-diffusion problems. Moreover, it offers an accurate control of total number of ions being added to the target and a predictable depth distribution in the target matrix determined by the incident ion-beam energy. Ion implantation in glassy structures provides the precipitation of metal colloids at a reasonably higher local concentration because of the large specific volume and more open structures of the glassy state relative to that of the crystalline counterpart.

The physical mechanisms governing cluster formation due to ion implantation have been reviewed comprehensively[30]. Both nuclear and electronic stopping have relative roles to the structural changes in materials. Perez et.al [31] considered a simple statistical model and described the role of a crystalline host matrix structure in the determination of the final compounds upon ion irradiation. Hosono [32] proposed a criterion based on physical and chemical considerations. The main defects produced during implantation in silica are oxygen - deficient centres, namely Si-Si bonds and a neutral oxygen monovacancies. The concentration of Si-Si bonds results ion specific, i.e , depending critically on the chemical interaction among the implanted elements, Si and oxygen atoms in the glass matrix. In case of strong chemical interactions, implanted ions (M) tend to form M-O bonds, leaving Si-Si bonds, whereas for weak chemical interactions, a large part of the implanted atoms do not react with oxygen atoms. Implanted (M) and Si ions make a competition for bonding oxygen, and cluster formation will occur when the chemical affinity of M ions for the oxygen is smaller than that of Si^{4+}. Clusterisation is expected when Gibb's free energy for an oxide formation with the implanted element M is greater than that for SiO_2. The chemistry considered here is the formation of M_xO_2 oxide starting from the element M in the metallic form and the molecular gaseous oxygen as a consequence of thermal spike phenomenon.

Glasses containing crystallites of metals exhibit an enhanced third-order susceptibility, whose real part is related to the intensity dependent refractive index η_2 defined in terms of the linear index η_0 and of the light intensity I, as expressed by Eq.(4). The technological interest in metal nanocluster -glass composites is strengthened by the general interest in strongly quantum-confined electronic systems, which exhibit peculiar effects deriving from the increased electronic density of states near the conduction-band edges. Metal nanoclusters in dielectric hosts exhibit several peculiar optical behaviors; depending on their size, mean energy-level spacing, and the local electronic and chemical environment. In contrast to semiconductor clusters, many more electronic energy levels are accessible for clusters of a given size and therefore, the problems of inhomogeneous broadening appear more severe. However, the nonlinearities in metal clusters also exhibit a strong quantum-size effect that can be exploited to enhance the magnitude of the third-order nonlinearity and the energy relaxation time may be substantially faster.

7. Characterizations of the Metal Nanocluster-Glass Composites

The study of metal nanocluster-glass composites is rapidly growing towards the novel aspects of their physical characteristics, so new techniques are investigated to characterize the composites from optical, structural, chemical and mechanical point of view. Optical absorption spectroscopy has been widely used on dielectric composites containing small metal colloids; because the location, amplitude and width of the SPR bands are an excellent zeroth-order diagnostic of the nanocluster species and their sizes. However, the resonant plasmon response exhibited in UV-Vis spectrum decreases in amplitude and broadens with increasing nanocluster size, making absorption spectroscopy a less useful tool precisely in the region where the nonlinear effects could be strongest. Cross-sectional transmission electron microscopy (X-TEM) and related x-ray energy dispersive spectroscopy (EDS) and energy-energy loss spectroscopy (EELS) give information on the depth distributions of cluster shape and size and also crystallinity of the nanostructures. Rutherford backscattering spectrometry (RBS) and secondary ion mass spectroscopy (SIMS) can provide the concentration-depth distributions of the implanted species.

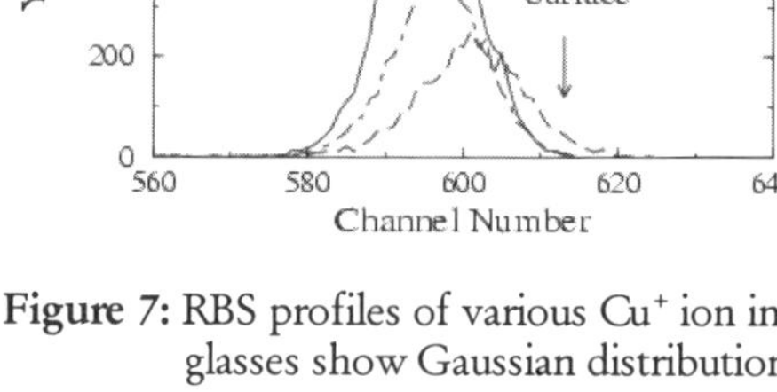

Figure 7: RBS profiles of various Cu^+ ion implanted glasses show Gaussian distributions

Figure 7 shows typical RBS profiles of copper implanted glass samples [34], showing Gaussian distributions in all the cases. In case of samples implanted with 100 keV Cu^+ ions, the implanted Cu distribution has been found to shift towards shallower depth with a projected range of about 50 nm for a dose of 3×10^{16} atoms/cm^2, as evidenced from Fig.(8). However, for a dose of 1×10^{17} atoms/cm^2, the distribution remains Gaussian with the projected ion range of about 94 nm. There have been quite a few studies reporting the out-diffusion of implanted Cu atoms during implantation with a higher beam current density. This probably has caused a splitting in the implanted Cu distribution, as seen in the figure, indicating the fractional concentrations of Cu near the surfaces. The distribution of Cu atoms in 200 keV Cu-implanted fused silica is nearly Gaussian with a projected ion range and longitudinal range straggling of 94 nm and 30 nm, respectively.

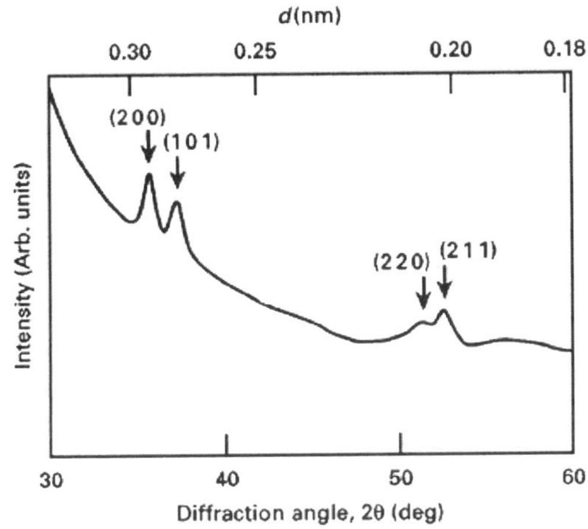

Figure 8: X-ray diffraction (XRD) of the Sn-nanoclusters in Sn^+ implanted silica glass

X-ray photoelectron spectroscopy (XPS) and Auger electron spectroscopy (AES) determine the chemical states of the implanted species. The XPS and AES spectra of copper-implanted silicate glasses [35] confirmed the presence of metallic copper inside the glass matrices. X-ray diffraction (XRD) peaks ensure the crystallinity of the nanoclusters, as evident from Figure 8.

7.1 Charactrizations for Nonlinear Optical Effects

Amongst the techniques directly sensitive to the third-order susceptibility, there are two widely used spectroscopic techniques to determine the size, speed and dephasing of the 3rd-order nonlinear optical effects. One, known as Z-scan [36], is conceptually derived from the far-field intensity distribution measurements, first described by Weaire et al [37]. The other one for measuring is the four-wave mixing in Degenerate Four-Wave Mixing (DFWM) method. The principles of these two nonlinear absorption spectroscopy techniques have been discussed in the following.

7.1.1 Z-Scan method:

The Z-scan is a simple and popular experimental technique to measure intensity- dependent nonlinear susceptibilities of materials. The excitation source used is a pulsed laser. Each laser pulse has a Gaussian spatial profile and is focused by a converging lens. The sample under investigation is placed near the waist of a focused Gaussian beam and moved in the direction of the propagation of the light (Z-axis). At the focal point of the lens the laser intensity is maximum. Therefore, at each point the sample faces different intensity of light as it moves in the Z-direction. The light intensity transmitted across the nonlinear material is measured in the far field (FF) as the sample is moved along the direction of the propagation of light, in the open z-scan mode of operation. When the sample moves towards the focus from negative Z, the laser power density intercepted by the sample increases, giving rise to 'self-focusing' [38]: owing to the Gaussian transverse intensity profile of the laser beam, the original plane wavefront gets progressively more distorted, in a way similar to that imposed by a positive lens (in the case of positive nonlinearity of the material), leading to a self-focusing that shifts the position of the actual focal point. A displacement of the focal point toward negative Z gives rise to an increased divergence of the output beam, and thus to a decrease in intensity at the detector. For positive values of the Z-position of the sample, the same effect gives rise to a decreased divergence, that is, to an increased intensity at the detector. When the sample is at Z = 0, the self-focusing effect does not affect the output. The same occurs when the sample is far from the focal point, because the power density is low and self-focusing effect is negligible. Figure 9 illustrates the principle of the Z-scan method based on self-focusing effect on a single beam allowing one to obtain the magnitude and the sign of the nonlinear refraction. A photodiode detector simultaneously measures the intensity of the transmitted beam through the sample.

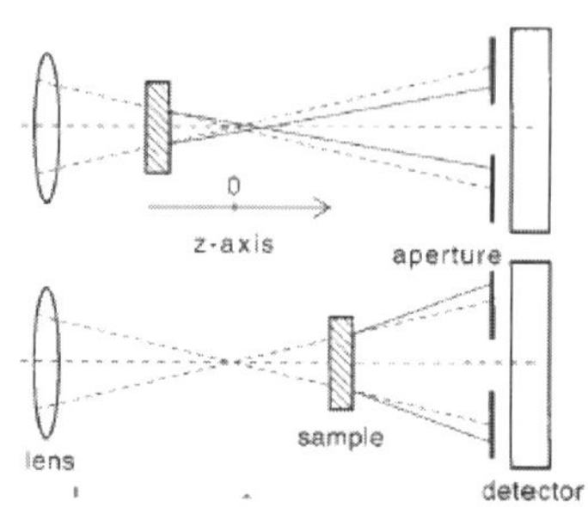

Figure 9: Schematic diagram of the principle of the Z-scan method

For open aperture Z-scan measurements, the analytical expression for the normalized transmittance can be written as [39]

$$T(z) = A(z)\frac{1}{q(z)\sqrt{\pi}}\int_{-\infty}^{\infty}\ln\left[1+q(z)\exp\left(-t^{2}\right)\right]dt \tag{25}$$

Where $A(z)=\exp(\alpha_o lI(z)/(I_s+I(z))$, $q(z)=\dfrac{\beta I_o L}{1+(\frac{z}{z_o})^2}$ and $L=\dfrac{1-\exp(-\alpha_o l)}{\alpha_o}$, I_o is the laser intensity at the focal point, l is the sample length, $z_O\,(=\pi\omega_O^2/\lambda)$ is the Rayleigh range, ω_0 is the minimum beam waist at the focal point. β and I_s are the fitting parameters used in Fig(13). The effective nonlinear absorption coefficient β_{eff} can be estimated from the graph of Fig.10 using the Z-scan theory [39]. The corresponding imaginary part of the third-order nonlinear susceptibility $\chi^{(3)}$ can be calculated from the value of β_{eff}. However, for open aperture Z-scan measurements two types of profiles are observed. One corresponds to the saturable absorption (Fig. 10) and the other to inverse saturable absorption. The phenomenon of inverse saturable absorption (not shown in the figure) can be easily understood. As the incident intensity increases while the sample moves towards the focal point, the absorption also increases within the sample, but in a nonlinear fashion, thus decreasing the transmitted intensity. Reverse happens when the sample is moved away from the focal point.

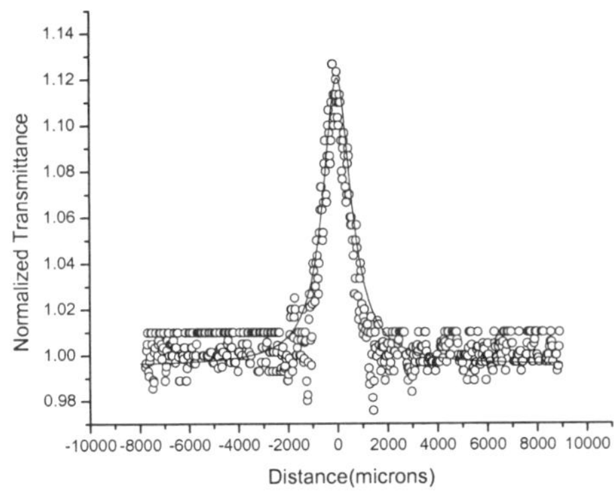

Figure 10: Open-aperture Z-scan for Cu^+ ion-implanted annealed glass

Exact mechanism of saturable absorption is not clear. However, the following processes can give some ideas. Electron dynamics in metal nanoparticles following laser excitation has been studied in detail [40]. Results of those investigations are relevant in the context of present measurements. When nanoparticles are irradiated, the electrons corresponding to the SPR band gets excited, which leads to the dipole and higher order (quadrupole, octopole etc) oscillations. These oscillations can couple with the applied electric field. Consequently, SPR frequency of excited atoms differs from that of unexcited atoms. The electrons to this SPR band cannot further absorb in the original SPR band region. This leads to SPR bleach [41]. The excited electrons have an energy higher than Fermi energy and hence, they are called hot electrons. These hot electrons get thermalized by dissipating excess energy through successive processes of electron-surface scattering, electron-electron scattering and electron-phonon scattering [40]. Exact time-scales of all these processes may vary according to the type and environment of nanoparticles. At the end of thermalization process, heat energy is transferred to the surrounding medium. This excess thermal energy will increase the temperature of the surrounding medium, which in turn influences the SPR. Full recovery of plasmon bleach is delayed and transient absorption is observed till thermalization of hot electrons is complete. The electron-electron scattering is very fast and occurs in sub-picosecond timescale. The component of transient absorption taking place in nanosecond timescale is mainly due to photochemical change induced absorption arising from photo-ejection of electrons from the sample on laser irradiation. Photo ejection of electrons is a multi photon process. The ejected electrons can change the nanoparticle surface electrically, leading to aggregation, which results in a transient state ($Cu^{+}e^{-}$) generated by photo induced intra-particle charge separation [40]. Besides, ejection of electrons creates holes, which can act as free carriers giving rise to free carrier absorption. Another kind of nonlinear absorption process could be due to nonlinear scattering causing optical limiting and possible particle size-selective excitations.

7.1.2. Four wave mixing method:

Principle:

When a nonlinear material is illuminated by a pair of coherent waves of equal magnitude traveling in different directions, they result in a phase grating within the material, which in turn can diffract a third wave (not necessarily coherent with the first two) incident on the material into a fourth one. The diffracted phase conjugated wave ids the fourth wave. This process is called 'four-wave mixing' and has been demonstrated in several crystals. When all the four waves involved in the process have the same frequency, it is known as the degenerate four wave mixing (DFWM). The effect can be understood in terms of the third-order nonlinear susceptibility $\chi^{(3)}$. Physically, one may understand this process by considering the individual interactions of the fields within a dielectric medium. The first incident field causes in the dielectric an oscillating polarization, which re-radiates with some phase shift determined by the damping of the individual dipoles; this is just traditional Raleigh scattering described by linear optics. The application of a second field will also drive the polarization of the dielectric, and the interference of the two waves will cause harmonics in the polarization at the sum and difference frequencies. Now, the application of a third field will also drive the polarization at the sum and difference frequencies. This beating with the sum and difference frequencies is what gives rise to the fourth field in four wave mixing. Since each of the beat frequencies produced can also act as new source fields, a bewildering number of interactions and fields may be produced from this basic process.

Experiment:

Figure 11 shows the forward and backward mixing arrangements. In both the cases, there are two pump beams and one probe beam that are split off from a single laser beam. The pump beams are arranged to be coincident in time in the sample volume, while the probe beam arrival time in the sample can be varied with respect to pump beams by an optical delay line. In both the cases, the nonlinear polarization radiates a signal field at ω in the phase-matched direction. The pump and the probe waves interfere giving rise to both amplitude gratings (resulting from the nonlinear loss due to two-photon absorption) and refractive index or phase gratings (due to the intensity dependent refractive index change), from which the waves scatter coherently. The output intensities of signal and probe are related to the nonlinear absorption from which one can be related to the nonlinear refractive index. Because the relative temporal duration of the laser-induced changes in the medium is measured directly varying the timing of the probe-and-pump beams, DFWM gives a direct indication of the speed at which the material relaxes, i.e. nonlinear response time.

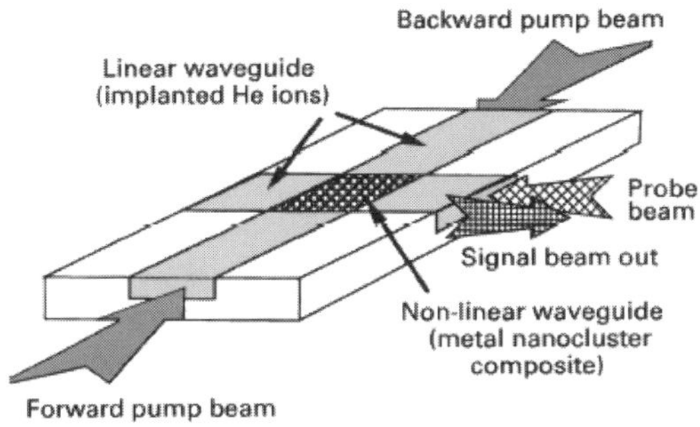

Figure 11: Forward and backward mixing arrangements of DFWM set up.

Measurements of the magnitude and the temporal response characteristics of $\chi^{(3)}$ are the most important part of the experiments. One can measure $\chi^{(3)}$ at wavelengths both on and off the SPR by forward DFWM in the geometry of Figure 11 [42]. DFWM experiments also give important information on the switching speed. $\chi^{(3)}$ for gold nanoclusters in glass was measured by Magruder *et al.* [44] by DFWM using a mode-locked Q-switched, frequency doubled Nd:YAG laser at a wavelength of 532 nm with a pulse repetition frequency of 10 Hz and a nominal pulse width of 35±5 ps. This wavelength is near the SPR of gold (530 nm). The two forward-going pump-beams and a weak probe-beam were arranged to intersect in the composite-layer containing the metal clusters. The probe was delayed with respect to the two pump beams using a computer-controlled optical delay line. The pump and probe beams interacted coherently via the third-order non-linear susceptibility to produce a phase-conjugated signal, detected in a photo-multiplier. Identical measurements were carried out on unimplanted samples; no measurable DFWM signal was observed. Figure 12 shows the intensity of the phase-conjugated signal as a function of pump probe delay time [44]. The symmetric shape of the time spectrum indicates that the third-order response is no longer than that of the pulse width (~35 ps). The measured values of $\chi^{(3)}$ were found to be 1.0×10^{-10} e.s.u and 1.7×10^{-10} e.s.u. for this nanocluster composite layer without and with heat treatment, respectively.

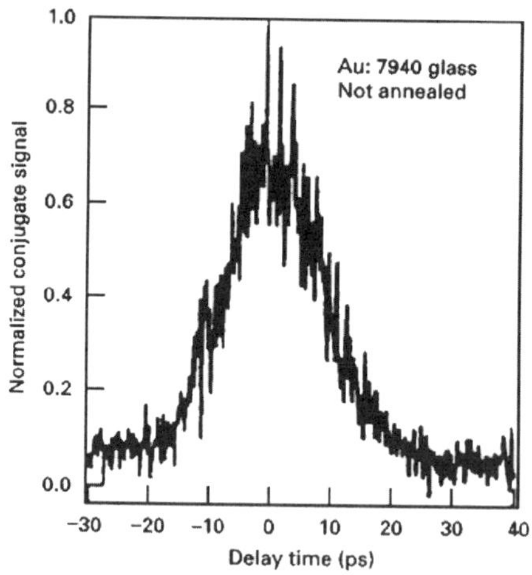

Figure 12: Intensity of the phase-conjugated beam in a DFWM measurement with cross-polarized pump-probe configuration as a function of pump-probe time in picoseconds.

There are other optical techniques, such as Raman scattering, femtosecond pump-and-probe spectroscopy, photon echo techniques, nondegenerate four-wave mixing, electro-absorption spectroscopy, magneto-absorption pump-probe spectroscopy, interferometry, etc. which have been developed in the last few years [30] for various promising applications in metal nanocluster-glass composites analysis. These techniques are not discussed in the present article. The main impediment to the practical exploitation of non-linear optics is the difficulty in understanding and developing materials that can be engineered to match device needs. New device concepts and material developments are two key factors for future success. The criteria used to define the figure of merit are conditional and so the selection of material is crucial. The best material for one application may be the worst for another. Although the researchers are dedicated in pursuing considerable amount of works in this direction globally, realization of the photonic materials is yet in the stage of infancy. The main bottleneck is in the optimization of the parameters used in the synthesis, probably because of the lack of systematics in the fabrication process. To our knowledge, third-order susceptibilities ($\chi^{(3)}$) having values greater than $\sim 3\times10^{-6}$ e.s.u (for Sn-nanoclusters of radii 2-10 nm in silica glass) [45] have not been reported so far. Furthermore, NLO materials prepared by metal ion implantation having responses in the less than picosecond time domain have also not been reported. It is therefore obvious that much more emphasis is still needed in the research towards a thorough and detailed investigation on the role of implantation energy, dose and annealing temperature and time for various sets of ion-target combinations.

References

1. M. Faraday, *Phil. Trans. R. Soc.* **147**, 145 (1857)
2. P. A. Franken, A. E. Hill, C. W. Peters, and G. Weinreich, *Phys. Rev Lett.* **7**, 118 (1961)
3. T. H. Maiman*, R. H. Hoskins*, I. J. D'Haenens, C. K. Asawa, and V. Evtuhov, *Phys. Rev.* **123**, 1151(1960)
4. P. Chakraborty, *Journal of Material Science*, **33** 2235 (1998)
5. Robert W. Boyd, *Nonlinear Optics*, Academic Press
6. C.N.Ironside, Ultrafast all-optical Switching, *Contemporary Physics*, **34** (1993)
7. Sheik-Bahae, M.Hutchings, D.C. Hagan, Dispersion of bound electronic nonlinear refracting solids, *IEEE J. Quant. Elect.* 1296
8. Y.Gao, X.Zhang,Y.Li, H. Liu, Y.Wang, Q.Chang, W.Jiao, Y. Song, *Opt. Commun.* **251,** 429, (2005)
9. E. Cattaruzza, G. Battaglin, F. Gonella, G. Mattei, P. Mazzoldi, R. Polloni and B. F. Scremin, *Appl Surf. Sci.* **247**, 390, (2005)
10. S. R. Friberg and P. W. Smith, *IEEE J. Quant. Elect* -**23** 2089 (1987)
11. Eva M. Vogel, Glass as Nonlinear Photonic Materials, *J. Am. Ceram. Soc*, **72**
12. Robert Adair, L. L. Chase, and Stephen A. Payne, *Phys. Rev B* **39**, 3337(1989)
13. J.C. Maxwell-Garnett, *Philos. Trans. R. Soc A* **203**, 385 (1904)
14. G.W.Arnold and J.A.Borders, *Journal of Appl. Phys*. **48**, 1488 (1977)
15. L.Yang, D.H. Osborne, R.F. Haglund, R.H. Magruder, III, C.W.White, R.A.Zuhr and H.Hosono, *Appl. Phys. A* **62**, 403 (1996)
16. W. C. Huang and J. T. Lue, *Phys. Rev B* **49**, 17, 179 (1994)
17. W. J. Doyle, *Phys. Rev.* **111** 1067(1958).
18. *Idem, ibid.* **42** 414 (1965).
19. H. Amekura, K. Kono, Y. Takada, N. Kishimoto, *Appl. Phys. Lett.* **87**, 153105, (2005)
20. G Mattei, G. De Marchi, C. Maurizio, P. Mazzoldi, C. Sada, V. Bello and G. Battaglin, Phys. Rev. Lett. **90**, 085502-1/4, (2003)
21. G. Battaglin, M. Catalano, E. Cattaruzza, F. D'Acapito, C. De Julian Fernandez, G. De Marchi, F. Gonella, G. Mattei, C. Maurizio, P. Mazzoldi, A. Miotello and C. Sada., *Nuclr. Instrum. Meth. In Phys. Res.* **B 178** ,176 (2001)
22. D. Stroud and P.M. Hui, *Phys. Rev B* **37**, 8719 (1988)

23. L. Ward, in "The Optical Constants of Bulk materials and Films," IOP Publishing, Bristol, (1988)

24. F. Hache, D. Richard and C. Girad, *Phys. Rev B.* **38**, 7990 (1988)

25. A.S. Grove, " Physics and Theory of Semiconductor Devices," Wiley, New York (1967)

26. S. Schmitt-Rink, D.A.B. Miller and D.S. Chemla, *Phys. Rev B* **35**, 8113 (1987)

27. W. P. Halperin, *Rev Mod. Phys.* **58**, 533 (1986)

28. F. Hache, D. Ricard, C. Flytzanis and U. Kreibig, *Appl. Phys. A* **47**, 347 (1988)

29. G De, G. Mattei, P. Mazzoldi, C. Sada, G. Battaglin, A. Quaranta, *Chem. Mater.* **12**, 2157 (2000)

30. F. Gonella and P. Mazzoldi, in : Handbook of Nanostructered Materilas and Nanotechnology (Ed: H.S. Nalwa) Vol. **4**, pp. 81, Academic Press, San Diego (2000), and references there in.

31. A. Perez, G. Marest, B. D. Sawicka, J.A. Sawicki and T. Tyliszczac, *Phys. Rev B* **28** 1227 (1983)

32. H. Hosono, *Jpn. J. Appl. Phys.* **32** 3892 (1993)

33. R. H. Magruder III, J. E. Witting and R. A. Zuhr, *J. Non-Cryst. Solids* **163** 162 (1993)

34. B. Ghosh, P. Chakraborty, S. Mohapatra, P. A. Kurian, C. Vijayan, P. C. Deshmukh and P. Mazzoldi, *Philosophical Magazine* (communicated)

35. P. Mazzoldi, F. Caccavale, E. Cattaruzza, P. Chakraborty, L. Tramontin, A. Boscolo-Boscoletto, R. Bertoncello, F. Trivillin, G. Battaglin and G.W. Arnold, *Nucl. Instrum. Meth.* B **91**, 505 (1994)

36. M. Sheik-Bahae, A. A. Said, T. H. Wei, D. J. Hagen, and E. W. van Stryland, *IEEE J. Quantum Electron.* **QE 26**, 760 (1990)

37. D. Weaire, B. X. Wherrett, D A B Miller, and S D Smith, *Opt. Lett.* **4**, 331 (1979)

38. Y. R. Shen, *The Principles of Nonlinear Optics* (Wiley, New York, 1984)

39. S. Link, C. Burda, Z. L. Wang and M. A. El-Sayed, *J. Chem. Phys.* **111,** 1255, (1999)

40. P V Kamat, M Flumiani and G V Hartland, *J. Phys. Chem* B **102,** 3123, (1998)

41. J Y Bigot, V Halte, J C Merle and A Daunois, *Chem. Phys.* **251,**181, (2000)

42. R. F. Haglund Jr, L. Yang, R. H. Magruder , C.W. White, R. A. Zuhr, Lina Yang, R. Dorsinville and R. R. Alfano, *Nucl. Instrum. Meth.* **B 91** 493 (1994)

43. R. C. Caro and R. C. Gower, *IEEE J. Quantum Electron.* **QE-18** 1375. (1982)

44. R. H. Magruder III, L. Yang, R. F. Haglund, C. W. White, Lina Yang, R. Dorsinville and R. R. Alfano, *Appl. Phys. ,Lett.* **62** 1730 (1993).

45. Y. Takeda, T. Hioki, T. Motohiro and S. Noda, *Nucl. Instrum. Methods B* **91** 515 (1994)

Study of atom-surface interaction using magnetic atom mirror

A. K. Mohapatra
Fundamental Interactions Lab (Gravitation Group)
Tata Institute of Fundamental Research
Homi Bhabha Road, Colaba
Mumbai - 400 005, India

Abstract

We describe the experimental study of the atom-surface interaction, especially the quantum electrodynamical van der Waals and Casimir-Polder force on atoms in ground state near metallic surface, employing the reflection of cold atoms from magnetic atom mirror. The coefficient of the van der Waals interaction is determined with an uncertainty of 15% using the novel technique.

1 Introduction

The attraction of neutral atoms to the conducting surface separated by more than the typical atomic size was predicted by van der Waals in 1881. The simple Lennard-Jones model, which was based on the electrostatic interaction of a fluctuating atomic dipole due to quantum fluctuation and it's image shows z^{-3} dependence law of the interaction potential. The full QED treatment of the van der Waals force leads to the famous long distance z^{-4} law known as the Casimir-Polder potential, which is a fundamental QED problem involving quantized electromagnetic field and the retardation effect [1]. The van der Waals force represents the universal interaction of importance in numerous field of physics, chemistry and biology. Its attractive nature is essential for cohesion in many chemical and bio-physical system. This force also plays an important role in atomic force microscopy. In the field of integrated atom-optics, the van der Waals and the Casimir-Polder forces cause the decoherence and limits the usefulness of mater-wave interferometry on the chip [2]. An accurate measurement of such force is also important in getting useful constraints on the hypothetical short range forces like non-Newtonian gravity at sub-micron length scale speculated in the context of higher dimensional particle physics theories [3].

Thus, the study and precision measurements of the atom-surface interactions are of importance to a variety of topics ranging from precision spectroscopy to nanoscience, and to even particle physics.

The interaction of atoms near a surface can be probed by measuring the spectral shifts of the atomic transition, modification of the atomic trajectories and phase shifts in the atom interferometers. The initial experiments to measure the van der Waals force between neutral atoms and a surface were done by passing a thermal atomic beam near a metallic cylinder. Other significant measurements include high resolution spectroscopy on Rydberg atoms inside a micron sized parallel plate metallic cavity and the verification of the energy level shift due to the van der Waals force, measurement of the Casimir-Polder force by measuring the transmission of atomic beams through a cavity, measurement of the van der Waals interaction of inert gases with Silicon Nitride by measuring the diffraction intensity of atoms through a transmission grating, measurement of the van der Waals force of alkali atoms with Silicon Nitride using atom interferometry and diffraction experiments, and the measurement of the large deflection of metastable noble gas atoms due to the change in their internal energy state arising from the quantum mechanical perturbation by the atom-surface van der Waals force, when atoms pass through a fine grating. There have also been some recent measurements on ultra-thin vapor cells. A short survey and relevant reference to the experiments and calculation are available in reference [1, 4].

2 Cold atoms as probe for atom-surface interaction

Neutral atoms cooled to a temperature of a few micro-Kelvin or even lower are the ideal tools for probing directly and precisely the effect of the short range interactions. Typical experiments probing short range interactions between surfaces and atoms measure the change in the trajectory, spectral frequencies or phase of the matter waves. Using the cold atoms improves the potential sensitivity by an enormous factor, as compared to the experiments with thermal beams. For example, the deflection of an atomic trajectory near a surface depends on the atom-surface force at a particular distance as

$$d \approx \frac{F(r)}{2m} t^2 \tag{1}$$

where m is the mass of the atom and t is the interaction time. $t \propto 1/v$, where v is the average velocity, which can be a factor of 10^3 lesser for laser cooled atoms. Hence, using cold atoms improves the sensitivity by factors exceeding a million. Similarly, in interferometry with atomic matter wave, the phase shift depends on the interaction energy and time as $\Delta\phi \approx Et$, and the measured shift in the fringes scales with the de Broglie wavelength as $\delta \approx \Delta\phi\lambda \approx 1/v^2$, as in the case of the deflection of the trajectory.

Advances in laser cooling and trapping of neutral atoms has opened up a wide range of other techniques to control and manipulate the atomic trajectories. It has become possible to control the atomic motions near surfaces, which has opened up a possibility to probe the atom-surface interaction with a better precision. Some notable experiments include the measurement of the van der Waals and the Casimir-Polder force from the modification of the reflectivity of cold atoms from a blue-detuned evanescent wave atom mirror [5], and from quantum reflection of cold Neon atoms from Silicon and BK7 surface at grazing incidence [6]. The recent high precision measurement by Harber et al [7] of the Casimir-Polder force using magnetically trapped ^{87}Rb Bose-Einstein condensate, by detecting the perturbation of the frequency of the center-of-mass oscillations of the condensate perpendicular to the surface, highlights the new possibilities in precision measurements in the field.

Recently, we have performed an experiment at TIFR to probe the atom-surface interaction employing laser cooled atoms. We performed a new measurement of the van der Waals force between a ground state atom and a conducting wall employing a novel technique involving reflection of cold atoms from a magnetic thin film atom mirror. The van der Waals coefficient was determined with an estimated 1σ error of 15% [8]. The details of the ideas and the experiment are reported here.

3 Principle of magnetic atom mirror

The principle of the magnetic atom mirror is based on spin dependent quantized Stern-Gerlach force. An atom in the magnetic field of magnitude B has a magnetic dipole interaction energy $U = -\mu_\varsigma B$, where μ_ς is the projection of its magnetic moment in the field direction. In the case of cold atom experiment, atoms moves slowly through the magnetic field. If zeroes of the magnetic field can be avoided, then the magnetic moment of the atoms follows the field direction adiabatically and the angle between the moment and the field remains constant. Hence, the potential energy of the atom becomes independent of the field direction and depends only on the magnitude of the field. In the linear Zeeman regime (less than $\sim$ 300 G), the interaction energy is

$$U = m_F g_F \mu_B B \tag{2}$$

where m_f is the magnetic quantum number, g_F is Lande's g factor and μ_B is Bohr magneton. Atoms in magnetic states of positive $m_F g_F$ are repelled by the Stern-Gerlach force (-∇U) arising from the region of increasing field magnitude. The magnetic atom mirror can be realized by making periodic magnetic structure. For a periodic magnetic structure, the magnitude of the magnetic field decreases exponentially above the mirror as

$$B(y) = B_0 \left(1 - e^{-kb}\right) e^{-ky} \tag{3}$$

where μ_0 is the permeability of vacuum, B_0 is the magnetic field at the mirror surface, b is the thickness of the magnetic layer and $k = 2\pi/a$ is the exponential decay constant, where a is the magnetization periodicity.

Previously, magnetic atom mirrors were realized experimentally by artificially making the magnetization alternate in sign on audio tape, floppy disk, video tape, using an array of permanent magnets, an array of current carrying wires or cobalt single crystal. The detail review of the magnetic atom mirror can be found in reference [9]. The typical magnetization periodicity of these mirrors are more than 10 μm and hence the closest approach of the atoms to the surface is of the same order. Hence the atom-surface interaction is negligible in these atom mirrors and are mostly used in modern atom optics where high reflectivity and smoothness are required.

4 Magnetic thin film atom mirror

We started the idea of using magnetic thin film as a mirror for cold atoms and experimentally realized in our group at TIFR. We used cobalt thin film as magnetic atom mirror. cobalt thin film possesses periodic stripe like domain structure, which was predicted by Kittel. The domain periodicity (λ) of the stripe like structure changes with the thickness of the film (d) and obeys the law $\lambda \propto \sqrt{d}$ [10]. The stripe like domain structure has been observed in the single crystal cobalt thin film grown by molecular beam epitaxy (MBE) [11]. For such a domain structure, the magnitude of the magnetic field above the surface reduces exponentially, with a decay constant related to the domain size. The typical domain size can be as small as 100 nm. So the closest approach of the atoms can be of the same order. At this distance, the atom-surface interaction becomes appreciable. Hence, there is a possibility to probe the atom-surface interaction in these systems. Since the domain periodicity depends on the thickness of the film, the magnetic interaction potential can be varied by using thin films of different thickness.

5 The magnetic atom-mirror experiment at TIFR

We used cobalt thin film as magnetic atom mirror in our experiment, which is ferro-magnetic at room temperature. Our cobalt thin films were prepared by dc magnetron sputtering. The domain structure of the film is imaged using

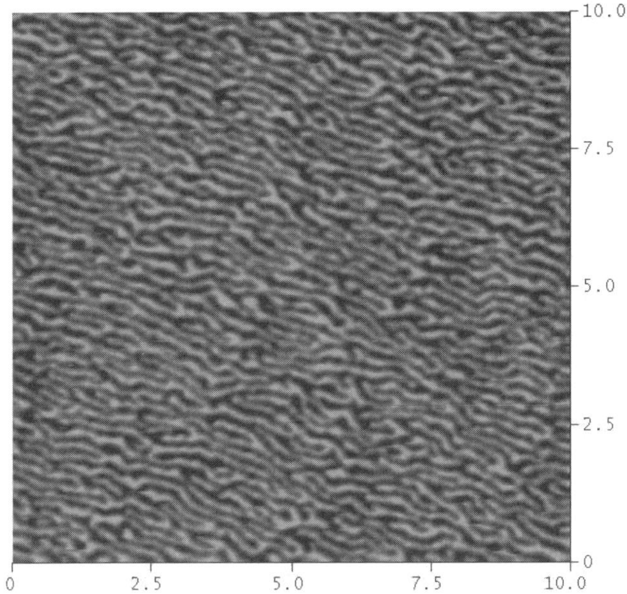

Figure 1: Magnetic force microscopy image of cobalt thin film of thickness 180 nm. The scanning area is 10 μm $\times$ 10 μm.

magnetic force microscopy (MFM), which is shown in figure 1. The pattern of the domains for these films are similar to the MBE grown thin films. Also the mean domain periodicity is same as the MBE grown thin films within 5%.

We used ^{85}Rb atoms for the experiments, cooled and trapped in a Magneto-Optical Trap (MOT) [12], formed in a SS octagonal chamber equipped with glass view ports. The MOT was loaded from Rb vapor from a heated getter source at a pressure less than 1×10^{-9} Torr for about 10 seconds to obtain the cold atomic cloud with about 10^7 atoms. Then the atoms were further cooled in the optical molasses [12] for 15 ms to a temperature of 10 μK (rms velocity of about 3 cm/sec). Atoms were then released on the cobalt thin film situated 17 mm below the MOT center. The rms size of the slowly expanding atom cloud grows to about 3 mm at the surface of the mirror with dimensions 1 cm $\times$ 1 cm. Thus, about 99% of the atoms interact with the thin film atom mirror.

The atoms were detected using a horizontal retroreflected probe laser beam kept between the MOT and the mirror, with uniform intensity over a rectangular area of 20 mm $\times$ 0.5 mm, resonant to the cooling transition line. The frequency of the probe was modulated by modulating the current to the laser at 17 KHz. The absorption of the probe intensity was detected by a photo-diode and a low noise amplifier feeding a lock-in amplifier with the reference derived from the

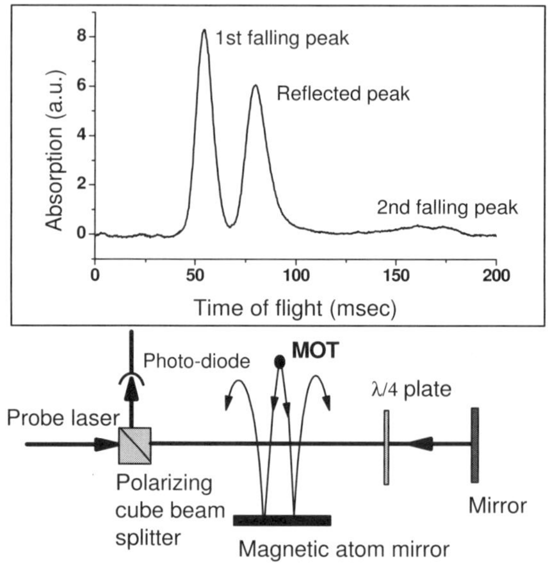

Figure 2: Schematic of the measurement (lower panel). The upper panel shows the time-of-flight signal of the spin polarized atomic cloud passing through the probe beam before and after reflection from the magnetic thin film.

laser current modulation. The sensitivity achieved is sufficient to detect a few hundred atoms in the probe beam using this technique [13]. The schematic of the experimental set up and the corresponding time-of-flight signal is shown in figure 2.

6 Measurement of the van der Waals force

To measure the van der Waals force, the reflectivities of unpolarized cold atoms from cobalt thin film of different thickness – 180 nm, 90 nm, 45 nm, and 20 nm – were measured from the observed time-of-flight signal. The van der Waals coefficient was determined by doing a least square fitting of the experimental data with the computed reflectivity function which was determined by modeling the effective potential (including the repulsive magnetic interaction and the van der Waals potential) experienced by the atoms.

To model the effective potential, the magnitude of the magnetic field above a cobalt thin film is approximated as

$$B(y) = B_0 \left(\tan^{-1} \left[e^{-2\pi y/\lambda(d)} \right] - \tan^{-1} \left[e^{-2\pi(y+d)/\lambda(d)} \right] \right) \tag{4}$$

where B_0 is the magnetic field at the film surface. λ is the domain periodicity

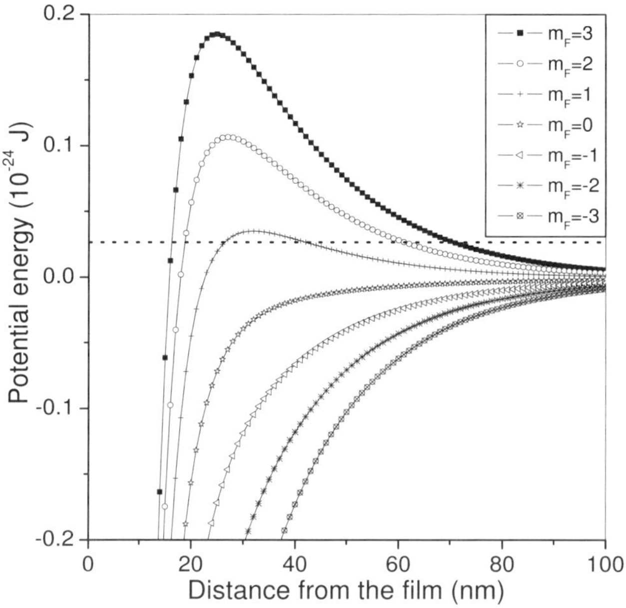

Figure 3: The effective potential experienced by the atoms in different Zeeman sub levels including the van der Waals interaction with the magnetic interaction potential energy. Dotted line is the kinetic energy of the atoms at the surface.

and obeys the law $\lambda \propto \sqrt{d}$, d is the thickness of the film. The surface field of cobalt is expected as a few kG. Hence, the quadratic Zeeman effect becomes important very close to the surface. The magnetic interaction potential for the atoms in various Zeeman sub levels is calculated including quadratic Zeeman effect. For the case of thin film atom mirrors with relatively small thickness, the atom-surface attractive interaction becomes significant, and the effective potential is the sum of the repulsive magnetic potential and the attractive van der Waals ($\frac{-C_3}{z^3}$) potential. The effective potential is shown in figure 3. Close to the surface (at 20-30 nm), the van der Waals attraction dominate over the magnetic repulsion. Hence, a barrier peak appears in the effective potential. Atoms in various Zeeman sub levels with less kinetic energy compared to the corresponding barrier height are reflected from the mirror. Atoms have well defined mean kinetic energy at the mirror surface, since they are dropped from a fixed height. The spread in kinetic energy of the atoms due to the temperature of the cloud is less than 1% of the mean kinetic energy gained due to gravity and hence is neglected. The distribution of atoms released from the optical molasses without spin-polarization in various Zeeman sub-levels is assumed as uniform. The reflectivity is computed by comparing the kinetic energy of the atoms in

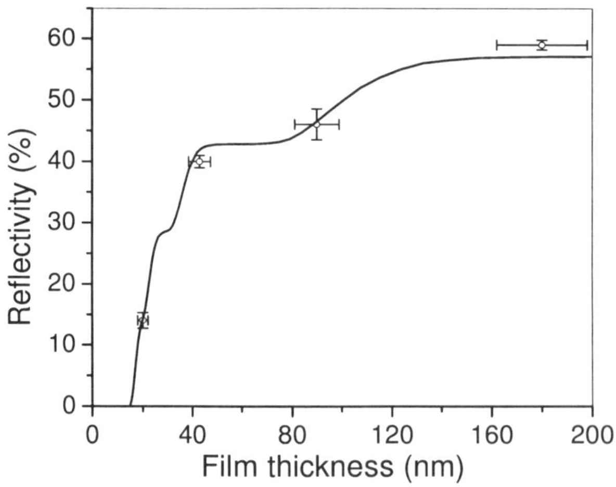

Figure 4: Reflectivity of the un-polarized cold atoms as a function of thickness of cobalt thin film. Circles are the experimental data. Solid line is the computed reflectivity using the van der Waals coefficient (C_3) and the surface field (B_0) determined from the least square fitting

various Zeeman sub-levels with the corresponding barrier heights. Since there are finite number of Zeeman sub-levels, the reflectivity as a function of the thickness of the film is a multi step (quantized) function. Including a small variance in the local surface field of the order of 10%, makes the steps in the reflectivity function smooth.

The computed reflectivity function was least square fitted with experimental data. The surface field (B_0) along with the van der Waals coefficient (C_3) were taken as the fitting parameters. The least square fitted curve along with the experimental data are shown in figure 4. The van der Waals coefficient is 1.75×10^{-48} Jm^3 and the surface field is 1.28 kG, which are determined from the least square fitting.

The errors on the parameters were determined by doing a Monte Carlo simulation. The error on the reflectivity was determined from repeated measurements. The error on the thickness of the films was determined from the profilometer measurement. Random numbers of Gaussian distribution around the mean value of the experimental data were generated by taking the errors on thickness and reflectivity as the Gaussian half width. Computed reflectivity was fitted with the generated points and the fitted values of the parameters were recorded. The distribution of the parameters generated from the Monte Carlo simulation is shown in figure 5. The 1σ estimate on the width of the distribution of the van der Waals coefficient is 15%.

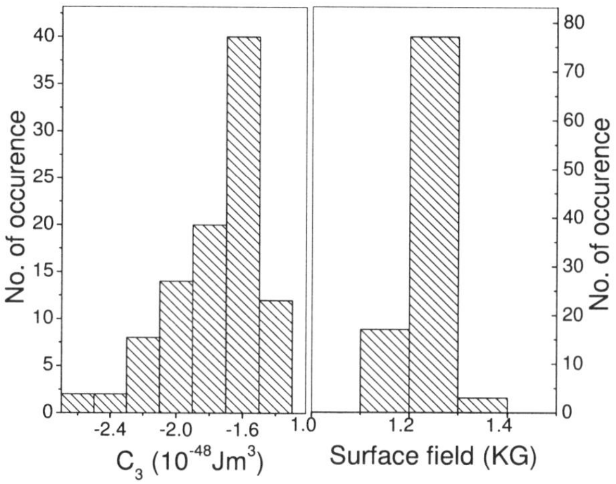

Figure 5: Distribution of the coefficient of the van der Waals potential (C_3) and the surface field (B_0) determined from the Monte Carlo simulation. The error on (C_3) is 15% and on (B_0) is 12%.

7 Conclusion

We have measured the van der Waals force between a ground state Rubidium atom and a conducting surface with an uncertainty of 15% employing a novel technique involving reflection of cold atoms from the magnetic thin film atom mirror. The measured van der Waals coefficient is 1.75×10^{-48} Jm^3. The theoretical value, after correction for the finite conductivity of cobalt, is 1.63×10^{-48} Jm^3, which is in good agreement with the value measured in our experiment. The error in the measurement can be reduced by increasing the number of data. So, conducting the experiment with variable kinetic energy would increase the precision of the measurement. The kinetic energy of the atoms can be varied over a range by controlling the position of the MOT or by using a moving molasses. Also new magnetic atom mirrors can be designed, such that the measurement will be sensitive at the retarded regime, which will enable to conduct a detailed and accurate study of Casimir-Polder interaction.

Acknowledgments: We would like to thank Prof. C. S. Unnikrishnan for his constant support in conducting the experiment. We also thank my colleagues of the group for their help in setting up the experiment. We thank G. Sheet, V Bagwe, S. C. Purndare and S. K. Gohil for help with thin film and P. G. Rodrigues for electronics.

References

[1] C. S. Unnikrishnan, ***Atoms in cavities and near surfaces***, in Lecture notes of the SERC school on Precision spectroscopy of Atoms, Molecules and Bose-Einstein Condensates, Eds. B. P. Das and V. Natarajan, (Allied, New Delhi, 2004)

[2] W. Hänsel, P. Hommelhoff, T.W. Hänsch and J. Richey, Nature **413**, 498 (2001)

[3] G. Rajalakshmi, ***Torsion balance investigation of the Casimir effect***, PhD thesis, Indian Institute of Astrophysics / Bangalore University (2004)

[4] A. K. Mohapatra, ***Study of atom-surface interaction using laser cooled atoms***, PhD thesis, TIFR (2005) and reference therein

[5] A. Landragin, Y. Courtios, G. Labeyrie, N. Vansteenkite, C.I. Westbrook and A. Aspect, Phys. Rev. Lett. **77**, 1464 (1996)

[6] F. Shimizu, Phys. Rev. Lett. **86**, 987 (2001)

[7] D. M. Harber, J. M. Obrecht, J. M. McGuirk and E. A. Cornell, Phys. Rev. A, **72**, 033610 (2005)

[8] A. K. Mohapatra and C. S. Unnikrishnan, Europhys. Lett., **73**, 839 (2006)

[9] E. A. Hinds, I. G. Hughes, J. Phys. D: Applied Physics, **32**, R119 (1999)

[10] C. Kittel, Phys. Rev.,**70**, 965 (1946)

[11] M. Hehn, S. Padovani, K. Ounadjela and J.P. Bucher, Phys. Rev. B **54**, 3428 (1996)

[12] Metcalf H J and Van Der Straten P 1999 Laser Cooling and Trapping(New York: Springer) and References therein

[13] A. K. Mohapatra and C. S. Unnikrishnan, ***High sensitivity probe absorption technique for time-of-flight measurements on cold atoms***, to be published in Pramana-J. of Physics